Graduate Texts in Physics

Graduate Texts in Physics publishes core learning/teaching material for graduate- and advanced-level undergraduate courses on topics of current and emerging fields within physics, both pure and applied. These textbooks serve students at the MS- or PhD-level and their instructors as comprehensive sources of principles, definitions, derivations, experiments and applications (as relevant) for their mastery and teaching, respectively. International in scope and relevance, the textbooks correspond to course syllabi sufficiently to serve as required reading. Their didactic style, comprehensiveness and coverage of fundamental material also make them suitable as introductions or references for scientists entering, or requiring timely knowledge of, a research field.

More information about this series at http://www.springer.com/series/8431

Pierre Meystre

Quantum Optics

Taming the Quantum

 Springer

Pierre Meystre
Department of Physics and James C. Wyant
College of Optics Sciences
University of Arizona
Tucson, AZ, USA

ISSN 1868-4513   ISSN 1868-4521   (electronic)
Graduate Texts in Physics
ISBN 978-3-030-76185-1   ISBN 978-3-030-76183-7   (eBook)
https://doi.org/10.1007/978-3-030-76183-7

Cover photo: Cavity cooling of a nanosphere to its quantum ground state of motion. Courtesy K. Dare,
Y. Coroli, M. Reisenbauer, Aspelmeyer group on Quantum Optics, Quantum Nanophotonics and
Quantum Information, Faculty of Physics, University of Vienna. https://aspelmeyer.quantum.at/.

This Springer imprint is published by the registered company Springer Nature Switzerland AG
The registered company address is: Gewerbestrasse 11, 6330 Cham, Switzerland

Pour Regina, Pierre-André et Lore

Preface

Quantum optics has witnessed remarkable developments in recent years, and finds itself at the center of what is sometimes called the "second quantum revolution." The central tenet of this "revolution" is the ability to control and manipulate individual quantum systems, with a profound impact in basic science and applications, from cosmology and tests of the foundations of physics to biology and geosciences, and from precision metrology to quantum information science and engineering.

These advances can be traced back to the inventions of the maser and the laser, and to the unique opportunities that they provide in controlling both the internal and the external state of atoms and molecules. This has led to the development of clocks of unfathomable precision, the cooling of atomic systems to unimaginably low temperatures, and the isolation of single atoms and molecules in tailored environments that allow for an exquisite control of their behavior, to mention just three of many remarkable breakthroughs.

All this is not the result of a grand plan developed by a government agency of some sort, but rather the consequence of the more or less random way in which scientific progress takes place. More often than not, ground-breaking advances come from the most unexpected places as a result of curiosity-driven research, and it is not unusual that a long time elapses before we fully realize their impact. Who would have expected when John Bell, a theoretical particle physicist, came up with his now famous inequalities, that AMO physics would provide an ideal testing ground to investigate them? And who would have thought when Peter Toschek first succeeded in trapping a single Barium ion that it would lead to ground-breaking developments in manipulating quantum entanglement, quantum steering, and to the recent advances in quantum information science and technology? I can't imagine that in his wildest dreams he ever thought of quantum computing at the time! And how about Edward Purcell's observation that spontaneous decay rates didn't seem to match their established values in some environments, and Dan Kleppner understanding then that this rate can be changed pretty much at will by controlling the electromagnetic environment of the atom? Or Vladimir Braginsky, who showed the way in understanding the standard quantum limit of quantum measurements and ways to circumvent it, when trying to figure out how to detect gravitational waves?

As scientists we are enormously lucky not just to have such brilliant minds as our colleagues, but also that there are folks with the vision and intelligence to understand the importance of curiosity-driven research, and the means and dedication to support it without being too worried about short-term financial or economic gains.

But I digress. This book is not a history book, and as such it is not the place to recount the successes, failures, wrong turns, brilliant insights, hard work, and lucky guesses that led to the current state of quantum optics. Rather, I have attempted to organize the building blocks that led us to that point in some kind of a logical fashion, starting from the simplest physical situations and moving to increasingly complex ones. As a result, the reader might be surprised to find topics such as quantum entanglement or measurement theory introduced quite early, before spontaneous emission or laser cooling, for example. My hope is that this perhaps unconventional approach will prove pedagogically appealing, while providing the reader with enough theoretical background to follow much of the current literature and start producing original research of their own.

After a review in Chap. 1 of a few basic elements of the semiclassical description of light-matter interaction that will be useful in later chapters, Chap. 2 presents an intuitively appealing approach to the quantization of the electromagnetic field based on an analogy with the simple harmonic oscillator, and reviews states of the field of particular importance in quantum optics, most importantly the coherent state, squeezed state and thermal field, and their descriptions in terms of quasiprobability distributions.

Atoms make their grand entrance in Chap. 3, which introduces the Jaynes-Cummings model, the linchpin of quantum optics, and its extension to the more general quantum Rabi model. It is quite remarkable that this model, which was introduced as a rather unrealistic toy model in the early 1970s, has become of increased importance over the years, due in large part to striking advances in experimental physics.

We then turn in Chap. 4 to a discussion of more general properties of composite systems, focusing on ways in which their quantum behavior can fundamentally differ from their classical counterparts. This brings us to the idea of quantum entanglement and to the demonstration of its profound significance in the violation of Bell's inequalities. We then turn to some of its properties, including entanglement monogamy and sharing and the no-cloning theorem, and their applications in quantum teleportation and quantum key distribution.

The next level of complication results from the observation that the small systems discussed so far can never be perfectly isolated from their environment and the need to deal with the implications of that coupling, most famously perhaps in quantum optics in the analysis of spontaneous emission. This is the topic of Chap. 5, which discusses several theoretical methods to describe the system-environment dynamics, and the onset of apparent irreversibility in a theory that is, at its most fundamental level, reversible.

These results lead us quite naturally to the challenging problem of quantum measurements, which we confront in Chap. 6. As eloquently stated by John Wheeler, "no phenomenon is a phenomenon until it is brought to a close by an irreversible act of amplification." That is, quantum measurements must somehow involve irreversibility. Without going into philosophical arguments, and at least from an operational point of view, this is achieved by coupling the system to be measured to an apparatus that provides this irreversibility. We start in the traditional way from von Neumann's projection postulate, followed by an analysis of measurement back action and the way to limit its effects through quantum non-demolition measurements. We then move to weak and to continuous measurements, and conclude with a short discussion of Zurek's pointer basis, which clarifies the role of the environment in the measurement process.

Not leaving the environment quite yet, we turn in Chap. 7 to ways to tailor and control it. This is the general topic of cavity QED, which we consider both in the resonant and the dispersive regimes. We also discuss the extension of these ideas to circuit QED, which uses Josephson junction-based artificial atoms coupled to transmission lines instead of real atoms and offers considerable promise for applications in quantum information science and technology. The chapter concludes with a brief discussion of the Casimir force, which presents the double advantage of illustrating what is arguably the simplest consequence of tailored electromagnetic environments, and also of introducing the idea of mechanical effects of light.

These mechanical effects are then analyzed in more detail beginning with Chap. 8, where we first identify the radiation pressure and gradient forces of light, and consider their effect both in the ray and wave optics regimes of atom optics. We show in particular how these forces can be exploited to trap atoms, and discuss several regimes of atomic diffraction and its application to atom interferometry. The importance of spontaneous emission and the associated random atomic recoil on these effects are also considered.

In addition to atom interferometry, the most important quantum optics application of the mechanical forces of light is arguably laser cooling, to which we turn in Chap. 9. We consider increasingly sophisticated approaches that permit to reach ever lower temperatures. Both the cooling of neutral atoms and the sideband cooling of trapped ions are described in some detail.

This takes us to atomic Bose-Einstein condensation, which is described in various situations in Chap. 10. Because in such systems the atoms are indistinguishable and subject to many-body effects, it is useful to treat them as a field. Following introductions to Schrödinger field quantization and to the mean-field Hartree approximation, we turn to ultracold atoms trapped in periodic lattices. This is perhaps the simplest example of a quantum simulator of a condensed matter problem, which we illustrate with the example of the Bose-Hubbard model. We also highlight the use of atomic microscopes to image and investigate these systems.

Chapter 11 then shifts to quantum optomechanics, which pairs optical and/or microwave resonators with massive mechanical oscillators. A sideband cooling technique directly adapted from the approach used for trapped ions permits to cool these objects down to their ground state of motion. This opens the way to determining and controlling the quantum state of truly macroscopic objects, with applications ranging from the development of extraordinarily sensitive force and acceleration sensors operating near or below the standard quantum limit to, possibly, a more profound understanding of quantum mechanics.

These various aspects of quantum optics demonstrate that in addition to its intrinsic scientific interest, it is also an enabling tool of considerable value for basic and applied science, engineering, and technology. Remarkably, it is also exceptionally positioned to help shed light on aspects of the physical world that are still a profound mystery to us. The final Chap. 12 elaborates on this point with a very brief overview of the role of quantum optics in testing the fundamental laws of nature, from quantum mechanics to relativity, and in exploring the nature of the particles and fields populating the Dark Sector, the 95% of the physical world that we still don't understand.

Tucson, AZ, USA   Pierre Meystre
2021

Acknowledgement

Physics is a team sport, and this book would not have been possible without the many students, postdocs, colleagues, and friends with whom I have had the pleasure to interact over the years. Special thanks to Markus Aspelmeyer, Dirk Bouwmeester, Qiongyi He, Jack Manley, Hal Metcalf, Oriol Romero-Isart, Keith Schwab, Ewan Wright, Keye Zhang, Weiping Zhang, Yakai Yang, and especially to Swati Singh and Kanupriya Sinha, who demonstrated infinite patience and tolerance in dealing with my endless questions and inquiries, and to my mentors Antonio Quattropani, Marlan Scully, and the late Herbert Walther. Of course all mistakes left in the book, hopefully not too many, are mine.

Contents

Chapter 1
Semiclassical Atom–Light Interaction

After a brief summary of the multipole expansion of the interaction between electromagnetic fields and charged particles and of the Lorentz atom, this chapter reviews a few aspects of the electric dipole interaction between two-level atoms and classical fields of particular relevance for the rest of this book, including semiclassical dressed states, the optical Bloch equations, Rabi oscillations, and relaxation mechanisms.

This chapter presents a brief review of selected aspects of the semiclassical interaction between an atom and an electromagnetic field that will be of use in this book. In subsequent chapters, we will deal mainly with quantized fields—with a number of notable exceptions, for instance, in some of the discussions of the mechanical effects of light and laser cooling. However, the atom–field interaction has the same physical origin, the Lorentz force, independently of whether the optical fields are treated classically or quantum mechanically. For this reason, this chapter starts by briefly reviewing how to exploit the multipole expansion of the Lorentz interaction, considering classical fields for simplicity. It then turns to a summary of selected key results in the semiclassical description of the electric dipole interaction between a two-level atom and a classical field. The reader is referenced to other texts, for instance, *Elements of Quantum Optics* by P. Meystre and M. Sargent III, for a more detailed and in-depth semiclassical discussion on these topics [1].

1.1 Multipole Expansion: A Brief Summary

Consider a test charge q of mass m and velocity $\mathbf{v}$ localized within an atom and acted upon by an external electromagnetic field with electric field $\mathbf{E}(\mathbf{r},t)$ and magnetic field $\mathbf{B}(\mathbf{r}, t)$, see Fig. 1.1. The Lorentz force acting on this charge is

$$\mathbf{F}(\mathbf{r}, t) = q\mathbf{E}(\mathbf{r}, t) + q\mathbf{v} \times \mathbf{B}(\mathbf{r}, t), \qquad (1.1)$$

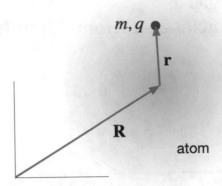

so that the electric and magnetic interaction energies between the charge and the
electromagnetic field are

$$V_e(t) = V_{E0}(t) - q \int_0^{\mathbf{r}} d\mathbf{s} \cdot \mathbf{E}(\mathbf{R} + \mathbf{s}, t),$$

$$V_m(t) = -q \int_0^{\mathbf{r}} d\mathbf{s} \cdot \mathbf{v} \times \mathbf{B}(\mathbf{R} + \mathbf{s}, t), \tag{1.2}$$

respectively, with all electric and magnetic fields considered as classical in the
chapter, as already indicated. These energies correspond to the work done by the
electric and magnetic components of the Lorentz force in first moving the charge
to a stationary origin of coordinates at a point $\mathbf{R}$ and then to a location $\mathbf{r}$ relative to
$\mathbf{R}$. Here, $V_{E0}(t)$ represents the energy of the charge when located at the reference
point $\mathbf{R}$. It may be expressed in terms of the electrostatic potential $\phi(\mathbf{R}, \mathbf{t})$ as
$V_{E0}(t) = +q\phi(\mathbf{R}, \mathbf{t})$. Because in electromagnetic waves the amplitude of the
electric field is of the order of c times that of the electromagnetic field, we have
that $V_m/V_e \approx v/c$. Considering further that the velocity of an electron in orbit
around a nucleus is of the order of $v \approx e^2/4\pi\epsilon_0\hbar$ gives

$$\frac{V_m}{V_e} \approx \frac{1}{4\pi\epsilon_0} \frac{e^2}{\hbar c} = \alpha, \tag{1.3}$$

where $\alpha \approx 1/137$ is the fine structure constant.

A Taylor series expansion of $V_e(t)$ and $V_m(t)$ about $\mathbf{r} = \mathbf{0}$ yields, see e.g. Refs. [2,
3],

$$V_e(t) = V_{E0}(t) - q \sum_{n=1}^{\infty} \frac{1}{n!} \left(\mathbf{r} \cdot \frac{\partial}{\partial \mathbf{R}} \right)^{n-1} \mathbf{r} \cdot \mathbf{E}(\mathbf{R}, t), \tag{1.4}$$

$$V_m(t) = -\frac{q\hbar}{m} \sum_{n=1}^{\infty} \frac{n}{(n+1)!} \left(\mathbf{r} \cdot \frac{\partial}{\partial \mathbf{R}} \right)^{n-1} \boldsymbol{\ell} \cdot \mathbf{B}(\mathbf{R}, t), \tag{1.5}$$

where $\hbar\boldsymbol{\ell} = \mathbf{r} \times \mathbf{p}$ is the angular momentum of the test charge relative to the coordinate origin $\mathbf{R}$. Note that the use of the mechanical momentum $\mathbf{p} = m\dot{\mathbf{r}}$ instead of the canonical momentum neglects the electromagnetic component of the momentum responsible for diamagnetic effects.

In addition to the electromagnetic interaction, electrons and nuclei are characterized by a spin magnetic moment $\mathbf{m}_s = (q\hbar/2m)g_s\mathbf{s}$, where $\mathbf{s}$ is the spin of the test charge and g_s its gyromagnetic factor, equal to $2.002\ldots$ for electrons. The factor $q\hbar/2m$ is the particle's magneton. The spin magnetic moment yields an additional term to the magnetic energy V_{m}, which becomes

$$V_{\mathrm{m}}(t) = -\frac{e\hbar}{2m}\sum_{n=1}^{\infty}\frac{1}{n!}\left(\mathbf{r}\cdot\frac{\partial}{\partial\mathbf{R}}\right)^{n-1}\left(\frac{2}{n+1}g_{\ell}\boldsymbol{\ell} + g_s\mathbf{s}\right)\cdot\mathbf{B}(\mathbf{R},t),\qquad(1.6)$$

where the orbital g-factor is $g_{\ell} = q/e$ and $g_{\ell} = -1$ for an electron. For an ensemble $\{\varsigma\}$ of particles of charges q_{ς} and masses m_{ς} in an atom, these expressions are to be summed over all particles. Thus, the electric energy becomes

$$V_{\mathrm{e}}(t) = V_{\mathrm{E0}}(t) + V_{\mathrm{E1}}(t) + V_{\mathrm{E2}}(t) + \ldots$$

$$-\sum_{\iota}q_{\alpha}\phi(\mathbf{R},t) - \sum_{i=1}^{3}\sum_{\iota}q_{\alpha}r_i(\varsigma)E_i(\mathbf{R},t)$$

$$-\frac{1}{2}\sum_{i,j=1}^{3}\left[\sum_{\varsigma}q_{\varsigma}r_i(\varsigma)r_j(\varsigma)\right]\frac{\partial}{\partial R_j}E_i(\mathbf{R},t) + \ldots,\qquad(1.7)$$

where $\phi(\mathbf{R},t)$ is the electrostatic potential, and the magnetic energy becomes

$$V_{\mathrm{m}}(t) \equiv V_{\mathrm{M1}}(t) + V_{\mathrm{M2}}(t) + \cdots$$

$$= \sum_{i=1}^{3}B_i(\mathbf{R},t)\times\sum_{\varsigma}\frac{e\hbar}{2m_{\varsigma}}[g_{\ell}(\varsigma)\ell_i(\varsigma) + g_s(\varsigma)s_i(\varsigma)]\qquad(1.8)$$

$$-\sum_{i,j=1}^{3}\frac{\partial B_i(\mathbf{R},t)}{\partial R_j}\sum_{\varsigma}\frac{e\hbar}{2m_{\varsigma}}\left[\frac{2}{3}g_{\ell}(\varsigma)\ell_i(\varsigma)r_j(\varsigma) + g_s(\varsigma)s_i(\varsigma)\right] + \cdots$$

For electromagnetic fields whose wavelength $\lambda = 2\pi/k$ is large compared with the size of interacting atom, only the first few terms in the Taylor expansions of $V_{\mathrm{e}}(t)$ and $V_{\mathrm{m}}(t)$ need to be retained, since the expansion factor

$$\mathbf{r}\frac{\partial}{\partial\mathbf{R}} \approx kr \ll 1.\qquad(1.9)$$

It is instructive to recast this condition in terms of generic order of magnitude atomic properties. Specifically, we know that the typical radius of an electron orbit around an atomic nucleus is given by the Bohr radius a_0 so that

$$r \approx a_0 = \frac{4\pi\epsilon_0\hbar^2}{me^2} = \frac{\hbar}{mc\alpha}. \tag{1.10}$$

Also, typical field frequencies ω are comparable to atomic transition frequencies, which are of the order of

$$\omega_0 \approx \frac{R_E}{\hbar} = \frac{mc^2\alpha^2}{\hbar}, \tag{1.11}$$

where $R_E = \frac{1}{2}mc^2\alpha^2$ is the Rydberg energy. It follows that in situations involving the interaction between atoms and optical fields, the product kr appearing in Eq. (1.9) is of the order of

$$kr \approx \hbar\alpha, \tag{1.12}$$

which shows that the expansion of $V_e(t)$ and $V_m(t)$ can be understood as an expansion in powers of the fine structure constant α. Armed with this insight, we now focus on the first two terms in the expansion of $V_e(t)$ and the first term only for $V_m(t)$.

Electric Dipole Interaction The first term $V_{E0}(t)$ of $V_e(t)$ is proportional to the net charge of the atom, and it vanishes for neutral atoms. The second term, $V_{E1}(t)$, is the electric dipole interaction energy. Introducing the electric dipole moment

$$\mathbf{d} = \sum_\varsigma q_\varsigma \mathbf{r}(\varsigma), \tag{1.13}$$

or $\mathbf{d} = \int d^3r\rho(\mathbf{r})\mathbf{r}$ for a charge distribution, this contribution to the interaction energy may be reexpressed as

$$V_{E1}(t) = -\mathbf{d} \cdot \mathbf{E}(\mathbf{R}, t). \tag{1.14}$$

This interaction dominates most quantum optical phenomena of interest in this book.

Electric Quadrupole Interaction The $V_{E2}(t)$ contribution to $V_e(t)$ describes electric quadrupole (E2) interactions. In terms of the quadrupole tensor

$$Q_{ij} = 3\int d^3r\rho(\mathbf{r})r_ir_j, \tag{1.15}$$

it becomes

$$V_{E2}(t) = -\frac{1}{6}\sum_{i,j=1}^{3} Q_{ij}\frac{\partial}{\partial R_i}E_j(\mathbf{R}, t). \tag{1.16}$$

Alternatively, electric quadrupole interactions can be expressed in terms of the traceless quadrupole tensor $Q_{ij}^{(2)} = \int d^3 r \hat{\rho}(\mathbf{r}) \times (3 r_i r_j - \delta_{ij} r^2)$. Electric quadrupole interactions are typically weaker than electric dipole interactions by a factor a_0/λ, where a_0 is the Bohr radius and λ is the wavelength of the transition. Since a_0/λ is very small for optical transitions, these interactions are typically neglected in quantum optics.

Magnetic Dipole Interaction The first term in the multipole expansion of the magnetic interaction is the magnetic dipole (M1) interaction of a magnetic moment $\mathbf{m}$ in a magnetic field

$$V_{\mathrm{M1}} = -\mathbf{m} \cdot \mathbf{B}(\mathbf{R}, t), \tag{1.17}$$

where

$$\mathbf{m} = \sum_\alpha \left(\frac{q_\alpha \hbar}{2 m_\alpha} \right) [g_\ell(\alpha) \ell(\alpha) + g_s(\alpha) \mathbf{s}(\alpha)] = -\mu_{\mathrm{B}}(\mathbf{L} + 2\mathbf{S}), \tag{1.18}$$

and we have used the fact that for electrons $g_\ell = -1$ and $g_s \simeq -2$. The Bohr magneton μ_{B} is

$$\mu_{\mathrm{B}} = \frac{e\hbar}{2mc} = \frac{\alpha e a_0}{2}, \tag{1.19}$$

where α is the fine structure constant. Thus, magnetic dipole interactions tend to be smaller than electric dipole interactions by a factor of order α. The connection between $\mathbf{m}$ and the angular momentum $\mathbf{J} = \ell + \mathbf{s}$ is $\mathbf{m} = \gamma \mathbf{J}$, where γ is the gyromagnetic ratio.

1.2 The Lorentz Atom

The Lorentz atom consists of a classical electron harmonically bound to a proton, see e.g. Refs. [1, 4, 5]. It provides a framework to understand a number of elementary aspects of the electric dipole interaction between a single atom and light. Assuming for now that the center of mass motion of the atom is unaffected by the field, a restriction that will be removed when we turn to the mechanical effects of light and laser cooling in Chaps. 8 and 9, and neglecting in addition magnetic effects, the equation of motion of the electron is

$$\left(\frac{d^2}{dt^2} + 2\gamma \frac{d}{dt} + \omega_0^2 \right) \mathbf{r} = -\frac{e}{m} \mathbf{E}(\mathbf{R}, t), \tag{1.20}$$

where ω_0 is the electron's natural oscillation frequency and γ represents a frictional decay rate that accounts for the effects of radiative damping. For the classical Lorentz atom, it is given by

$$\gamma = \omega_0^2 r_0/3c\,, \qquad (1.21)$$

where

$$r_0 = \frac{1}{4\pi\epsilon_0}\left(\frac{e^2}{mc^2}\right) \qquad (1.22)$$

is the classical electron radius. This damping arises physically from the radiation reaction of the field radiated by the atom on itself, as will be analyzed in detail in Sect. 5.1. In the electric dipole approximation, the electric field is evaluated at the location $\mathbf{R}$ of the atomic center of mass.

The study of light-matter interactions is simplified by the introduction of complex variables [6–8]; for example, an electric field

$$\mathbf{E}(\mathbf{R}, t) = \sum_{n,\mu} \vec{\epsilon}_\mu \mathcal{E}_n \cos(\omega_n t)\,, \qquad (1.23)$$

where $\vec{\epsilon}_\mu$ is the polarization vector of the Fourier component of the field at frequency ω_n, is expressed as

$$\mathbf{E}(\mathbf{R}, t) = \mathbf{E}^+(\mathbf{R}, t) + \mathbf{E}^-(\mathbf{R}, t)\,, \qquad (1.24)$$

where the *positive frequency part* of the field is

$$\mathbf{E}^+(\mathbf{R}, t) = \frac{1}{2}\sum_{n,\mu} \vec{\epsilon}_\mu \mathcal{E}_n \exp[i(\mathbf{k}_n \cdot \mathbf{R} - \omega_n t)]\,. \qquad (1.25)$$

Due to the linearity of Eq. (1.20), it is sufficient to study the response of the Lorentz atom to a plane monochromatic electric field of frequency ω, complex amplitude $\mathcal{E}$, and polarization $\vec{\epsilon}$. Introducing the *complex dipole moment*

$$\mathbf{d} = -e\mathbf{r} = \vec{\epsilon}\,\alpha(\omega)\mathcal{E}\exp[i(\mathbf{k}\cdot\mathbf{R} - \omega t)] + \text{c.c.}\,, \qquad (1.26)$$

where $\alpha(\omega)$ is the *complex polarizability*, one finds readily

$$\alpha(\omega) = \frac{e^2/m}{\omega_0^2 - \omega^2 - i\gamma\omega}\,. \qquad (1.27)$$

Beer's Law Combining Eq. (1.25) with the Maxwell wave equation

$$\left(\nabla^2 - \frac{1}{c^2}\frac{\partial^2}{\partial t^2}\right)\mathbf{E}(\mathbf{R}, t) = \frac{1}{\epsilon_0 c^2}\frac{\partial^2 \mathbf{P}(\mathbf{R}, t)}{\partial t^2}, \tag{1.28}$$

where $\mathbf{P}(\mathbf{R}, t)$ is the electric polarization, given by the electric dipole density of the medium as

$$\mathbf{P} \equiv N\mathbf{d} = -Ne\mathbf{r} = N\vec{\epsilon}\,\alpha(\omega)\mathcal{E}\exp[i(\mathbf{k}\cdot\mathbf{R} - \omega t)] + \text{c.c.}, \tag{1.29}$$

N being the atomic density, the plane wave dispersion relation is easily found to be

$$k^2 = \frac{\omega^2}{c^2}n^2(\omega), \tag{1.30}$$

where the index of refraction $n(\omega)$ is

$$n(\omega) = \sqrt{1 + \frac{N\alpha(\omega)}{\epsilon_0}}. \tag{1.31}$$

Since the polarizability (1.27) is normally complex, so is the index of refraction. Its real part leads to dispersive effects and its imaginary part to absorption. Specifically, $\text{Re}[n(\omega)] - 1$ has the form of a standard dispersion curve, positive for $\omega - \omega_0 < 0$ and negative for $\omega - \omega_0 > 0$, while $\text{Im}[n(\omega)]$ is a Lorentzian peaked at $\omega = \omega_0$. The intensity absorption coefficient $a(\omega)$ is

$$a(\omega) = 2\text{Im}\,[n(\omega)]\,\omega/c = \frac{2\omega}{c}\text{Im}\left[1 + \left(\frac{Ne^2}{m\epsilon_0}\right)\frac{(\omega_0^2 - \omega^2) + i\gamma\omega}{(\omega_0^2 - \omega^2)^2 + \gamma^2\omega^2}\right]^{1/2}. \tag{1.32}$$

For atomic vapors, the corrections to the vacuum index of refraction are normally small, so that the square root in Eq. (1.32) can be expanded to first order, giving

$$a(\omega) = \left(\frac{Ne^2}{\epsilon_0 mc}\right)\frac{\gamma\omega^2}{(\omega_0^2 - \omega^2)^2 + \gamma^2\omega^2}, \tag{1.33}$$

see Fig. 1.2. The intensity of a monochromatic field propagating along the z direction through a gas of Lorentz atoms is therefore attenuated according to *Beer's law*,

$$I(\omega, z) = I(\omega, 0)e^{-a(\omega)z}. \tag{1.34}$$

Fig. 1.2 Real part (dashed curve) and imaginary part (solid curve) of the complex polarizability $\alpha(\omega)$ as a function of $\omega_0^2 - \omega^2$, in arbitrary units. The dispersive real part of $\alpha(\omega)$ results in changes in index of refraction of the medium, while its nearly Lorentzian imaginary part results in the absorption of the light field by the atomic medium

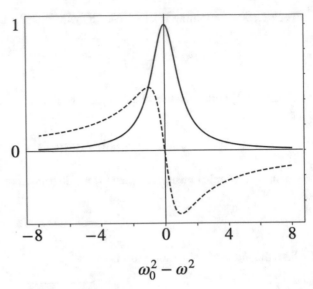

$$\omega_0^2 - \omega^2$$

If the index of refraction at a given frequency becomes purely imaginary, no electromagnetic wave can propagate inside the medium. This is the case for field frequencies smaller than the plasma frequency

$$\omega_{\mathrm{p}} = \sqrt{\frac{Ne^2}{m\epsilon_0}}. \tag{1.35}$$

While the Lorentz atom model gives an adequate description of absorption and dispersion in weakly excited absorbing media, it fails to predict important phenomena such as saturation and light amplification. This is because, in this model, the phase of the induced atomic dipoles with respect to the incident field is always such that the polarization field adds destructively to the incident field. The description of light amplification requires a quantum treatment of the medium, which gives a greater flexibility to the possible relative phases between the incident and polarization fields. Despite its important limitations, though, the Lorentz model often provides valuable intuition. One such example is the cavity cooling of atoms that will be discussed in Sect. 9.4.

Slowly Varying Envelope Approximation Light-matter interactions often involve quasi-monochromatic fields. Their electric field, taken for concreteness to propagate along the z-axis, can be expressed in the form

$$\mathbf{E}(\mathbf{R}, t) = \frac{1}{2}\vec{\epsilon}\,\mathcal{E}^+(\mathbf{R}, t)e^{\mathrm{i}(kz-\omega t)} + \text{c.c.}, \tag{1.36}$$

with

$$\left|\frac{\partial \mathcal{E}^+}{\partial t}\right| \ll \omega \left|\mathcal{E}^+\right| , \quad \left|\frac{\partial \mathcal{E}^+}{\partial z}\right| \ll k \left|\mathcal{E}^+\right| . \tag{1.37}$$

It is consistent within this approximation to assume that the polarization (1.29) takes then the form

$$\mathbf{P}(\mathbf{R}, t) = \frac{1}{2} \vec{\epsilon} \, \mathcal{P}^+(\mathbf{R}, t) e^{i(kz - \omega t)} + \text{c.c.} , \tag{1.38}$$

with

$$\left|\frac{\partial \mathcal{P}^+}{\partial t}\right| \ll \omega \left|\mathcal{P}^+\right| . \tag{1.39}$$

Under these conditions, known as the *slowly varying envelope approximation*, Maxwell's wave equation reduces to

$$\left(\frac{\partial}{\partial z} + \frac{1}{c}\frac{\partial}{\partial t}\right) \mathcal{E}^+(z, t) = -\frac{k}{2i\epsilon_0} \mathcal{P}^+(z, t). \tag{1.40}$$

Hence, in the slowly varying envelope approximation, we ignore the backward propagation of the field [9]. The slowly varying amplitude and phase approximation is essentially the same, except that it expresses the electric field envelope in terms of a real amplitude and phase.

The slowly varying polarization $\mathcal{P}^+(z, t)$ associated with the classical Lorentz oscillator is readily obtained by expressing $x(t)$ as

$$x(z, t) = \frac{1}{2} X(z, t) e^{-i\omega t} + \text{c.c.} , \tag{1.41}$$

where z should be understood for now as a parameter labeling the position of the oscillating dipole in the electric field $E(z, t)$. Later in this book, in Chaps. 8 and following, we will discuss the mechanical effects of light on atoms, in particular how light can be exploited to cool them. The oscillators will then be allowed to move under the influence of optical forces, and z will become a dynamical variable. In this sense, the current description essentially assumes that the dipole has an infinite mass.

Substituting $x(t)$ into Eq. (1.20) and neglecting the small quantities $\ddot{X}$ and $\gamma \dot{X}$, this gives readily

$$\frac{dX(z, t)}{dt} = -\left[\gamma + i(\omega_0^2 - \omega^2)/2\omega\right] X - \frac{ie\mathcal{E}(z, t)}{2\omega m} . \tag{1.42}$$

In steady state, we have therefore

$$X(z) = -\left(\frac{ie\mathcal{E}(z)}{2m\omega}\right)\frac{1}{\gamma + i(\omega_0^2 - \omega^2)/2\omega}, \tag{1.43}$$

or, for

$$\omega_0^2 - \omega^2 \approx 2\omega(\omega_0 - \omega), \tag{1.44}$$

an approximation valid for detunings $\Delta = \omega_0 - \omega$ small compared to ω and ω_0,

$$X(z) = -\left(\frac{ie\mathcal{E}(z)}{2m\omega}\right)\frac{1}{\gamma + i(\omega_0 - \omega)}, \tag{1.45}$$

and

$$x(z, t) = X(z, t)e^{i\omega t} + \text{c.c.} \tag{1.46}$$

The slowly varying polarization $\mathcal{P}(z)$ is then

$$\mathcal{P}(z) = -N(z)eX(z), \tag{1.47}$$

with $N(z)$ the number of dipoles per unit volume.

It will occasionally prove useful to decompose the slowly varying dipole oscillation amplitude $X(t)$ as the sum of its real and imaginary parts as

$$X(z, t) = U(z, t) - iV(z, t), \tag{1.48}$$

with equations of motion

$$\frac{dU}{dt} = -(\omega_0 - \omega)V - \gamma U$$

$$\frac{dV}{dt} = (\omega_0 - \omega)U - \gamma V + e\mathcal{E}/2m\omega, \tag{1.49}$$

which gives in steady state

$$U(z) = \left(\frac{-e\mathcal{E}(z)}{2m\omega}\right)\frac{\omega_0 - \omega}{(\omega_0 - \omega)^2 + \gamma^2},$$

$$V(z) = \left(\frac{-e\mathcal{E}(z)}{2m\omega}\right)\frac{\gamma}{(\omega_0 - \omega)^2 + \gamma^2}. \tag{1.50}$$

These equations will be encountered again in Sect. 1.3.2, where the classical oscillator will be replaced by a two-level atom. In that case, the atomic dynamics will acquire a third component $W(z)$, which describes the difference in populations of the excited and ground atomic state, or inversion. The resulting equations of motion for the vector $\mathbf{U} = (U, V, W)$ will become the optical Bloch equations, from which the dynamics of the Lorentz oscillator results from assuming that $W = -1$, that is, the atom remains in its lower level.

1.3 Two-Level Atoms

A large number of optical phenomena can be understood by considering the interaction between a quasi-monochromatic field of central frequency ω and a two-level atom, which simulates a dipole-allowed transition in an atom, a molecule, or an artificial atom.[1] This approximation, which is well justified for near-resonant interactions, $\omega \simeq \omega_0$, is central to the discussion of a wide range of phenomena in quantum optics [1, 5, 7, 8, 10, 11]. It is also of particular importance in quantum information science, where the two-level atom changes name to become a qubit, the quantum mechanical version of the bit familiar from classical information science. This section discusses the model Hamiltonian for this system in the *semiclassical approximation* where the electromagnetic field can be described classically. It will then be revisited at considerably more length for the case of quantized fields in subsequent chapters.

1.3.1 Hamiltonian

In the absence of dissipation mechanisms, the dipole interaction between a quasi-monochromatic classical field and a two-level atom is

$$\hat{H} = \hbar\omega_e |e\rangle\langle e| + \hbar\omega_g |g\rangle\langle g| - \mathbf{d} \cdot \mathbf{E}(\mathbf{R}, t), \qquad (1.51)$$

where $|e\rangle$ and $|g\rangle$ label the upper and lower atomic levels, of frequencies ω_e and ω_g, respectively, with $\omega_e - \omega_g = \omega_0$, and $\mathbf{R}$ is the location of the center of mass of the atom. The electric dipole operator that couples the excited and ground levels may be expressed as

$$\mathbf{d} = \vec{\epsilon}_d d \left(|e\rangle\langle g| + |g\rangle\langle e| \right), \qquad (1.52)$$

---

[1] A dipole-allowed transition between two atomic levels $|g\rangle$ and $|e\rangle$ is a transition for which the matrix element $\langle e|V_{E1}|g\rangle \neq 0$. Since successive terms in the multipole expansion of the atom–field interaction scale with increasing powers of $1/\alpha$, it is therefore usually—but not always—sufficient to ignore higher order terms in that case.

where $\vec{\epsilon}_d$ is a unit vector in the direction of the dipole and d is the matrix element of the electric dipole operator between the ground and excited states, which we take to be real for simplicity. We also neglect the vector character of d and $\mathbf{E}(\mathbf{R}, t)$ in the following, assuming, for example, that both $\vec{\epsilon}_d$ and $\vec{\epsilon}$ are parallel to the x-axis. The Hamiltonian (1.51) may then be expressed as

$$\hat{H} = \hbar\omega_e |e\rangle\langle e| + \hbar\omega_g |g\rangle\langle g| - d\left(|e\rangle\langle g| + |g\rangle\langle e|\right)\left[E^+(\mathbf{R}, t) + E^-(\mathbf{R}, t)\right].$$
(1.53)

It is often convenient to introduce the matrix representation of the atomic level

$$|e\rangle = \begin{pmatrix} 1 \\ 0 \end{pmatrix} \quad ; \quad |g\rangle = \begin{pmatrix} 0 \\ 1 \end{pmatrix}$$
(1.54)

and the *Pauli spin operators*

$$\hat{\sigma}_x = \begin{pmatrix} 0 & 1 \\ 1 & 0 \end{pmatrix} \quad ; \quad \hat{\sigma}_y = \begin{pmatrix} 0 & -i \\ i & 0 \end{pmatrix} \quad ; \quad \hat{\sigma}_z = \begin{pmatrix} 1 & 0 \\ 0 & -1 \end{pmatrix}.$$
(1.55)

With Eqs. (1.54), these matrices can readily be written in terms of $|e\rangle$ and $|g\rangle$, for example,

$$\hat{\sigma}_z = \left(|e\rangle\langle e| - |g\rangle\langle g|\right).$$
(1.56)

Introducing in addition the spin raising and lowering operators

$$\hat{\sigma}_+ \equiv \tfrac{1}{2}\left(\hat{\sigma}_x + i\hat{\sigma}_y\right) = |e\rangle\langle g| = \hat{\sigma}_-^\dagger$$
(1.57)

and redefining the zero of atomic energy result in the commonly used form of the Hamiltonian

$$\hat{H} = \frac{1}{2}\hbar\omega_0\hat{\sigma}_z - d\left(\hat{\sigma}_+ + \hat{\sigma}_-\right)\left[E^+(\mathbf{R}, t) + E^-(\mathbf{R}, t)\right].$$
(1.58)

Rotating Wave Approximation   Under the influence of a monochromatic electromagnetic field of frequency ω, atoms undergo transitions between their lower and upper states by interacting with both the positive and the negative frequency parts of the field. The corresponding contributions to the atomic dynamics oscillate at frequencies $\pm(\omega_0 - \omega)$ and $\pm(\omega_0 + \omega)$, respectively, and their contributions to the probability amplitudes involve denominators containing these same frequency dependences. For near-resonant atom–field interactions, $|\omega_0 - \omega| \ll \omega$, the rapidly oscillating contributions lead to small corrections, the first-order one being the Bloch–Siegert shift, whose value near resonance $\omega \simeq \omega_0$ is

$$\delta\omega_{eg} = -\frac{\left(d|E^+|/\hbar\right)^2}{4\omega}$$
(1.59)

to lowest order in $d\mathcal{E}/\hbar\omega$. The neglect of these terms is the *rotating wave approximation* (RWA). Note that it is often (but not always) inconsistent to regard an atom as a two-level system and not to perform the RWA, a point further discussed in Sect. 3.6.

In the RWA, the atomic system is described by the Hamiltonian

$$\hat{H} = \frac{1}{2}\hbar\omega_0\hat{\sigma}_z - d\left[\hat{\sigma}_+E^+(\mathbf{R},t) + \hat{\sigma}_-E^-(\mathbf{R},t)\right], \tag{1.60}$$

or, in a frame rotating at the frequency ω of the field,

$$\hat{H} = \frac{1}{2}\hbar\Delta\hat{\sigma}_z - \frac{1}{2}d\left(\hat{\sigma}_+\mathcal{E}e^{i\mathbf{k}\cdot\mathbf{R}} + \text{h.c.}\right)$$

$$= \frac{1}{2}\hbar\Delta\hat{\sigma}_z - \frac{1}{2}\hbar\Omega_r\left(\hat{\sigma}_+e^{i\mathbf{k}\cdot\mathbf{R}} + \text{h.c.}\right), \tag{1.61}$$

where

$$\Omega_r \equiv d\mathcal{E}/\hbar \tag{1.62}$$

is called the *resonant Rabi frequency* for a reason that will soon be apparent, and

$$\Delta = \omega_0 - \omega \tag{1.63}$$

is the atom–light detuning.[2] In the rest of this chapter, we consider atoms at rest and located at the origin, $\mathbf{R} = 0$.

Rabi Frequency The dynamics of the two-level atom is conveniently expressed in terms of its density operator $\hat{\rho}$, whose evolution is given by the Schrödinger equation

$$\frac{d\hat{\rho}}{dt} = -\frac{i}{\hbar}[\hat{H}, \hat{\rho}]. \tag{1.64}$$

Its diagonal elements $\hat{\rho}_{ee} = \langle e|\hat{\rho}|e\rangle$ and $\rho_{gg} = \langle g|\hat{\rho}|g\rangle$ are the upper and lower state populations, respectively, while the off-diagonal matrix elements $\hat{\rho}_{eg} = \langle e|\hat{\rho}|g\rangle = \hat{\rho}_{ge}^\star$ are called the atomic coherences, or simply coherences, between levels $|e\rangle$ and $|g\rangle$. These coherences, which are proportional to the expectation value of the electric dipole operator, play a key role in much of optical physics and quantum optics.

Equation (1.64) readily gives the equations of motion

---

[2]Note that the alternative detuning definition $\delta = \omega - \omega_0$ is also frequently used in the quantum optics and laser spectroscopy literature, as there is usually no obvious reason to prefer one over the other. It is therefore important to always check which definition is used when comparing results from different publications.

$$\frac{d\rho_{ee}}{dt} = -\frac{i\Omega_r}{2}\left(\rho_{eg} - \rho_{ge}\right),\tag{1.65a}$$

$$\frac{d\rho_{gg}}{dt} = \frac{i\Omega_r}{2}(\rho_{eg} - \rho_{ge}),\tag{1.65b}$$

$$\frac{d\rho_{eg}}{dt} = -i\Delta\rho_{eg} - \frac{i\Omega_r}{2}(\rho_{ee} - \rho_{gg}).\tag{1.65c}$$

As we will see more explicitly in the discussion of the optical Bloch equations of Sect. 1.3.2, the evolution of the atomic populations $P_g(t)$ and $P_e(t) = 1 - P_g(t)$ is characterized by oscillations at the *generalized Rabi frequency*

$$\Omega = \left(\Omega_r^2 + \Delta^2\right)^{1/2}.\tag{1.66}$$

Specifically, assuming that the atom is initially in its ground state $|g\rangle$, the probability that it is in the excited state $|e\rangle$ at a subsequent time t is given by Rabi's formula

$$\rho_{ee}(t) = (\Omega_r/\Omega)^2 \sin^2(\Omega t/2).\tag{1.67}$$

At resonance $\Delta = 0$, the generalized Rabi frequency Ω reduces to the Rabi frequency Ω_r.

Semiclassical Dressed States Instead of using as a basis set the eigenstates $|e\rangle$ and $|g\rangle$ of the Hamiltonian $\hbar\omega_0\hat{\sigma}_z$ of non-interacting atoms, the so-called *bare states*, their dynamics can alternatively be described in terms of the *dressed states* basis of the eigenstates of the full Hamiltonian (1.60), see in particular Ref. [12]. By convention, the state $|1\rangle$ is the one with the greatest energy. They are conveniently expressed in terms of the bare states via the Stückelberg angle θ as

$$|1\rangle = \sin\theta|g\rangle + \cos\theta|e\rangle,$$
$$|2\rangle = \cos\theta|g\rangle - \sin\theta|e\rangle,\tag{1.68}$$

where $\sin(2\theta) = -\Omega_r/\Omega$ and $\cos(2\theta) = \Delta/\Omega$. The corresponding eigenenergies are

$$E_1 = +\frac{1}{2}\hbar\Omega \quad ; \quad E_2 = -\frac{1}{2}\hbar\Omega\tag{1.69}$$

and are plotted in Fig. 1.3. These dressed levels repel each other and form an anticrossing at resonance $\omega = \omega_0$. As the detuning Δ varies from positive to negative values, state $|1\rangle$ passes continuously from the excited state $|e\rangle$ to the bare ground state $|g\rangle$, with both bare states having equal weights at resonance. The distances between the perturbed levels and their asymptotes for $|\Delta| \gg \Omega_1$ represent the ac Stark shifts, or light shifts, of the atomic states when coupled to the optical

Fig. 1.3 Semiclassical
dressed states as a function of
the detuning $\Delta = \omega_0 - \omega$.
The upper part of the figure
shows for reference the
interaction picture bare
energies of the atom in the
absence of field, taking the
energy $\hbar\omega_g$ of the ground
state $|g\rangle$ to be constant. The
bottom part illustrates that the
energy separation of the
dressed states is the Rabi
frequency Ω and reaches its
lowest value Ω_r at the
avoided crossing point $\Delta = 0$

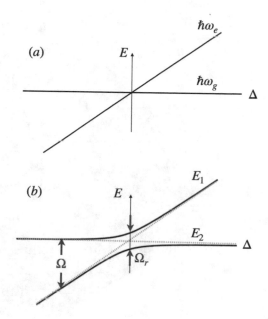

field. The ac Stark shift of $|g\rangle$ is positive for $\Delta < 0$ and negative for $\Delta > 0$, while
the $|e\rangle$ state shift is negative for $\Delta < 0$ and positive for $\Delta > 0$.

1.3.2   Optical Bloch Equations

Introducing the real quantities

$$U = \rho_{eg}e^{i\omega t} + \text{c.c.},$$
$$V = i\rho_{eg}e^{i\omega t} + \text{c.c.},$$
$$W = \rho_{ee} - \rho_{gg}, \tag{1.70}$$

their equations of motion may be expressed with Eq. (1.65) as

$$\frac{dU}{dt} = -\Delta V$$

$$\frac{dV}{dt} = \Delta U + \Omega_1 W$$

$$\frac{dW}{dt} = -\Omega_1 V. \tag{1.71}$$

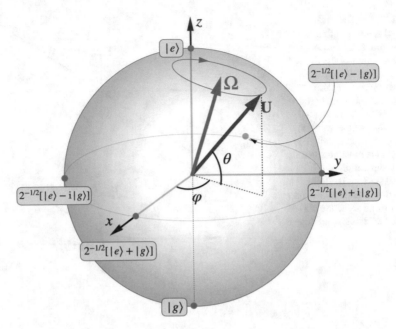

Fig. 1.4 Schematic of the Bloch sphere. In the absence of dissipation, the Bloch vector $\mathbf{U} = (U, V, W)$ processes on the surface of that sphere about the vector $\mathbf{\Omega} \equiv (-\Omega_r, 0, \Delta)$. The state of the two-level system at the locations of the blue points on the axes of the sphere is also indicated

These are the *optical Bloch equations*. Physically, U describes the component of the atomic coherence in phase with the driving field, V the component in quadrature with the field, and W the atomic inversion.

The optical Bloch equations have a simple geometrical interpretation offered by thinking of U, V, and W as the three components of a vector called the Bloch vector $\mathbf{U}$ whose equation of motion is

$$\frac{d\mathbf{U}}{dt} = \mathbf{\Omega} \times \mathbf{U}, \tag{1.72}$$

where we have introduced the vector $\mathbf{\Omega} \equiv (-\Omega_r, 0, \Delta)$. Thus, $\mathbf{U}$ processes about the vector $\mathbf{\Omega}$, of length Ω, while conserving its length, as illustrated in Fig. 1.4. The evolution of a two-level atom driven by a monochromatic field is thus a rotation on the surface of the Bloch sphere,

$$|\psi(t)\rangle = e^{-i\hat{H}t}|\psi(0)\rangle, \tag{1.73}$$

about the "Hamiltonian vector"

$$\hat{H} = -\Omega_r \hat{\sigma}_x + \Delta \hat{\sigma}_z. \tag{1.74}$$

Mathematically, this motion is equivalent to that of a spin-$\frac{1}{2}$ system in two magnetic fields $\mathbf{B}_0$ and $2\mathbf{B}_1 \cos \omega t$ that are parallel to the z- and x-axis, respectively, and whose amplitudes are such that the Larmor spin precession frequencies around them are ω and $2\Omega_1 \cos \omega t$.

1.3.3 Relaxation Mechanisms

In addition to their coherent interaction with light fields, atoms suffer incoherent relaxation mechanisms, whose origin can be as diverse as elastic and inelastic colli-sions or spontaneous emission. This will be considered in detail in the discussion of system–reservoir interactions in Chap. 5. For now, it is sufficient to remark that one advantage of describing the atomic state in terms of the density operator $\hat{\rho}$ is that the physical interpretation of its matrix elements allows us to add phenomenologically various relaxation terms directly to them.

Relaxation Toward Unobserved Levels If the relaxation mechanisms transfer populations or atomic coherences toward uninteresting or unobserved levels, their description can normally be given in terms of a Schrödinger equation, but with an effective non-Hermitian Hamiltonian. Specifically, the evolution of the atomic density operator restricted to the levels of interest is of the general form

$$\frac{\mathrm{d}\hat{\rho}}{\mathrm{d}t} = -\frac{\mathrm{i}}{\hbar} \left(\hat{H}_{\mathrm{eff}}\hat{\rho} - \hat{\rho}\hat{H}_{\mathrm{eff}}^{\dagger} \right) , \qquad (1.75)$$

where

$$\hat{H}_{\mathrm{eff}} = \hat{H} + \hat{\Gamma} , \qquad (1.76)$$

$\hat{H}$ being the atom–field Hamiltonian and $\hat{\Gamma}$ a non-Hermitian relaxation operator defined by its matrix elements

$$\langle n|\hat{\Gamma}|m\rangle = \frac{\hbar}{2\mathrm{i}} \gamma_n \delta_{nm} . \qquad (1.77)$$

Both inelastic collisions and spontaneous emission to unobserved levels can be described by this form of evolution.

Relaxation Toward Levels of Interest As will be discussed in detail in Chap. 5, a description of the atomic dynamics in terms of a so-called *master equation*, or alternatively of quantum Langevin equations, is necessary when all involved levels are observed. This is also the case when a proper account of the coupling of the atom to the full electromagnetic field is needed and/or atomic collisions are involved. For now we limit ourselves to stating without proof an equation that describes the dynamics of a two-level atom subject to upper to lower level spontaneous decay and to elastic or soft collisions, that is, collisions that change the separation of energy levels during the collision but leave the level populations unchanged. In that case, the atomic evolution is no longer governed by a Hermitian Hamiltonian or the simple non-Hermitian Hamiltonian of Eq. (1.75), but rather by a master equation of the form

$$\frac{d\hat{\rho}}{dt} = -\frac{i}{\hbar}[\hat{H}, \hat{\rho}] - \frac{\Gamma}{2}\left(\hat{\sigma}_+\hat{\sigma}_-\hat{\rho} + \hat{\rho}\hat{\sigma}_+\hat{\sigma}_- - 2\hat{\sigma}_-\hat{\rho}\hat{\sigma}_+\right) - \frac{1}{2}\gamma_{ph}\hat{\rho} + 2\gamma_{ph}\hat{\sigma}_z\hat{\rho}\hat{\sigma}_z .$$

$$(1.78)$$

The first term on the right-hand side of this equation accounts for the Hamiltonian dynamics of the system, while the term proportional to Γ accounts for the fact that the atom is coupled to a continuum of modes of the electromagnetic field, which are responsible for spontaneous emission and the irreversible decay of the atom from the excited state $|e\rangle$ to the ground state $|g\rangle$. The free space spontaneous decay rate Γ, whose determination requires a detailed quantum electrodynamics (QED) analysis that will be given in Sect. 5.1, is

$$\Gamma = \frac{1}{4\pi\epsilon_0}\frac{4d^2\omega_0^3}{3\hbar c^3} , \qquad (1.79)$$

Finally, the terms proportional to γ_{ph} account for additional decay rate(s) that the atoms may be subject to, oftentimes as a result of elastic collisions, as discussed, for instance, in Ref. [1].

Optical Bloch Equations with Decay In general, the optical Bloch equations cannot be generalized straightforwardly to cases where relaxation mechanisms are present. There are, however, two notable exceptions corresponding to situations where

1. the upper level spontaneously decays to the lower level only, while the atom undergoes only elastic collisions and
2. spontaneous emission between the upper and lower levels can be ignored in comparison with decay to unobserved levels, which occurs at equal rates $\gamma_e = \gamma_g = 1/T_1$.

Under these conditions, the Bloch equations generalize to

$$\frac{dU}{dt} = -U/T_2 - \Delta V$$

$$\frac{dV}{dt} = -V/T_2 + \Delta U + \Omega_r W$$

$$\frac{dW}{dt} = -(W - W_{eq})/T_1 - \Omega_r V , \qquad (1.80)$$

where we have introduced the longitudinal and transverse relaxation times T_1 and T_2, with $T_1 = 1/\Gamma$ and $T_2 = (1/2T_1 + \gamma_{ph})^{-1}$ in the first case and $T_2 = (1/T_1 + \gamma_{ph})^{-1}$ in the second case. The equilibrium inversion W_{eq} is equal to zero in the second case since the decay is to unobserved levels.

1.3.4 Density Matrix Equations

In the general case, it is necessary to revert to the master equation (5.31) that results from a proper analysis of the coupling of the two-level atom to a reservoir instead of the optical Bloch equations (1.80). The equations of motion for the matrix elements of $\hat{\rho}$ become then, for the general case of a complex Rabi frequency Ω_r,

$$\frac{d\rho_{ee}}{dt} = -\gamma_e \rho_{ee} - \frac{1}{2}\left(i\Omega_r^* \rho_{eg} + \text{c.c.}\right)$$

$$\frac{d\rho_{gg}}{dt} = -\gamma_g \rho_{gg} + \frac{1}{2}\left(i\Omega_r^* \rho_{eg} + \text{c.c.}\right)$$

$$\frac{d\rho_{eg}}{dt} = -(\gamma + i\Delta)\rho_{eg} - i\frac{\Omega_r}{2}\left(\rho_{ee} - \rho_{gg}\right), \qquad (1.81)$$

where $\gamma = (\gamma_e + \gamma_g)/2 + \gamma_{ph}$ and $\tilde{\rho}_{eg} = \rho_{eg}\, e^{i\omega t}$. In the case of spontaneous decay from the upper to the lower level, these equations become

$$\frac{d\rho_{ee}}{dt} = -\Gamma\rho_{ee} - \frac{1}{2}\left(i\Omega_r^* \rho_{eg} + \text{c.c.}\right)$$

$$\frac{d\rho_{gg}}{dt} = +\Gamma\rho_{ee} + \frac{1}{2}\left(i\Omega_r^* \rho_{eg} + \text{c.c.}\right)$$

$$\frac{d\rho_{eg}}{dt} = -(\gamma + i\Delta)\rho_{eg} - i\frac{\Omega_r}{2}\left(\rho_{ee} - \rho_{gg}\right), \qquad (1.82)$$

where $\gamma = \Gamma/2 + \gamma_{ph}$.

Rate Equation Approximation If the coherence decay rate γ is dominated by elastic collisions and hence is much larger than the population decay rates γ_e and γ_g, ρ_{eg} can be adiabatically eliminated from the equations of motion (1.81) and (1.82) to obtain the rate equations

$$\frac{d\rho_{ee}}{dt} = -\gamma_e \rho_{ee} - R\left(\rho_{ee} - \rho_{gg}\right),$$

$$\frac{d\rho_{gg}}{dt} = -\gamma_g \rho_{gg} + R\left(\rho_{ee} - \rho_{gg}\right), \qquad (1.83)$$

and

$$\frac{d\rho_{ee}}{dt} = -\Gamma\rho_{ee} - R(\rho_{ee} - \rho_{gg}),$$

$$\frac{d\rho_{gg}}{dt} = +\Gamma\rho_{gg} + R(\rho_{ee} - \rho_{gg}), \tag{1.84}$$

respectively, where the transition rate is

$$R = |\Omega_r|^2 \mathcal{L}(\Delta)/(2\gamma), \tag{1.85}$$

and we have introduced the dimensionless Lorentzian

$$\mathcal{L}(\Delta) = \frac{\gamma^2}{\gamma^2 + \Delta^2}. \tag{1.86}$$

The transitions between the upper and lower states are thus described in terms of simple rate equations.

Adding phenomenological pumping rates Λ_e and Λ_g on the right-hand side of these pairs of equations provides a description of the excitation of the upper and lower levels from some distant levels, as would be the case in a laser. The equations then form the basis of conventional, single-mode laser theory.

Steady State In the absence of additional external processes, often referred to as pump mechanisms, that repopulate the atomic levels, the populations ρ_{ee} and ρ_{gg} eventually decay away for the case of decay to unobserved levels, while for the case of upper to lower level decay, they reach a steady state with corresponding inversion

$$W_{st} = -\frac{\Gamma}{\Gamma + 2R} = -\frac{1}{1 + s}, \tag{1.87}$$

where s is the *saturation parameter*. In the case of pure radiative decay, $\gamma_{ph} = 0$, s is given by

$$s = \frac{\Omega_r^2/2}{\Gamma^2/4 + \Delta^2}. \tag{1.88}$$

The inversion W_{st}, which equals -1 for $\Omega_r = 0$, first increases quadratically and asymptotically approaches $W_{st} = 0$ as $\Omega_r \to \infty$. At this point, where the upper and lower state populations are equal, the transition is said to be saturated, and the medium becomes effectively transparent or bleached. The inversion is always negative, which means in particular that no steady-state light amplification can be achieved in this system. This is one reason why external pump mechanisms are required in lasers.

In steady state, the other two components of the Bloch vector **U** are given by

$$U_{\text{st}} = -\frac{2\Delta}{\Omega_r}\left(\frac{s}{1+s}\right) \tag{1.89}$$

and

$$V_{\text{st}} = \frac{\Gamma}{\Omega_r}\left(\frac{s}{1+s}\right). \tag{1.90}$$

U_{st} varies as a dispersion curve as a function of the detuning Δ, while V_{st} is a Lorentzian of power-broadened half-width at half maximum $\left(\Gamma^2/4 + \Omega_r^2\right)^{1/2}$. As the intensity of the driving field, or Ω_r^2, increases, U_{st} and V_{st} first increase linearly with Ω_r, reach a maximum, and finally tend to zero as $\Omega_r \to \infty$.

Einstein's A and B Coefficients When atoms interact with broadband radiation instead of the monochromatic fields considered so far, the rate equations still apply, but the rate R becomes

$$R \to B_{eg}\varrho(\omega), \tag{1.91}$$

where $\varrho(\omega)$ is the spectral energy density of the inducing radiation. Einstein's A and B coefficients apply to an atom in thermal equilibrium with the field, which is described by Planck's blackbody radiation

$$\varrho(\omega) = \frac{\hbar\omega^3}{\pi^2 c^3}\frac{1}{e^{\hbar\omega/k_B T} - 1}, \tag{1.92}$$

where T is the temperature of the source and k_B is Boltzmann's constant. Invoking the principle of detailed balance, which states that in thermal equilibrium the average number of transitions $|i\rangle \to |k\rangle$ between arbitrary states $|i\rangle$ and $|k\rangle$ must be equal to the number of transitions $|k\rangle \to |i\rangle$, one finds

$$\frac{A_{eg}}{B_{eg}} = \frac{\hbar\omega^3}{\pi^2 c^3}, \tag{1.93}$$

where A_{eg} is the rate of spontaneous emission Γ from $|e\rangle$ to $|g\rangle$ of Eq. (1.79), in the notation traditionally used when discussing the Einstein A and B coefficients.

Problems

Problem 1.1 The fine structure constant $\alpha = e^2/4\pi\epsilon_0\hbar c$ is one of the most important constants in physics. In addition to having a rich history, it has several physical interpretations that are worth thinking about. With all this in mind, consider how it is related to (a) the classical electron radius r_e, the Compton wavelength of the electron λ and the Bohr radius a_0, (b) to the electrostatic of two electrons a distance d apart and the energy of a photon of wavelength $2\pi d$, (c) to the velocity of an electron on its lowest energy orbit in the Bohr model of the atom and the velocity of light, and (d) to the elementary charge e and the Planck charge.

Problem 1.2 Derive the slowly varying expression for the polarization $\mathcal{P}(t)$ of the Lorentz atom to obtain the full expression of the slowly varying Maxwell wave equation (1.40).

Problem 1.3 Derive the rate equations (1.83) and (1.84), and solve them in steady state in terms of the transition rate R and the saturation parameter s.

Problem 1.4 (Lagrangian and Hamiltonian Formulations)
This and the next three problems address important aspects of the Lagrangian and Hamiltonian formulations of the interaction between charges and electromagnetic fields and their connection to the minimum coupling Hamiltonian. This topic is discussed pedagogically in the text by Cohen-Tannoudji, Dupont-Roc, and Grynberg [2], which gives an excellent discussion of the Hamiltonian approach to electrodynamics, the electric dipole interaction, and the $\mathbf{A} \cdot \mathbf{p}$ vs. $\mathbf{E} \cdot \mathbf{r}$ forms of the electric dipole interaction.

The classical Lagrangian describing the coupling of the electromagnetic field to a collection of charges $\{q_\alpha\}$ of masses $\{m_\alpha\}$ at locations $\mathbf{r}_\alpha$ and with velocities $\dot{\mathbf{r}}_\alpha$ is

$$\mathcal{L} = \frac{1}{2} \sum_\alpha m_\alpha \dot{\mathbf{r}}_\alpha^2 + \frac{\epsilon_0}{2} \int d^3r \left\{ \left[\mathbf{E}(\mathbf{r},t)^2 - c^2 \mathbf{B}(\mathbf{r},t)^2 \right] \right\}$$

$$+ \int d^3r \left\{ \mathbf{J}(\mathbf{r},t) \cdot \mathbf{A}(\mathbf{r},t) - \rho(\mathbf{r},t) U(\mathbf{r},t) \right\}, \tag{1.94}$$

where the first and second terms are the free particle and free field Lagrangian, respectively, and the third term describes their coupling. Here, $\mathbf{A}(\mathbf{r}_\alpha, t)$ is the vector potential, $U(\mathbf{r}, t)$ is the scalar potential, and $\mathbf{E}(\mathbf{r}, t)$ and $\mathbf{B}(\mathbf{r}, t)$ are the electric and magnetic fields, with

$$\mathbf{E}(\mathbf{r},t) = \frac{\partial \mathbf{A}(\mathbf{r},t)}{\partial t} - \nabla U(\mathbf{r},t), \tag{1.95}$$

$$\mathbf{B}(\mathbf{r},t) = \nabla \times \mathbf{A}(\mathbf{r},t). \tag{1.96}$$

Finally, the total charge is

$$\rho(\mathbf{r},t) = \sum_\alpha q_\alpha \delta(\mathbf{r} - \mathbf{r}_\alpha(t)), \tag{1.97}$$

and the current is

$$J(\mathbf{r}, t) = \sum_\alpha \dot{\mathbf{r}}_\alpha(t)\delta(\mathbf{r} - \mathbf{r}_\alpha(t)). \tag{1.98}$$

Show that when applied to this Lagrangian, the Euler–Lagrange equations of motion

$$\frac{\partial \mathcal{L}}{\partial F_i} - \nabla \frac{\partial \mathcal{L}}{\partial (\nabla F_i)} - \frac{d}{dt}\frac{\partial \mathcal{L}}{\partial \dot{F}_i} = 0, \tag{1.99}$$

where the components F_i of the field $\mathbf{F}$ are the vector potential $\mathbf{A}$ and the scalar potential U, yield the Maxwell equations

$$\nabla \times \mathbf{B} = \frac{1}{c^2}\frac{\partial \mathbf{E}}{\partial t} + \mu_0 \mathbf{J}, \tag{1.100}$$

$$\nabla \cdot \mathbf{E} = \frac{\rho}{\epsilon_0}. \tag{1.101}$$

Show also that the Euler–Lagrange equations of motion for a particle,

$$\frac{\partial \mathcal{L}}{\partial \mathbf{r}_\alpha} - \frac{d}{dt}\frac{\partial \mathcal{L}}{\partial \dot{\mathbf{r}}_\alpha} = 0, \tag{1.102}$$

yield the Lorentz equations of motion

$$m_\alpha \ddot{\mathbf{r}}_\alpha = q_\alpha[\mathbf{E}(\mathbf{r}_\alpha, t) + \dot{\mathbf{r}}_\alpha \times \mathbf{B}(\mathbf{r}_\alpha, t)]. \tag{1.103}$$

Hint: Use the vector identity

$$\nabla(\mathbf{A} \cdot \mathbf{B}) = (\mathbf{B} \cdot \nabla) \cdot \mathbf{A} + (\mathbf{A} \cdot \nabla) \cdot \mathbf{B} + \mathbf{B} \times (\nabla \times \mathbf{A}) + \mathbf{A} \times (\nabla \times \mathbf{B}). \tag{1.104}$$

Problem 1.5 Starting from the Lagrangian (1.94) of Problem (1.4), show that the conjugate momenta $\mathbf{p}_\alpha$ of $\mathbf{r}_\alpha$ and $\mathbf{\Pi}$ of $\mathbf{A}$ are

$$\mathbf{p}_\alpha = \frac{\partial \mathcal{L}}{\partial \dot{\mathbf{r}}_\alpha} = m_\alpha \dot{\mathbf{r}}_\alpha + q_\alpha \mathbf{A}(\mathbf{r}_\alpha, t) \tag{1.105}$$

and

$$\mathbf{\Pi} = \frac{\partial \mathcal{L}}{\partial \dot{\mathbf{A}}} = \epsilon_0 \mathbf{E}(\mathbf{r}, t). \tag{1.106}$$

What is the conjugate momentum of the scalar potential U?

From these results, show that the Hamiltonian corresponding to that Lagrangian is the minimum coupling Hamiltonian

$$H = \sum_{\alpha} [\mathbf{p}_\alpha - q_\alpha \mathbf{A}(\mathbf{r}_\alpha, t)]^2 + \sum_{\alpha} q_\alpha U(\mathbf{r}_\alpha, t)$$

$$+ \epsilon_0 \int d^3 r [\mathbf{E}^2(\mathbf{r}, t) + c^2 \mathbf{B}^2(\mathbf{r}, t)] + \mathbf{E}(\mathbf{r}, t) \cdot \nabla U(\mathbf{r}, t), \quad (1.107)$$

In the Coulomb gauge

$$U(\mathbf{r}, t) = 0 \qquad\qquad\qquad (1.108)$$

$$\nabla \cdot \mathbf{A}(\mathbf{r}, t) = 0, \qquad\qquad\qquad (1.109)$$

commonly used in quantum optics, it simplifies to

$$H = \sum_{\alpha} \int d^3 r \left\{ \frac{1}{2m_\alpha} [\mathbf{p}_\alpha - q_\alpha \mathbf{A}(\mathbf{r}_\alpha, t)]^2 \delta(\mathbf{r} - \mathbf{r}_\alpha) \right.$$

$$\left. + \epsilon_0 \int d^3 r [\mathbf{E}^2(\mathbf{r}, t) + c^2 \mathbf{B}^2(\mathbf{r}, t)] \right\}, \qquad (1.110)$$

where the second line accounts for the free field part, which is essential to obtain Maxwell's equations.

Problem 1.6 Consider the form of the minimum coupling Hamiltonian (1.110), and use the Hamilton–Jacobi equations of motion

$$\dot{\mathbf{p}} = \frac{\partial H}{\partial \mathbf{r}},$$

$$\dot{\mathbf{r}} = -\frac{\partial H}{\partial \mathbf{p}}, \qquad\qquad\qquad (1.111)$$

to prove that the motion of a particle of charge q is governed by the Lorentz equation

$$m\ddot{\mathbf{r}} = q \left(-\nabla U - \frac{\partial \mathbf{A}}{\partial t} \right) + q\dot{\mathbf{r}} \times (\nabla \times \mathbf{A}), \qquad (1.112)$$

or, with $\mathbf{E} = -\partial \mathbf{A}/\partial t$, $\mathbf{B} = \nabla \times \mathbf{A}$ and $U = 0$,

$$m\ddot{\mathbf{r}} = q\mathbf{E} + q(\dot{\mathbf{r}} \times \mathbf{B}). \qquad\qquad (1.113)$$

Hint: Use the same vector identity as in Problem 1.4.

Problem 1.7 Quantizing the motion of a charge is achieved by promoting its position $\mathbf{r}$ and canonical momentum $\mathbf{p}$ to operators satisfying the canonical commutation relation

$$[\hat{r}_i, \hat{p}_j] = i\hbar\delta_{ij} . \tag{1.114}$$

With the coordinate representation form of the canonical momentum $\hat{\mathbf{p}} = -i\hbar\nabla$, and the Hamiltonian (1.110), the Schrödinger equation describing the dynamics of a single charge bound to a nucleus by a potential $\hat{V}(\hat{\mathbf{r}})$ is then

$$i\hbar\frac{\partial\psi(\mathbf{r},t)}{\partial t} = -\frac{\hbar^2}{2m}\left[\nabla - \frac{iq}{\hbar}\mathbf{A}(\mathbf{R},t)\right]^2 + \hat{V}(\mathbf{r}) . \tag{1.115}$$

Show by introducing the new wave function $\phi(\mathbf{r},t)$ via the gauge transformation

$$\psi(\mathbf{r},t) = \exp[(-iq\mathbf{r}/\hbar)\cdot\mathbf{A}(\mathbf{R},t)]\phi(\mathbf{r},t) , \tag{1.116}$$

we find that $\phi(\mathbf{r},t)$ obeys the Schrödinger equation

$$i\hbar\frac{\partial\phi(\mathbf{r},t)}{\partial t} = [\hat{H}_0 - q\hat{\mathbf{r}}\cdot\mathbf{E}(\mathbf{R},t)]\phi(\mathbf{r},t) , \tag{1.117}$$

where $\hat{H}_0 = \hat{\mathbf{p}}^2/2m + V(\hat{\mathbf{r}})$ is the unperturbed Hamiltonian of the charge. This shows that in the electric dipole approximation, the interaction between the electron and the electromagnetic field is described by the electric dipole interaction Hamiltonian (1.14),

$$\hat{V}_{\text{EI}} = -q\,\hat{\mathbf{r}}\cdot\mathbf{E}(\mathbf{R},t),$$

where $\mathbf{R}$ is the position of the center of the mass of the atom. Hint: Remember that in the coordinate representation $[p, f(x)] = -i\hbar f'(x)$ and in the Coulomb gauge, the electric field and the potential vector are related by

$$\mathbf{E}(\mathbf{R},t) = -\frac{\partial\mathbf{A}(\mathbf{R},t)}{\partial t} . \tag{1.118}$$

Problem 1.8 Consider an ensemble of two-level atoms, 30% of which are in the state $1/\sqrt{2}(|e\rangle + |g\rangle)$, 50% are in the state $1/\sqrt{10}(|e\rangle - 3|g\rangle)$, and 20% are in the state $|g\rangle$. Find the density matrix $\hat{\rho}$ of this system, and determine the probability for the atoms to be in the ground state $|g\rangle$. Is this state a pure state or a mixed state? Why?

Problem 1.9 Quantum mechanically, the von Neumann entropy is defined as

$$S = -k_B \text{Tr}\{\hat{\rho}\ln\hat{\rho}\} , \tag{1.119}$$

where $\hat{\rho}$ is the density operator of the system and k_B is Boltzmann's constant. Show that S vanishes for a pure state. What does this mean physically?

Problem 1.10 Calculate $|C_e(t)|^2$ for an atom interacting with a resonant but incoherent light source characterized by the intensity fluctuation function $P(I)$ given by

$$P(I) = \frac{1}{I_0} e^{-I/I_0}$$

in the absence of relaxation mechanisms, that is, assuming that $\gamma_e = \gamma_g = 0$. Assume also that the field phase vanishes, so that $E = \sqrt{I}$. Discuss the results as $t \to \infty$.

Problem 1.11 Consider a two-level atom subject to the Hamiltonian (1.60) and initially in its ground state $|g\rangle$, with a Rabi frequency $\Omega_r(t)$ that is constant for a time τ and zero otherwise and some detuning Δ. Under what condition(s) will the system evolve to (a) the state $|\psi\rangle = (1/\sqrt{2})[|g\rangle + |e\rangle]$ and (b) the state $|\psi\rangle = |e\rangle$. A pulse that permits to achieve the first state is called a $\pi/2$-pulse, and a pulse that permits to reach the second state is called a π-pulse.[3] Why?

Problem 1.12 (a) Derive the rate equations (1.83) by adiabatically eliminating ρ_{eg} from the density operator equations (1.81), and solve these equations in steady state. (b) Do the same for the rate equations (1.84). (c) Discuss physically the difference in the steady-state solutions for these two systems.

References

1. P. Meystre, M. Sargent III, *Elements of Quantum Optics*, 4th edn. (Springer, Berlin, 2007)
2. C. Cohen-Tannoudji, J. Dupont-Roc, G. Grynberg, *Photons and Atoms: Introduction to Quantum Electrodynamics* (Wiley-Interscience, New York, 1989)
3. B. Shore, *The Theory of Coherent Atomic Excitation* (Wiley-Interscience, New York, 1990)
4. J.D. Jackson, *Classical Electrodynamics*, 3rd edn. (Wiley, New York, 1999)
5. L. Allen, J.H. Eberly, *Optical Resonance and Two-Level Atoms* (Dover, New York, 1987)
6. M. Born, E. Wolf, *Principles of Optics* (Pergamon, Oxford, 1970)
7. L. Mandel, E. Wolf, *Optical Coherence and Quantum Optics* (Cambridge University Press, Cambridge, 1995)
8. S. Zubairy, M.O. Scully, *Quantum Optics* (Cambridge University Press, Cambridge, 1997)
9. Y.R. Shen, *Principles of Nonlinear Optics* (Wiley, New York, 1984)

---

[3] π-pulses and $\pi/2$ pulses play an important role in quantum optics. In particular, we will see in Chap. 8 how they can be exploited in the design of atom interferometers.

10. M. Orszag, *Quantum Optics: Including Noise Reduction, Trapped Ions, Quantum Trajectories, and Decoherence* (Springer, Berlin, 2016)
11. W.P. Schleich, *Quantum Optics in Phase Space* (Wiley VCH, Weinheim, 2001)
12. C. Cohen-Tannoudji, J. Dupont-Roc, G. Grynberg, *Atom-Photon Interactions: Basic Processes and Applications* (Wiley-Interscience, New York, 1992)

Chapter 2
Electromagnetic Field Quantization

Following a brief review of the quantization of the simple harmonic oscillator starting from the Lagrangian formalism, we use a formal analogy to quantize the single-mode electromagnetic field along the same lines. We then extend the analysis to multimode fields and consider a number of states of the field of importance for quantum optics before discussing two important ways to characterize them, photocounting and homodyne detection. The chapter concludes with a discussion of quasiprobability distribution function descriptions of the field.

This chapter starts with a brief review of the quantization of the simple harmonic oscillator, starting from its Lagrangian and introducing conjugate variables to derive the corresponding Hamiltonian and proceed with its canonical quantization. We then exploit this result to carry out the quantization first of a single-mode electromagnetic field and then of a multimode field, relying on a formal analogy between their classical Hamiltonians. We also clarify a small but important difference between field quantization in terms of running and standing modes. We then analyze a few specific examples of states of the field of particular interest in quantum optics, most importantly thermal fields, coherent states, and squeezed states. This is followed by a review of two important models of photodetection, photocounting and balanced homodyne detection. They illustrate in particular the physical importance of normally ordered field correlation functions and of the field quadratures. We conclude with a discussion of three useful descriptions of the electromagnetic field in terms quasiprobability functions, the P-representation, the Husimi function, and the Wigner distribution.

2.1 Quantum Harmonic Oscillator

The harmonic oscillator is a central ingredient of physics in general and of quantum optics in particular, where it appears in a number of guises, including in particular the description of electromagnetic fields, phonon modes, and Schrödinger

© The Author(s), under exclusive license to Springer Nature Switzerland AG 2021   29
P. Meystre, *Quantum Optics*, Graduate Texts in Physics,
https://doi.org/10.1007/978-3-030-76183-7_2

matter wave fields. Because of that central role, it is perhaps useful to review its quantization starting from its classical Lagrangian, rather than from its classical Hamiltonian as is sometimes the case. This may help in refreshing our memory on the standard way in which conjugate momenta and canonical commutation relations are introduced and will also prove useful when we carry out the quantization of LC circuits and transmission line resonators in Chap. 6.

The Lagrangian of the simple harmonic oscillator with position q, mass m, and spring constant k is the difference between its kinetic and potential energies,

$$\mathcal{L} = \frac{1}{2}m\dot{q}^2 - \frac{1}{2}kq^2 , \tag{2.1}$$

from which the Euler–Lagrange equation

$$\frac{\partial \mathcal{L}}{\partial q} - \frac{d}{dt}\frac{\partial \mathcal{L}}{\partial \dot{q}} = 0 \tag{2.2}$$

gives the familiar oscillator equation of motion

$$\ddot{q} + \omega^2 q = 0, \tag{2.3}$$

where $\omega = \sqrt{k/m}$. The corresponding system Hamiltonian H is obtained by first determining the momentum conjugate to q via

$$p = \frac{\partial \mathcal{L}}{\partial \dot{q}} = m\dot{q} , \tag{2.4}$$

and H is then defined as

$$H = p\dot{q} - \mathcal{L} = \frac{p^2}{2m} + \frac{1}{2}m\omega^2 q^2 , \tag{2.5}$$

with Hamilton equations of motion

$$\dot{q} = \frac{\partial H}{\partial p} = \frac{p}{m} , \tag{2.6}$$

$$\dot{p} = -\frac{\partial H}{\partial q} = -m\omega^2 q = -kq . \tag{2.7}$$

The quantization of the oscillator is completed by promoting q and p to quantum operators with canonical commutation relation

$$[\hat{q}, \hat{p}] = i\hbar \tag{2.8}$$

and associated Heisenberg uncertainty relation

$$\Delta p \Delta q \geq 1/2 |\langle [\hat{p}, \hat{q}] \rangle| = \hbar/2 \tag{2.9}$$

between their standard deviations $\Delta p = \sqrt{\langle \hat{p}^2 \rangle - \langle \hat{p} \rangle^2}$ and $\Delta q = \sqrt{\langle \hat{q}^2 \rangle - \langle \hat{q} \rangle^2}$, so that

$$\hat{H} = \frac{\hat{p}^2}{2m} + \frac{1}{2} m \omega^2 \hat{q}^2 . \tag{2.10}$$

In the coordinate representation, we have

$$\hat{p} = -i\hbar \frac{d}{d\hat{q}} , \tag{2.11}$$

so that $\hat{H}$ takes the form

$$\hat{H} = -\frac{\hbar^2}{2m} \frac{d^2}{d\hat{q}^2} + \frac{m}{2} \omega^2 \hat{q}^2. \tag{2.12}$$

Its eigenfunctions $u_n(q)$ are well known. They may be expressed in terms of Hermite polynomials $H_n(\alpha q)$ as

$$u_n(q) = \sqrt{\frac{\alpha}{\sqrt{\pi} 2^n n!}} H_n(\alpha q) \exp(-\alpha^2 q^2/2) , \tag{2.13}$$

where $\alpha = \sqrt{m\omega/\hbar}$, see Fig. 2.1, with corresponding eigenenergies

$$\hbar \omega_n = \hbar \omega (n + \tfrac{1}{2}) , \ n = 0, 1, 2, \dots \tag{2.14}$$

Fig. 2.1 First four eigenstates $n = 0 \dots 3$ of the one-dimensional harmonic oscillator, with eigenenergies $E_n = (n + \frac{1}{2})\hbar\omega$

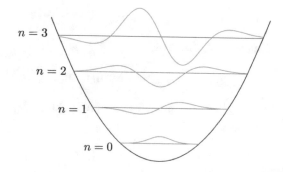

We now introduce two new operators $\hat{a}$ and $\hat{a}^\dagger$ defined as

$$\hat{a} = \sqrt{\frac{m\omega}{2\hbar}}\left(\hat{q} + \frac{\mathrm{i}}{m\omega}\hat{p}\right),\tag{2.15}$$

$$\hat{a}^\dagger = \sqrt{\frac{m\omega}{2\hbar}}\left(\hat{q} - \frac{\mathrm{i}}{m\omega}\hat{p}\right),\tag{2.16}$$

which are readily seen from Eq. (2.8) to obey the boson commutation relation

$$[\hat{a}, \hat{a}^\dagger] = 1.\tag{2.17}$$

Inverting these expressions, we find the position and momentum

$$\hat{q} = \sqrt{\frac{\hbar}{2m\omega}}(\hat{a} + \hat{a}^\dagger) \equiv q_{\mathrm{zpf}}(\hat{a} + \hat{a}^\dagger),\tag{2.18}$$

$$\hat{p} = \mathrm{i}\sqrt{\frac{\hbar m\omega}{2}}(\hat{a} - \hat{a}^\dagger) \equiv \mathrm{i}p_{\mathrm{zpf}}(\hat{a} - \hat{a}^\dagger),\tag{2.19}$$

where q_{zpf} and q_{zpf} are called the zero-point position and momentum, so that in terms of $\hat{a}$ and $\hat{a}^\dagger$, $\hat{H}$ becomes

$$\hat{H} = \hbar\omega\left(\hat{a}^\dagger\hat{a} + \tfrac{1}{2}\right).\tag{2.20}$$

In the Heisenberg picture, the time evolution of $\hat{a}$ is easily found to be

$$\frac{\mathrm{d}\hat{a}}{\mathrm{d}t} = \frac{\mathrm{i}}{\hbar}[\hat{H}, \hat{a}] = -\mathrm{i}\omega\hat{a}\tag{2.21}$$

with solution

$$\hat{a}(t) = \hat{a}(0)\,\mathrm{e}^{-\mathrm{i}\omega t}.\tag{2.22}$$

Similarly, we find that

$$\hat{a}^\dagger(t) = \hat{a}^\dagger(0)\,\mathrm{e}^{\mathrm{i}\omega t}.\tag{2.23}$$

Consider now an energy eigenstate $|H\rangle$ of the harmonic oscillator with eigenvalue $\hbar\Omega$,

$$\hat{H}|H\rangle = \hbar\Omega|H\rangle,\tag{2.24}$$

and evaluate the energy of the state $|H'\rangle = \hat{a}|H\rangle$. From Eq. (2.17), we have $\hat{H}\hat{a} = \hat{a}\hat{H} - \hbar\omega\hat{a}$ so that

$$\hat{H}\hat{a}|H\rangle = \hat{H}|H\rangle - \hbar\omega\hat{a}|H\rangle = \hbar(\Omega - \omega)\hat{a}|H\rangle ; \tag{2.25}$$

that is, $\hat{a}|H\rangle$ is again an eigenstate of the Hamiltonian, but with eigenenergy $\hbar\omega$ lower than $|H\rangle$. Because $\hat{a}$ lowers the energy, it is called an *annihilation operator*. Repeating the operation m times, we find

$$\hat{H}\hat{a}^m|H\rangle = \hbar(\Omega - m\omega)\hat{a}^m|H\rangle. \tag{2.26}$$

We can see that the lowest of these eigenvalues is positive as follows. For an arbitrary normalized vector $|\phi\rangle$, the expectation value of $\hat{H}$ is

$$\hbar\omega\langle\phi|a^\dagger\hat{a} + \tfrac{1}{2}|\phi\rangle = \hbar\omega\langle\phi'|\phi'\rangle + \hbar\omega/2 > 0, \tag{2.27}$$

where $|\phi'\rangle = \hat{a}|\phi\rangle$. Calling the lowest eigenvalue $\hbar\omega_0$ with eigenstate $|0\rangle$, we have

$$\hat{a}|0\rangle = 0 \tag{2.28}$$

and

$$\hat{H}|0\rangle = \hbar\omega(\hat{a}^\dagger\hat{a} + \tfrac{1}{2})|0\rangle = \hbar\omega_0|0\rangle = \tfrac{1}{2}\hbar\omega|0\rangle , \tag{2.29}$$

that is, the lowest energy eigenvalue $\hbar\omega_0 = \hbar\omega/2$.
Similarly, we find

$$\hat{H}\hat{a}^\dagger|0\rangle = [\hat{a}^\dagger\hat{H} + \hbar\omega\hat{a}^\dagger]|0\rangle = \hbar\omega(1 + \tfrac{1}{2})\hat{a}^\dagger|0\rangle ; \tag{2.30}$$

that is, the eigenstate $|1\rangle$ has eigenvalue $\hbar\omega(1 + \tfrac{1}{2})$. Because $\hat{a}^\dagger$ raises the energy, it is called a *creation operator*. Substituting successively higher eigenstates into this equation, we find

$$\hat{H}(\hat{a}^\dagger)^n|0\rangle = \hbar\omega(n + \tfrac{1}{2})(\hat{a}^\dagger)^n|0\rangle, \tag{2.31}$$

and hence the eigenstate $|n\rangle \propto (\hat{a}^\dagger)^n|0\rangle$ has the eigenvalue $\hbar\omega(n + \tfrac{1}{2})$. To find the constant of proportionality, we note that

$$\hat{a}|n\rangle = s_n|n - 1\rangle , \tag{2.32}$$

where s_n is some scalar. This implies that

$$\langle n|\hat{a}^\dagger\hat{a}|n\rangle = |s_n|^2\langle n - 1|n - 1\rangle = |s_n|^2 . \tag{2.33}$$

Since $\hat{a}^\dagger \hat{a}|n\rangle = n|n\rangle$, this gives $s_n = \sqrt{n}$. Thus,

$$\hat{a}|n\rangle = \sqrt{n}|n-1\rangle$$
$$\hat{a}^\dagger|n\rangle = \sqrt{n+1}|n+1\rangle, \tag{2.34}$$

yielding the normalized eigenstates

$$|n\rangle = \frac{1}{\sqrt{n!}}(\hat{a}^\dagger)^n|0\rangle . \tag{2.35}$$

Since $\hat{a}^\dagger \hat{a}|n\rangle = n|n\rangle$, $\hat{a}^\dagger a$ is called the *number operator*. It gives the number of quanta of excitation of the harmonic oscillator in the states $\{|n\rangle\}$, which are called number states or Fock states.

We can obtain the coordinate representation $u_0(q)$ of the ground state $|0\rangle$ by substituting the definition (2.15) of $\hat{a}$ into Eq. (2.28) to find

$$(m\omega\hat{q} + i\hat{p})u_0(q) = 0. \tag{2.36}$$

With $\hat{p} = -i\hbar(d/d\hat{q})$, Eq. (2.11), this gives

$$\frac{d}{dq}u_0(q) = -\left(\frac{m\omega}{\hbar}\right)q\, u_0(q), \tag{2.37}$$

which has the normalized solution

$$u_0(q) = \left(\frac{m\omega}{\pi\hbar}\right)^{1/4} e^{-(m\omega/2\hbar)q^2}. \tag{2.38}$$

Similarly, we have

$$u_n(q) = \frac{1}{\sqrt{n!}}(a^\dagger)^n u_0(q) = \frac{1}{\sqrt{n!}}\left(\frac{m\omega}{2\hbar}\right)^{n/2}\left(q - \frac{\hbar}{m\omega}\frac{d}{dq}\right)^n u_0(q), \tag{2.39}$$

which yields the wave function (2.13).

2.2   Electromagnetic Field Quantization

2.2.1   Single-Mode Field

To quantize the electromagnetic field, we consider a cavity of volume V closed by perfectly reflecting mirrors, see Fig. 2.2. For problems in free space, we can then

Fig. 2.2 One-dimensional cavity of length L closed by end mirrors of reflectivity $R = 1$, supporting field modes of wavelength $\lambda = 2L/n$, with n integer

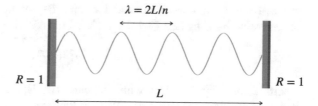

take this volume to be infinite at the end of the calculation. The electromagnetic field satisfies Maxwell's equations in a vacuum,

$$\nabla \cdot \mathbf{B} = 0, \tag{2.40}$$

$$\nabla \cdot \mathbf{E} = 0, \tag{2.41}$$

$$\nabla \times \mathbf{E} = -\frac{\partial \mathbf{B}}{\partial t}, \tag{2.42}$$

$$\nabla \times \mathbf{B} = \mu_0 \epsilon_0 \frac{\partial \mathbf{E}}{\partial t}, \tag{2.43}$$

where $\mathbf{E}$ is the electric field, $\mathbf{B}$ is the magnetic field, and μ_0 and ϵ_0 are the permeability and the permittivity of free space, respectively. Alternatively, it is useful to substitute c^2 for $1/\mu_0\epsilon_0$, where c is the speed of light in vacuum. The electric and magnetic components of a classical monochromatic, single-mode electromagnetic field propagating along z and polarized in the $\vec{x}$ direction can conveniently be cast in the form

$$\mathbf{E}(z, t) = \vec{x}\sqrt{\frac{2\omega^2}{\epsilon_0 V}}q(t)\sin(Kz), \tag{2.44}$$

$$\mathbf{B}(z, t) = \frac{\vec{y}}{c^2 K}\sqrt{\frac{2\omega^2}{\epsilon_0 V}}\dot{q}(t)\cos(Kz), \tag{2.45}$$

where ω is the single-mode field oscillation frequency, $K = \omega/c$ is the wave number ω/c, $q(t)$ is a measure of the field amplitude, and $\vec{x}$ and $\vec{y}$ are unit vectors in directions x and y perpendicular to the field propagation direction z. The classical electromagnetic energy density is given by

$$U = \frac{1}{2}\left[\epsilon_0 E^2 + B^2/\mu_0\right], \tag{2.46}$$

with corresponding classical Hamiltonian

$$H = \frac{1}{2}\int_V dV[\epsilon_0 E^2 + B^2/\mu_0], \tag{2.47}$$

where dV is a volume element and E and B are the magnitudes of $\mathbf{E}$ and $\mathbf{B}$, respectively. Inserting Eqs. (2.44) and (2.45) into H gives

$$H = \tfrac{1}{2}(\omega^2 q^2 + p^2), \tag{2.48}$$

which is formally identical with the classical Hamiltonian (2.5) for a simple harmonic oscillator of unit mass. We can therefore immediately quantize a single mode of the electromagnetic field by applying the results of the previous section on the quantization of the simple harmonic oscillator. In terms of the annihilation and creation operators $\hat{a}$ and $\hat{a}^\dagger$, the single-mode electromagnetic field Hamiltonian is therefore

$$\hat{H}_\omega = \hbar\omega(\hat{a}^\dagger\hat{a} + \tfrac{1}{2}). \tag{2.49}$$

The corresponding eigenstates $|n\rangle$ of the field satisfy

$$\hat{H}|n\rangle = \hbar\omega(n + \tfrac{1}{2})|n\rangle, \qquad n = 0, 1, 2, \ldots, \tag{2.50}$$

where n may be loosely interpreted as the "number of photons" in the state $|n\rangle$, and the ground state $|0\rangle$ of the field mode is referred to as the *vacuum state*. General state vectors of the field are linear superpositions of these energy eigenstates

$$|\psi\rangle = \sum_n c_n |n\rangle. \tag{2.51}$$

Substituting the expression (2.18) of the position operator in terms of creation and annihilation operators into Eq. (2.44) with $m = 1$ gives for the electric and magnetic fields

$$\hat{E}(z, t) = \mathcal{E}_\omega(\hat{a} + \hat{a}^\dagger)\sin Kz, \tag{2.52}$$

$$\hat{B}(z, t) = i\frac{\mathcal{E}_\omega}{c}(\hat{a} - \hat{a}^\dagger)\cos Kz, \tag{2.53}$$

where we have introduced the "electric field per photon"

$$\mathcal{E}_\omega \equiv [\hbar\omega/\epsilon_0 V]^{1/2}. \tag{2.54}$$

In complete analogy with classical fields, see Eq. (1.36), it is often convenient to decompose $\hat{E}(z, t)$ into its positive and negative frequency components as

$$\hat{E}(z, t) = \hat{E}^+(z, t) + \hat{E}^-(z, t) \tag{2.55}$$

with

$$\hat{E}^+(z,t) = \mathcal{E}_\omega \hat{a} e^{-i\omega t} \sin Kz \,, \tag{2.56}$$

$$\hat{E}^-(z,t) = (\hat{E}^+)^\dagger = \mathcal{E}_\omega \hat{a}^\dagger e^{i\omega t} \sin Kz \,, \tag{2.57}$$

and where we have used Eqs. (2.22) and (2.23) for the free time evolution of $\hat{a}$ and $\hat{a}^\dagger$, respectively.

2.2.2 Multimode Field

The generalization of the single-mode analysis to a multimode electromagnetic field is straightforward. Considering for concreteness linearly polarized plane wave modes, the classical electric field becomes

$$E(z,t) = \sum_s \vec{\epsilon}_s \sqrt{\frac{2\omega_s^2}{\epsilon_0 V}} q_s(t) \sin(K_s z) \,, \tag{2.58}$$

where $\vec{\epsilon}_s$ is the polarization of mode s, $\omega_s = cK_s$, $K_s = s\pi/L, s = 1, 2, 3, \ldots$, and L is the length of the cavity in the $\hat{z}$ direction. The Hamiltonian $\hat{H}$ of such a multimode field is simply the sum of the Hamiltonians $\hat{H}_s$ of the individual modes

$$\hat{H} = \sum_s \hat{H}_s \,, \tag{2.59}$$

where $\hat{H}_s$ is given in terms of the single-mode annihilation and creation operators $\hat{a}_s$ and $\hat{a}_s^\dagger$ of Eqs. (2.15) and (2.16) by

$$\hat{H}_s = \hbar \omega_s (\hat{a}_s^\dagger \hat{a}_s + \tfrac{1}{2}) \,, \tag{2.60}$$

with

$$[\hat{a}_s, \hat{a}_{s'}^\dagger] = \delta_{s,s'} \,. \tag{2.61}$$

The single-mode electric field operator generalizes then to

$$\hat{E}(z,t) = \sum_s \mathcal{E}_s (\hat{a}_s + \hat{a}_s^\dagger) \sin(K_s z) \,, \tag{2.62}$$

where

$$\mathcal{E}_s = [\hbar \omega_s / \epsilon_0 V]^{1/2} \tag{2.63}$$

is the electric field per photon of mode s for standing wave quantization.

The eigenstates of the multimode Hamiltonian (2.60) are products of the single-mode eigenstates

$$|n_1 n_2 \ldots n_s \ldots\rangle \equiv |\{n\}\rangle \tag{2.64}$$

and have eigenvalues given by the eigenvalue equation

$$\hat{H}|\{n\}\rangle = \hbar \sum_s \omega_s \left(n_s + \tfrac{1}{2}\right) |\{n\}\rangle . \tag{2.65}$$

When acting on the state $|\{n\}\rangle$, the creation and annihilation operators $\hat{a}_s^\dagger$ and $\hat{a}_s$ of the sth mode give

$$\hat{a}_s^\dagger |\{n\}\rangle = \sqrt{n_s + 1}|n_1 n_2 \ldots n_s + 1 \ldots\rangle ,$$
$$\hat{a}_s |\{n\}\rangle = \sqrt{n_s}|n_1 n_2 \ldots n_s - 1 \ldots\rangle ; \tag{2.66}$$

for example,

$$\hat{a}_1 |n_1, n_2\rangle = \sqrt{n_1}|n_1 - 1, n_2\rangle \quad ; \quad \hat{a}_2 |n_1, n_2\rangle = \sqrt{n_2}|n_1, n_2 - 1\rangle . \tag{2.67}$$

The general state vector of a multimode field is a linear superposition of states such as $|\{n\}\rangle$,

$$|\psi\rangle = \sum_{n_1} \sum_{n_2} \cdots \sum_{\{n_s\}} c_{n_1 n_2 \ldots n_s \ldots} |n_1 n_2 \ldots n_s \ldots\rangle \equiv \sum_{\{n\}} c_{\{n\}} |\{n\}\rangle . \tag{2.68}$$

Note that this form is more general than the factorized state

$$|\psi\rangle = |\psi_1\rangle|\psi_2\rangle \ldots |\psi_s\rangle \ldots \tag{2.69}$$

where the $|\psi_s\rangle$ are state vectors for the individual modes.

For the field ground state $|\{n\}\rangle = |\{0\}\rangle$, Eq. (2.65) reduces to

$$\hat{H}|\{0\}\rangle = \frac{1}{2}\hbar \sum_s \omega_s |\{0\}\rangle , \tag{2.70}$$

so that the ground state energy of the field, the so-called *zero-point energy*, is $\sum_s \frac{1}{2}\hbar\omega_s$. Since this energy is a constant for a given mode configuration, it is often possible to just ignore it. However, this is not always the case, and we will encounter in Sect. 7.5 an important example of its effect in the discussion of the attractive Casimir force between two perfectly conducting plates in a vacuum. As we shall see, the origin of that force is this zero-point energy, more precisely the radiation pressure exerted by the vacuum field on the cavity end plates. While this force is minute and can be ignored in most everyday situations, this is not always the case. In particular, it plays an increasing role in a number of nanophotonics and nanoscience applications.

Standing Waves Versus Traveling Waves Both standing and traveling wave modes are routinely used as basis field modes in quantum theories of electrodynamics. To appreciate the difference between them, consider the positive frequency field operator

$$\hat{E}^+(z, t) = \mathcal{E}_r[\hat{a}_1 e^{iKz} + \hat{a}_2 e^{-iKz}]e^{-i\omega t} . \tag{2.71}$$

Here, $\mathcal{E}_r$ is the running wave electric field per photon, and $\hat{a}_i$ and $\hat{a}_i^\dagger$, $i = 1, 2$, are the annihilation operators for two oppositely running wave modes, obeying the boson commutation relations $[\hat{a}_i, \hat{a}_j^\dagger] = \delta_{ij}$. Their action on the state $|n_1, n_2\rangle_r$ describing the two running waves is

$$\hat{a}_1|n_1, n_2\rangle_r = \sqrt{n_1}|n_1 - 1, n_2\rangle_r; \quad \hat{a}_2|n_1, n_2\rangle_r = \sqrt{n_2}|n_1, n_2 - 1\rangle_r . \tag{2.72}$$

Alternatively, in terms of the operators $\hat{a}_c = (\hat{a}_1 + \hat{a}_2)/\sqrt{2}$ and $\hat{a}_s = (\hat{a}_1 - \hat{a}_2)/\sqrt{2}$, the electric field operator (2.71) is given by

$$\hat{E}^+(z, t) = \sqrt{2}\,\mathcal{E}_r[\hat{a}_c \cos(Kz) + i\hat{a}_s \sin(Kz)]e^{-i\omega t} , \tag{2.73}$$

where $\sqrt{2}\,\mathcal{E}_r$ is just the electric field per photon $\mathcal{E}_s$ of Eq. (2.63). The operators $\hat{a}_c$ and $\hat{a}_s$ also obey boson commutation relations but act on standing wave rather than traveling wave modes. Importantly, then, if one chooses to expand the field in terms of running wave modes instead of standing wave modes as done here, the "electric field per photon" $\mathcal{E}_r$ is reduced by a factor of $\sqrt{2}$ compared to its standing wave value,

$$\mathcal{E}_r = \frac{\mathcal{E}_s}{\sqrt{2}} = \sqrt{\frac{\hbar\omega}{2\epsilon_0 V}} . \tag{2.74}$$

The choice of using a standing or running wave quantization scheme is one of mathematical convenience, and both approaches predict the same single-photon transition rates, such as those for spontaneous emission and photoionization, provided that the contributions of all modes are accounted for. In practice, though, there are cases in which the two approaches are not equivalent. Standing wave modes are the natural choice when a field is contained within a two-mirror cavity. In such situations, one can then oftentimes use a single-mode description, but it is then important to keep in mind that one of the two field modes needed to provide a full correspondence with running wave quantization, either the $\sin(Kz)$ or the $\cos(Kz)$ mode, is implicitly assumed to never be excited, due to boundary conditions imposed, say, by a cavity with perfectly reflecting end mirrors. Conversely, traveling waves are for instance the natural choice when the field consists of counter-propagating waves in a three-mirror ring cavity.

In such situations, one can expect differences between the physics described by a single standing mode field and by two counter-propagating running waves of equal frequencies and amplitudes, in contrast with the classical intuition. In particular, for small photon numbers, atoms are diffracted differently by the true standing wave than by the superposition of two waves.[1] We can understand this result intuitively from the following argument: with running waves, it is possible in principle to know which running wave exchanges a unit of momentum with the atom. In contrast, a standing wave is an inseparable quantum unit with zero average momentum. This unity is imposed by the fixed mirrors that establish the standing wave and that act as infinite sinks and sources of momentum. Quantum mechanics forbids one *even in principle* to determine, via a field measurement, "which traveling wave" exchanges momentum with the atom, and hence one expects interference phenomena. Indeed, the atomic diffraction patterns reflect this fundamental difference between a "true" standing wave and a superposition of two running waves [1].

2.3 States of the Field

2.3.1 Single-Mode Field in Thermal Equilibrium

A thermal single-mode field is a field from which we only know the average energy $\langle \hat{H} \rangle$

$$\langle \hat{H} \rangle = \text{Tr}\{\hat{\rho} \hat{H}\} . \tag{2.75}$$

Much as would be the case in classical physics, we can find its state by invoking the maximum entropy principle. Quantum mechanically, this amounts to determining the density operator $\hat{\rho}$ by using the method of Lagrange multipliers to maximize the von Neumann entropy of the mode subject to the constraint (2.75).

Von Neumann Entropy Before proceeding, it may be useful to first give a very brief refresher on the von Neumann entropy. We recall that this entropy is the quantum extension of the classical Gibbs entropy $S_{\text{classical}}$ of a system,

$$S_{\text{classical}} = -k_B \sum_{\ell} P_{\ell} \ln P_{\ell} , \tag{2.76}$$

where P_{ℓ} is the probability of finding the system in the state ℓ. In quantum systems, P_{ℓ} is replaced by the density operator $\hat{\rho}$, and the sum is replaced by the trace,

---

[1] We will discuss atomic diffraction by optical fields in some detail in Chap. 8.

resulting in the *von Neumann entropy*

$$S = -k_B \text{Tr}\{\hat{\rho}\ln\hat{\rho}\} \quad \text{or} \quad S = -\text{Tr}\{\hat{\rho}\log_2\hat{\rho}\}. \tag{2.77}$$

The quantum equivalent of the normalization condition $\Sigma_\ell P_\ell = 1$ is $\text{Tr}\{\hat{\rho}\} = 1$.

Note that the von Neumann entropy (2.77) is defined sometimes with and sometimes without the Boltzmann factor k_B, depending on the context of the problem. In statistical physics and quantum thermodynamics, S usually includes the Boltzmann factor, but in quantum information science it does not, as we shall see in some detail in Chap. 4. Like the Gibbs entropy in the classical world, the von Neumann entropy has an important interpretation in terms of information content: it characterizes the missing information about the system. For this reason, in quantum information science, it is computed on a $\log_2$ basis and measured in bits.

Maximizing the Entropy We now proceed to determine the density operator of a single-mode field in thermal equilibrium by maximizing S under the two constraints (2.75) and $\text{Tr}\hat{\rho} = 1$. We use here its statistical mechanics form with the Boltzmann factor included. Without constraints, we have

$$\delta S \sim -\text{Tr}\left[\frac{\partial}{\partial\hat{\rho}}(\hat{\rho}\ln\hat{\rho})\right]\delta\hat{\rho} = -\text{Tr}\{(1+\ln\hat{\rho})\delta\hat{\rho}\}. \tag{2.78}$$

The constraint (2.75) leads to the extra contribution $\text{Tr}\{\hat{H}\delta\hat{\rho}\} = 0$ and the normalization of $\hat{\rho}$ to $\text{Tr}\{\delta\hat{\rho}\} = 0$, with β and λ the associated Lagrange multipliers. Inserting these into Eq. (2.78) gives

$$\delta S = -\text{Tr}\{(1+\ln\hat{\rho}+\lambda+\beta\hat{H})\delta\hat{\rho}\}. \tag{2.79}$$

Maximizing the entropy requires that $\delta S = 0$ for any $\delta\hat{\rho}$, which yields

$$1 + \ln\hat{\rho} + \lambda + \beta\hat{H} = 0, \tag{2.80}$$

so that

$$\hat{\rho} = e^{-(1+\lambda)}e^{-\beta\hat{H}}. \tag{2.81}$$

We still have to determine the two Lagrange multipliers λ and β. With $\text{Tr}\{\hat{\rho}\} = 1$, Eq. (2.81) gives

$$e^{1+\lambda} = \text{Tr}\{\exp(-\beta\hat{H})\} \equiv Z, \tag{2.82}$$

where Z is the so-called *partition function* of the system. Substituting this definition into Eq. (2.81), then, we have

$$\hat{\rho} = \frac{\exp(-\beta\hat{H})}{\text{Tr}\{\exp(-\beta\hat{H})\}} = \frac{\exp(-\beta\hat{H})}{Z}. \tag{2.83}$$

From classical statistical mechanics, we recognize

$$\beta \equiv 1/k_B T \tag{2.84}$$

as the Boltzmann coefficient, which we use as a definition of the temperature T.

So far, our result is quite general, and the density operator (2.83) describes the thermal equilibrium state of any system with Hamiltonian $\hat{H}$ and subject to the mean energy constraint (2.75).[2] We now specialize it to the case of the simple harmonic oscillator Hamiltonian. We proceed by redefining the zero of the energy scale by removing the "zero-point energy" $\hbar\omega/2$ from $\hat{H}$. Then, the density operator (2.81) becomes

$$\hat{\rho} = \frac{e^{-\beta\hbar\omega\hat{a}^\dagger\hat{a}}}{\text{Tr}\{e^{-\beta\hbar\omega\hat{a}^\dagger\hat{a}}\}}. \tag{2.85}$$

In general, we can expand the field density operator on any complete set of states, in particular on the number states $|n\rangle$ where $\hat{\rho}$ takes the form

$$\hat{\rho} = \sum_{n,m} m|n\rangle\langle n|\hat{\rho}|m\rangle\langle m| = \sum_{n,m} \rho_{nm}|n\rangle\langle m|. \tag{2.86}$$

Noting that $\langle n|\hat{a}^\dagger\hat{a}|m\rangle = n\delta_{nm}$, we obtain the photon number expansion

$$\rho_{nm} = e^{-n\beta\hbar\omega}\Big[\sum_n e^{-n\beta\hbar\omega}\Big]^{-1}\delta_{nm} = e^{-n\beta\hbar\omega}\Big[1 - e^{-\beta\hbar\omega}\Big]\delta_{nm}, \tag{2.87}$$

where we have used the identity

$$\sum_{n=0}^{\infty}\exp(-nx) = 1 - \exp(-x).$$

Hence, we see that the thermal distribution has a diagonal expansion in terms of the photon number states. This diagonality causes the expectation value of the

---

[2]Additional constraints can be accounted in the same way with the introduction of additional Lagrange multipliers. We will encounter such an example in the analysis of Bose–Einstein condensation of Chap. 10, where the conserved mean number of particles will be an additional constraint and the chemical potential μ the associated Lagrange multiplier.

electric field to vanish in thermal equilibrium. Equation (2.86) also shows that the probability p_n that the field has n photons, the so-called *photon statistics*, is given by the Maxwell–Boltzmann distribution

$$p_n = \rho_{nn} = [1 - e^{-\beta\hbar\omega}]e^{-n\beta\hbar\omega} . \tag{2.88}$$

The density operator (2.85) permits us to compute the expectation value of any observable of interest, such as, e.g. the mean energy in the field

$$\langle \hat{H} \rangle = \text{Tr}(\hat{\rho}\hat{H}) = \frac{1}{Z}\text{Tr}\left(\hbar\omega\hat{a}^\dagger\hat{a}\,e^{-\beta\hbar\omega\hat{a}^\dagger\hat{a}}\right) = \frac{1}{Z}\sum_n \hbar\omega n\,e^{-\hbar\omega n/k_B T} . \tag{2.89}$$

Expanding the partition function Z of Eq. (2.82) on a number states basis and differentiating it with respect to the temperature gives

$$\frac{dZ}{dT} = \frac{1}{k_B T^2}\sum_n n\hbar\omega\,e^{-n\hbar\omega/k_B T} . \tag{2.90}$$

Comparing this result with Eq. (2.89), we find

$$\langle \hat{H} \rangle = k_B T^2 \frac{1}{Z}\frac{dZ}{dT} = \frac{\hbar\omega}{e^{\hbar\omega/k_B T} - 1} , \tag{2.91}$$

or reintroducing the field zero-point energy $\hbar\omega/2$,

$$\langle \hat{H} \rangle = \frac{\hbar\omega}{2} + \frac{\hbar\omega}{e^{\hbar\omega/k_B T} - 1} . \tag{2.92}$$

At absolute zero, the oscillator is in its ground state, with zero-point energy $\langle \hat{H} \rangle = \hbar\omega/2$. This is to be contrasted to the classical oscillator energy, which is $2 \cdot \frac{1}{2}k_B T$ in thermal equilibrium and hence vanishes as $T \to 0$. At high temperatures, $k_B T \gg \hbar\omega$, $\langle \hat{H} \rangle \to k_B T$ so that the quantum and classical oscillators approach the same mean energy. Note that in Eq. (2.92), it was important to keep the zero-point energy of the oscillator since we wish to compare the quantum to the classical energy and therefore need to use the same reference point in both cases. As already mentioned, though, it is often—but not always—appropriate to redefine the energy of the quantum oscillator such that its zero-point energy is zero, a convenient typographical simplification.

The mean energy calculation also permits us to find the mean number of photons $\langle n \rangle$ in a mode in thermal equilibrium at temperature T,

$$\langle n \rangle = \sum_n n\rho_{nn} = \frac{1}{e^{\hbar\omega/k_B T} - 1} , \tag{2.93}$$

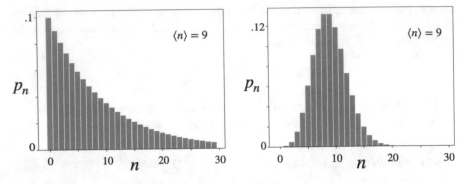

Fig. 2.3 Left: photon statistics p_n of the single-mode thermal field (2.85) with mean photon number $\langle n \rangle = 9$. Right: corresponding photon statistics (2.107) for a coherent field $|\alpha\rangle$ with $\langle n \rangle = |\alpha|^2 = 9$

which allows us to reexpress the photon statistics (2.88) as

$$p_n = \frac{1}{\langle n \rangle + 1} \left(\frac{\langle n \rangle}{\langle n \rangle + 1} \right)^n . \tag{2.94}$$

Figure 2.3 shows an example of a single-mode thermal field photon statistics, with $\langle n \rangle = 9$.

2.3.2 Coherent States

We now turn to a class of states of the simple harmonic oscillator that play a central role in the quantum theory of radiation and in quantum optics. These are the states that minimize the Heisenberg uncertainty relation (2.9), $\Delta p \Delta q \geq \hbar/2$. Since only the product of uncertainties is bound by the Heisenberg uncertainty, one can minimize it with a smaller uncertainty for one conjugate variable at the expense of increasing the uncertainty of the other. Such states, called "squeezed states," are discussed in Sect. 2.3.3. One subclass of particular importance is the so-called coherent states [2], which simultaneously minimize the variance in both the position and momentum operators.

There are several ways to introduce coherent states. Here, we choose a method that emphasizes their nearly classical character. Specifically, we seek pure states of the harmonic oscillator with mean energy equal to the classical energy. We proceed by applying Ehrenfest's theorem, which states that in the cases of free particles, particles in uniform fields, or particles in quadratic potentials (harmonic oscillators), the motion of the center of the quantum wave packet obeys precisely the laws of classical mechanics

$$\langle \psi | \hat{q}(t) | \psi \rangle = q_c(t) \,,$$

$$\langle \psi | \hat{p}(t) | \psi \rangle = p_c(t) \,,$$

where we use the index c to label the classical variables. Inserting these into Eq. (2.10), we obtain the classical energy

$$H_c = \frac{p_c^2}{2} + \frac{\omega^2 q_c^2}{2} = \frac{1}{2}[\langle \psi | \hat{p}(t) | \psi \rangle^2 + \omega^2 \langle \psi | \hat{q}(t) | \psi \rangle^2] \tag{2.95}$$

or, with Eqs. (2.15) and (2.16),

$$H_c = \hbar\omega \langle \psi | \hat{a}^\dagger | \psi \rangle \langle \psi | \hat{a} | \psi \rangle \,. \tag{2.96}$$

The corresponding quantum mechanical oscillator has the energy

$$\langle \hat{H} \rangle = \langle \psi | \hat{H} | \psi \rangle = \hbar\omega \langle \psi | \hat{a}^\dagger \hat{a} | \psi \rangle \,, \tag{2.97}$$

where we have shifted the zero of energy by the zero-point energy $\hbar\omega/2$.

The requirement that the classical energy be equal to the quantum mechanical energy for the coherent state, denoted $|\alpha\rangle$, leads therefore to the factorization condition

$$\langle \alpha | \hat{a}^\dagger | \alpha \rangle \langle \alpha | \hat{a} | \alpha \rangle = \langle \alpha | \hat{a}^\dagger \hat{a} | \alpha \rangle \,. \tag{2.98}$$

This implies that coherent states are eigenstates of the annihilation operator,

$$\hat{a} | \alpha \rangle = \alpha | \alpha \rangle \,. \tag{2.99}$$

It is readily seen by direct substitution that Eq. (2.99) satisfies the condition (2.98). To show conversely that Eq. (2.98) implies (2.99), we note that it may be rewritten as

$$|\langle \alpha | \hat{a}^\dagger | \alpha \rangle|^2 = \langle \alpha | \hat{a}^\dagger \hat{a} | \alpha \rangle \,. \tag{2.100}$$

The Gram–Schmidt orthogonalization procedure tells us that starting with $|\alpha\rangle$, we can construct a complete orthonormal basis set consisting of $|\alpha\rangle$ and an infinite complementary set of vectors $\{|R\rangle\}$. Writing this statement in terms of the identity operator, we have

$$\hat{I} = |\alpha\rangle\langle\alpha| + \sum_R |R\rangle\langle R| \,. \tag{2.101}$$

Inserting this expression between the $\hat{a}^\dagger$ and $\hat{a}$ operators on the right-hand side of Eq. (2.100), we have

$$\langle\alpha|\hat{a}^\dagger\hat{a}|\alpha\rangle = \langle\alpha|\hat{a}^\dagger|\alpha\rangle\langle\alpha|\hat{a}|\alpha\rangle + \sum_R\langle\alpha|\hat{a}^\dagger|R\rangle\langle R|\hat{a}|\alpha\rangle$$

$$= |\langle\alpha|\hat{a}^\dagger|\alpha\rangle|^2 + \sum_R|\langle R|\hat{a}|\alpha\rangle|^2 , \qquad (2.102)$$

and equating this result with the left-hand side of Eq. (2.100) gives

$$\sum_R|\langle R|\hat{a}|\alpha\rangle|^2 = 0 . \qquad (2.103)$$

Since every term in this sum is positive definite, we must have $\langle R|\hat{a}|\alpha\rangle = 0$ for all $|R\rangle$, which implies that $\hat{a}|\alpha\rangle$ must be orthogonal to any $|R\rangle$, that is, proportional to $|\alpha\rangle$, namely,

$$\hat{a}|\alpha\rangle = \alpha|\alpha\rangle . \qquad (2.104)$$

This concludes the proof that all states satisfying the factorization property (2.98) must be eigenstates of the annihilation operator.

Number States Representation To obtain an explicit form for $|\alpha\rangle$ in terms of the number states $|n\rangle$, we write

$$|\alpha\rangle = \sum_n|n\rangle\langle n|\alpha\rangle = \sum_n|n\rangle\langle 0|\frac{\hat{a}^n}{\sqrt{n!}}|\alpha\rangle = \langle 0|\alpha\rangle\sum_n\frac{\alpha^n}{\sqrt{n!}}|n\rangle . \qquad (2.105)$$

Using the normalization condition $\langle\alpha|\alpha\rangle = 1$, we find that $|\langle 0|\alpha\rangle|^2 = e^{-|\alpha|^2}$, so that choosing a unit phase factor gives finally

$$|\alpha\rangle = e^{-|\alpha|^2/2}\sum_n\frac{\alpha^n}{\sqrt{n!}}|n\rangle . \qquad (2.106)$$

This expression immediately gives the probability of finding n photons in the coherent state as the Poisson distribution, with corresponding *photon statistics*

$$p_n = |\langle n|\alpha\rangle|^2 = e^{-|\alpha|^2}\frac{|\alpha|^{2n}}{n!} , \qquad (2.107)$$

see Fig. 2.3, from which the mean photon numbers $\langle n\rangle$ is the field is readily found to be

$$\langle n \rangle = e^{-|\alpha|^2} \sum_{n=0}^{\infty} n \frac{|\alpha|^{2n}}{n!} = |\alpha|^2 . \tag{2.108}$$

The corresponding variance is likewise $|\alpha|^2$,

$$\sigma_n^2 \equiv \langle n^2 \rangle - \langle n \rangle^2 = |\alpha|^2 . \tag{2.109}$$

It is also useful to express the coherent states in terms of the vacuum state $|0\rangle$. We find, with $(\hat{a}^\dagger)^n |0\rangle = \sqrt{n!}|n\rangle$,

$$|\alpha\rangle = e^{-|\alpha|^2/2} \sum_{n} \frac{(\alpha \hat{a}^\dagger)^n}{n!} |0\rangle = e^{-|\alpha|^2/2} e^{\alpha \hat{a}^\dagger} |0\rangle$$

$$= e^{-|\alpha|^2/2} e^{\alpha \hat{a}^\dagger} e^{-\alpha^* \hat{a}} |0\rangle , \tag{2.110}$$

where we have used the fact that $\hat{a}|0\rangle = 0$ to perform the last step. Using the Baker–Hausdorff relation

$$e^{\hat{A}+\hat{B}} = e^{\hat{A}} e^{\hat{B}} e^{-[\hat{A},\hat{B}]/2} , \tag{2.111}$$

which holds if the operators $\hat{A}$ and $\hat{B}$ commute with their commutator

$$[\hat{A}, [\hat{A}, \hat{B}]] = [\hat{B}, [\hat{A}, \hat{B}]] = 0 ,$$

we can rewrite this expression as

$$|\alpha\rangle = \hat{D}(\alpha)|0\rangle , \tag{2.112}$$

where

$$\hat{D}(\alpha) \equiv e^{\alpha \hat{a}^\dagger - \alpha^* \hat{a}} \tag{2.113}$$

is the *displacement operator*. As such, we can call the coherent states $|\alpha\rangle$ displaced states of the vacuum.

The coherent states will prove very useful in describing a number of electromagnetic fields, although they have the complication of being overcomplete, with

$$\frac{1}{\pi} \int d(\text{Re}\,\alpha) d(\text{Im}\,\alpha) |\alpha\rangle \langle \alpha| = \frac{1}{\pi} \int d^2\alpha |\alpha\rangle \langle \alpha| = 1 , \tag{2.114}$$

as well as nonorthogonal, since

$$\langle \alpha | \beta \rangle = \exp\left[-\frac{1}{2}(|\alpha|^2 + |\beta|^2 - 2\alpha^* \beta) \right] \tag{2.115}$$

does not vanish for $\alpha \neq \beta$. However, squaring this expression, we have that

$$|\langle\alpha|\beta\rangle|^2 = \exp(-|\alpha - \beta|^2),\tag{2.116}$$

which shows that the states become increasingly orthogonal if α differs sufficiently from β.

Minimum Uncertainty States We now show that as advertised, the coherent states $|\alpha\rangle$ are minimum uncertainty states of the harmonic oscillator. Since $\hat{a}$ and $\hat{a}^\dagger$ are not self-adjoint operators and therefore not subject to the Heisenberg uncertainty relations, this requires returning to the position and momentum operators $\hat{q}$ and $\hat{p}$ instead, see Eqs. (2.18) and (2.19). From $\hat{a}|\alpha\rangle = \alpha|\alpha\rangle$, we find readily

$$\langle\alpha|\hat{a} + \hat{a}^\dagger|\alpha\rangle = (\alpha + \alpha^*),$$

$$\langle\alpha|(\hat{a} + \hat{a}^\dagger)^2|\alpha\rangle = (\alpha + \alpha^*)^2 + 1,$$

$$\langle\alpha|(\hat{a} - \hat{a}^\dagger)^2|\alpha\rangle = (\alpha - \alpha^*)^2 - 1.$$

Substituting these expressions into the expressions for $\hat{q}$, $\hat{p}$, and $\hat{q}^2$ and $\hat{p}^2$ then gives the variances

$$\sigma_q^2 = \langle\hat{q}^2\rangle - \langle\hat{q}\rangle^2 = \hbar/2\omega\tag{2.117}$$

$$\sigma_p^2 = \langle\hat{p}^2\rangle - \langle\hat{p}\rangle^2 = \hbar\omega/2\tag{2.118}$$

so that

$$\sigma_q^2 \sigma_p^2 = \hbar^2/4.\tag{2.119}$$

Since $[\hat{q}, \hat{p}] = i\hbar$, this is precisely the minimum product of uncertainties between two arbitrary observables $\hat{A}$ and $\hat{B}$ permitted by the Heisenberg uncertainty principle

$$\sigma_A \sigma_B \geq \frac{1}{2}|\langle[\hat{A}, \hat{B}]\rangle|.\tag{2.120}$$

Interestingly, the vacuum state $|0\rangle$ is itself a minimum uncertainty state, since it is an eigenstate of the annihilation operator with $\hat{a}|0\rangle = 0$.

2.3.3 Squeezed States

As already indicated, the Heisenberg uncertainty principle has a built-in degree of freedom: one can "squeeze" the standard deviation of one observable provided one "stretches" that for the conjugate observable [3–5]. For example, while the standard deviations in position and momentum obey the uncertainty relation $\Delta x \Delta p \geq \hbar/2$, all quantum mechanics requires is that the *product* be bounded from below: one can in principle squeeze Δx to an arbitrarily small value at the expense of increasing the standard deviation Δp accordingly. As we have seen, the electric and magnetic fields form a pair of observables analogous to the position and momentum of a simple harmonic oscillator. Hence, they obey a similar uncertainty relation

$$\sigma_E \sigma_B \geq \text{(constant)} \hbar/2, \tag{2.121}$$

and we can likewise squeeze the variance σ_E^2 at the expense of stretching σ_B^2, or vice versa. Such squeezing of the electromagnetic field is of considerable importance in the context of precision quantum measurements, where it offers the promise of achieving quantum noise reduction beyond the "standard shot noise limit," as will be discussed in Chap. 6.

Field Quadratures However, as a monochromatic electromagnetic field oscillates in time, its energy is transferred between E and B each quarter period. As a result, if we initially squeeze σ_E, it will then spread for a quarter of a cycle, then return to the squeezed value at the half cycle, and so on. This is in contrast to the result for coherent states: a displaced ground state of the simple harmonic oscillator of the correct width oscillates back and forth with unchanging width. Looking at the mean and standard deviation of the electric field vector in the complex α plane, the coherent state appears as in Fig. 2.4a, while a squeezed state appears as in Fig. 2.4b.

To observe the squeezing in σ_E, we must therefore somehow select its active quadratures from the general electromagnetic oscillation. Given a field described by the annihilation operator $\hat{a}$, we proceed by forming two Hermitian conjugate

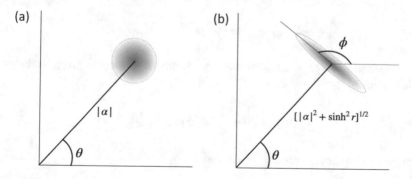

Fig. 2.4 (a) Amplitude vector of length $|\alpha|$ of a coherent state $|\alpha e^{i\theta}\rangle$ and its variance. (b) Amplitude vector of length $[|\alpha|^2 + \sinh^2 r]^{1/2}$ and variance of a squeezed coherent state $|\alpha, \zeta\rangle$, that is, a coherent state $|\alpha e^{i\theta}\rangle$ that has been squeezed by the operator $S(\zeta)$ of Eq. (2.127) at an angle ϕ with respect to the real axis. See Eq. (2.145) for a discussion of the length of its state vector

quadrature operators as

$$\hat{d}_1(\phi) = \frac{1}{2}(\hat{a}\,e^{i\phi} + \hat{a}^\dagger e^{-i\phi})$$

$$\hat{d}_2(\phi) = \frac{1}{2i}(\hat{a}\,e^{i\phi} - \hat{a}^\dagger e^{-i\phi})\,, \tag{2.122}$$

with $[\hat{d}_1, \hat{d}_2] = i/2$, so that the product of their variances is $\sigma_{d_1}^2\, \sigma_{d_2}^2 \geq 1/4$. These operators would correspond to position and momentum in the case of a mechanical oscillator.

The quadrature operators are of considerable importance in quantum optics. In particular, they are the observables that are measured in homodyne and heterodyne detection, two detection methods that multiply the signal to be measured with a reference sine wave called the local oscillator and are frequently exploited for precision field measurements near the limits imposed by quantum mechanics. Sect. 2.4.2 will show explicitly for the specific example of balanced homodyne detection how one can examine any quadrature of the signal by varying its phase relative to the phase of the local oscillator. We will also see in Sect. 11.4 how these quadratures play a central role in characterizing the noise of optical interferometers and in establishing their standard quantum limit, that is, the minimum level of quantum noise that is achievable under normal circumstances, and how squeezed light permits to circumvent that limit.

Consider then for a moment a quantum state such that the expectation value of the electric field is zero, $\langle \hat{a} \rangle = \langle \hat{a}^\dagger \rangle = \langle \hat{d}_i \rangle = 0$. This reduces the variance $\sigma_{d_1}^2$ to

$$\sigma_{d_1}^2 = \langle \hat{d}_1^2 \rangle - \langle \hat{d}_1 \rangle^2 = \frac{1}{4}\left[\langle \hat{a}^\dagger \hat{a} \rangle + \langle \hat{a}\hat{a}^\dagger \rangle + (\langle \hat{a}^2 \rangle e^{2i\phi} + \text{c.c.}) \right]$$

$$= \frac{1}{4} + \frac{1}{2}\langle \hat{a}^\dagger \hat{a} \rangle + \frac{1}{2}\text{Re}\,\{\langle \hat{a}^2 \rangle e^{2i\phi}\}\,. \tag{2.123}$$

For a given set of expectation values, the minimum variance is given by the phase ϕ that yields $\langle \hat{a}^2 \rangle\, e^{2i\phi} + \text{c.c.} = -2|\langle \hat{a}^2 \rangle|$, that is,

$$\sigma_{d_1}^2 = \frac{1}{4} + \frac{1}{2}\langle \hat{a}^\dagger \hat{a} \rangle - \frac{1}{2}|\langle \hat{a}^2 \rangle|\,, \tag{2.124}$$

with the conjugate variance $\sigma_{d_2}^2$ given then by

$$\sigma_{d_2}^2 = \frac{1}{4} + \frac{1}{2}\langle \hat{a}^\dagger \hat{a} \rangle + \frac{1}{2}|\langle \hat{a}^2 \rangle|\,, \tag{2.125}$$

which is greater than or equal to $\sigma_{d_1}^2$. These equations satisfy the Heisenberg uncertainty

$$\sigma_{d_1}\sigma_{d_2} = \frac{1}{4}\sqrt{[1 + 2\langle\hat{a}^\dagger\hat{a}\rangle - 2|\langle\hat{a}^2\rangle|][1 + 2\langle\hat{a}^\dagger\hat{a}\rangle + 2|\langle\hat{a}^2\rangle|]} \geq \frac{1}{4}, \qquad (2.126)$$

which corresponds geometrically to the equation for an ellipse. For a coherent state $|\alpha\rangle$, this gives $\sigma_{d_1}^2 = \sigma_{d_2}^2 = 1/4$. Squeezing occurs for the quadrature $\hat{d}_1$ if its standard deviation σ_{d_1} becomes smaller than $\frac{1}{2}$, that is, is squeezed below the minimum uncertainty product value for a coherent state. Again, this does not violate the uncertainty principle, since $\hat{d}_2$ then has a correspondingly increased variance.

Squeeze Operator In the present example, it is the $|\langle\hat{a}^2\rangle|$ term that leads to squeezing. A way to obtain such squeezing formally is to "squeeze" the state vector with the *squeeze operator*

$$\hat{S}(\zeta) = e^{\zeta\hat{a}^{\dagger 2} - \zeta^*\hat{a}^2}, \qquad (2.127)$$

which converts the circular variance of a coherent state illustrated in Fig. 2.4a into a rotated ellipse of Fig. 2.4b. That this is the case can be seen by calculating the standard deviations σ_{d_1} and σ_{d_2} in the state $\hat{S}(\zeta)|\alpha\rangle$. We proceed by first evaluating the expectation values of $\hat{a}$, $\hat{a}^\dagger$, $\hat{a}^2$, and $\hat{a}^{\dagger 2}$, which involve the operator products $S^\dagger(\zeta)\hat{a}\hat{S}(\zeta)$ and $\hat{S}^\dagger(\zeta)\hat{a}^\dagger\hat{S}(\zeta)$. We can express these products in terms of simple powers of $\hat{a}$ and $\hat{a}^\dagger$ by using the Baker–Hausdorff operator identity

$$e^{\hat{B}}\hat{X}e^{-\hat{B}} = \hat{X} + [\hat{B}, \hat{X}] + \frac{1}{2!}[\hat{B}, [\hat{B}, \hat{X}]] + \ldots + \frac{1}{n!}[\hat{B}, [\hat{B}, \ldots [\hat{B}, \hat{X}]\ldots]]] + \ldots, \qquad (2.128)$$

where we take $\exp(-\hat{B}) = \hat{S}$ and note that $\hat{S}^\dagger(\zeta) = \hat{S}^{-1}(\zeta)$, that is, $\hat{S}(\zeta)$ is a unitary operator. In working with the operator $\hat{S}$, it is convenient to write ζ in polar coordinates as

$$\zeta = \frac{1}{2}re^{-2i\phi}. \qquad (2.129)$$

With the relations

$$[\hat{a}, \hat{a}^{\dagger m}] = m\hat{a}^{\dagger(m-1)} \quad ; \quad [\hat{a}^\dagger, \hat{a}^m] = -m\hat{a}^{m-1}, \qquad (2.130)$$

which can easily be proven by induction, we have that

$$[\hat{B}, \hat{a}e^{i\phi}] = [\zeta^*\hat{a}^2 - \zeta\hat{a}^{\dagger 2}, \hat{a}e^{i\phi}] = \zeta e^{i\phi}[\hat{a}, \hat{a}^{\dagger 2}] = r\hat{a}^\dagger e^{-i\phi}, \qquad (2.131)$$

where we included the factor $e^{i\phi}$ since it simplifies the derivation. The adjoint of this equation is

$$[\hat{B}, \hat{a}^\dagger e^{-i\phi}] = r\hat{a}e^{i\phi}. \tag{2.132}$$

Using these commutators repeatedly, we obtain the series

$$\hat{S}^\dagger(\zeta)\hat{a}\, e^{i\phi}\hat{S}(\zeta) = \hat{a}e^{i\phi} + r\hat{a}^\dagger e^{-i\phi} + \frac{r^2}{2!}\hat{a}\, e^{i\phi} + \frac{r^3}{3!}\hat{a}^\dagger e^{-i\phi} + \cdots$$

$$= \hat{a}\, e^{i\phi}\cosh r + \hat{a}^\dagger e^{-i\phi}\sinh r, \tag{2.133}$$

which has the adjoint

$$\hat{S}^\dagger(\zeta)\hat{a}^\dagger e^{-i\phi}\hat{S}(\zeta) = \hat{a}^\dagger e^{-i\phi}\cosh r + \hat{a}e^{i\phi}\sinh r. \tag{2.134}$$

The corresponding squeezed versions of the Hermitian operators $\hat{d}_i$ are therefore

$$\hat{S}^\dagger(\zeta)\hat{d}_1\hat{S}(\zeta) = \frac{1}{2}\hat{d}_1[\cosh r + \sinh r] = \hat{d}_1 e^r,$$

$$\hat{S}^\dagger(\zeta)\hat{d}_2\hat{S}(\zeta) = \hat{d}_2 e^{-r}, \tag{2.135}$$

and likewise

$$\hat{S}^\dagger(\zeta)\hat{d}_1^2\hat{S}(\zeta) = \hat{S}^\dagger(\zeta)\hat{d}_1\hat{S}(\zeta)\hat{S}^\dagger(\zeta)\hat{d}_1\hat{S}(\zeta) = \hat{d}_1^2 e^{2r},$$

$$\hat{S}^\dagger(\zeta)\hat{d}_2^2\hat{S}(\zeta) = \hat{d}_2^2 e^{-2r}. \tag{2.136}$$

Hence, the unitary transformation (2.127) has indeed the effect of squeezing and stretching the operators $\hat{d}_1$ and $\hat{d}_2$, as advertised.

Squeezed Coherent States These results provide all the pieces required to calculate the standard deviations σ_{d_1} and σ_{d_2} in the squeezed coherent state $\hat{S}(\zeta)|\alpha\rangle$. With Eqs. (2.135), we have

$$\langle\alpha|\hat{S}^\dagger(\zeta)\hat{d}_1\hat{S}(\zeta)|\alpha\rangle = e^r\langle\alpha|\hat{d}_1|\alpha\rangle = \frac{1}{2}e^r(\alpha e^{i\phi} + \alpha^* e^{-i\phi}),$$

$$\langle\alpha|\hat{S}^\dagger(\zeta)\hat{d}_2\hat{S}(\zeta)|\alpha\rangle = \frac{1}{2}ie^r(\alpha e^{i\phi} - \alpha^* e^{-i\phi}), \tag{2.137}$$

while the expressions in Eq. (2.136) give

$$\langle\alpha|\hat{S}^\dagger(\zeta)\hat{d}_1^2\hat{S}(\zeta)|\alpha\rangle = \frac{1}{4}e^{2r}\langle\alpha|\hat{a}^2 e^{2i\phi} + \hat{a}^{\dagger2}e^{-2i\phi} + 2\hat{a}^\dagger\hat{a} + 1|\alpha\rangle$$

$$= \frac{1}{4}e^{2r}(\alpha^2 e^{2i\phi} + \alpha^{*2}e^{-2i\phi} + 2\alpha^*\alpha + 1). \tag{2.138}$$

Combining these expressions gives the standard deviation

$$\sigma_{d_1} = \frac{1}{2}e^r \qquad (2.139)$$

and similarly

$$\sigma_{d_2} = \frac{1}{2}e^{-r}. \qquad (2.140)$$

Equations (2.139) and (2.140) reveal that the squeezed state

$$|\alpha, \zeta\rangle \equiv \hat{S}(\zeta)|\alpha\rangle = \hat{S}(\zeta)\hat{D}(\alpha)|0\rangle, \qquad (2.141)$$

where we have used Eq. (2.112), is a minimum uncertainty state, or *squeezed coherent state*, since $\sigma_{d_1}\sigma_{d_2} = \frac{1}{4}$ independently of r and ϕ.

The standard deviation of the field quadrature at the angle ϕ with respect to the real and imaginary α axes is stretched, and that of the field quadrature at the angle $\phi + \frac{1}{2}\pi$ is squeezed. The angle ϕ is determined by the squeezing parameter ζ, and the angle θ that the phasor α makes with respect to its real and imaginary axes is in general independent of ϕ. The state with $\phi = \theta$ is called a phase squeezed state, and the state with $\phi = \theta + \frac{1}{2}\pi$ is called an amplitude squeezed state. They are illustrated in Fig. 2.5b, c, respectively.

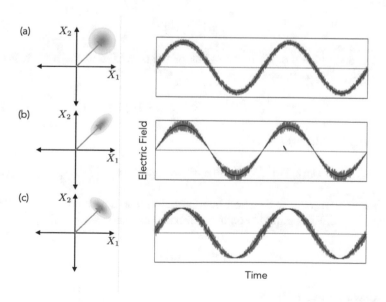

Fig. 2.5 Electric field of a monochromatic light wave versus time, for (**a**) a coherent state, (**b**) a phase squeezed state, and (**c**) amplitude squeezed state. Arbitrary units

More generally, squeezed states exist that yield variances less than the average minimum uncertainty for one quadrature, but whose uncertainty product exceeds the minimum uncertainty value of $\frac{1}{4}$. It is also easily shown that both the magnitude and the mean photon numbers of the squeezed states increase with squeezing.

Squeezed Vacuum A squeezed state of particular importance is the *squeezed vacuum*

$$|\zeta\rangle \equiv \hat{S}(\zeta)|\alpha\rangle = \hat{S}(\zeta)|0\rangle \,, \tag{2.142}$$

as it permits as we shall see in Sect. 11.4 to reduce the noise of interferometers below the standard quantum limit. With Eqs. (2.133) and (2.134), we find that the mean photon number in that state is

$$\langle n\rangle = \langle \zeta|\hat{a}^\dagger\hat{a}|\zeta\rangle = \langle 0|\hat{S}^\dagger(\zeta)\hat{a}^\dagger\hat{S}(\zeta)\hat{S}^\dagger(\zeta)\hat{a}\hat{S}(\zeta)|0\rangle = \sinh^2 r \,. \tag{2.143}$$

The mean photon number in the squeezed vacuum $|\zeta\rangle$ is therefore not equal to zero, much like the mean photon number $\langle n\rangle = |\alpha|^2$, or the coherent state, or displaced vacuum, is larger than zero as well. Indeed, combining these two results, the mean photon number $\langle n\rangle$ of the squeezed coherent state (2.141)

$$|\alpha, \zeta\rangle \equiv \hat{S}(\zeta)\hat{D}(\alpha)|0\rangle \tag{2.144}$$

is readily found to be

$$\langle n\rangle = |\alpha|^2 + \sinh^2 r \,. \tag{2.145}$$

We can determine the photon statistics of the squeezed vacuum by first noting that since $\hat{S}(\zeta)\hat{a}|0\rangle = 0$, we have

$$\hat{S}(\zeta)\hat{a}\hat{S}^\dagger(\zeta)\hat{S}(\zeta)|0\rangle = \hat{S}(\zeta)\hat{a}\hat{S}^\dagger(\zeta)|\zeta\rangle = 0 \,, \tag{2.146}$$

so that, with Eq. (2.134),

$$\hat{a}\cosh r + \hat{a}^\dagger e^{-2i\phi}\sinh r = \mu\hat{a} + \nu\hat{a}^\dagger = 0 \,, \tag{2.147}$$

where we have introduced the short-hand notation $\mu \equiv \cosh r$ and $\nu \equiv e^{-2i\phi}\sinh r$ for convenience. Expanding $|\zeta\rangle$ on the number states basis as

$$|\zeta\rangle = \sum_n c_n|n\rangle, \tag{2.148}$$

we have, with Eq. (2.147),

$$\left(\mu\hat{a} + \nu\hat{a}^\dagger\right)\sum_n c_n|n\rangle = 0 \tag{2.149}$$

so that

$$c_{n+1} = -\frac{\nu}{\mu}\sqrt{\frac{n}{n+1}}c_{n-1}. \tag{2.150}$$

It follows that only even number states are populated. This is of course not surprising, since we start from the vacuum and $\hat{S}(\zeta)$ adds or subtracts excitations to or from the field two photons at a time. Determining the probability amplitudes c_{2m} iteratively from c_0 and requiring that the state be properly normalized, one finds finally

$$|\zeta\rangle = \frac{1}{\sqrt{\cosh r}}\sum_{n=0}^{\infty}(-1)^m \frac{\sqrt{2n!}}{2^n n!}e^{-2in\phi}\tanh^n r|2n\rangle, \tag{2.151}$$

with corresponding photon statistics

$$p_{2m} = |\langle 2m|\zeta\rangle|^2 = \frac{2m!}{2^{2m}(m!)^2}\frac{\tanh^{2m} r}{\cosh r},$$

$$p_{2m+1} = |\langle 2m+1|\zeta\rangle|^2 = 0, \tag{2.152}$$

as illustrated in Fig. 2.6.

Two-Mode Squeezing Instead of considering a single mode of the electromagnetic field, one can also extend these considerations to multimode fields and, in particular, to two-mode fields, in which case squeezed uncertainties can be achieved in the combined quadratures of the two modes. The *two-mode squeeze operator* is defined as

$$\hat{S}_2(\zeta) = e^{\zeta\hat{a}_1^\dagger\hat{a}_2^\dagger - \zeta^*\hat{a}_1\hat{a}_2}, \tag{2.153}$$

Fig. 2.6 Photon statistics of the squeezed vacuum state $|\zeta\rangle = \hat{S}(\zeta)|0\rangle$ with $r = 1.8185$ and mean photon number $\langle n\rangle = \sinh^2 r = 9$, illustrating that $p_n = 0$ for n odd. Compare to Fig. 2.3

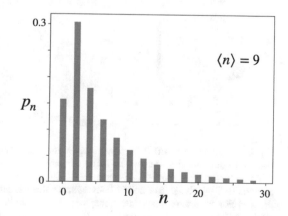

where $\hat{a}_i^{\dagger}$ and $\hat{a}_i$, $i = 1, 2$, are the creation and annihilation operators of the two modes. (Note that since field modes are fully characterized not just by their frequency but also by their wave vector and polarization, the two modes could have the same frequency, $\omega_1 = \omega_2$.) With

$$\zeta = re^{-2i\phi}, \tag{2.154}$$

we find, see Problem 2.12, that the corresponding two-mode squeezed vacuum sate $|TMSV\rangle$ is

$$|TMSV\rangle \equiv \hat{S}_2(\lambda)|0, 0\rangle = \frac{1}{\cosh r} \sum_{n=0}^{\infty} (-e^{-2i\phi} \tanh r)^n |n, n\rangle. \tag{2.155}$$

Importantly, when tracing over one of the modes, say mode 1, we find, see Problem 2.13, that the remaining mode is left in a thermal state

$$\mathrm{Tr}_1|TMSV\rangle\langle TMSV| = \frac{1}{\cosh^2 r} \sum_{n=0}^{\infty} \tanh^{2n}(r)|n_2\rangle\langle n_2|, \tag{2.156}$$

with mean photon number $\langle n_2 \rangle = \sinh^2 r$.

We will encounter two-mode squeezing again in the discussion of optomechanics of Chap. 11. In particular, Sect. 11.3.1 will show explicitly that the two-mode squeeze operator permits to squeeze the variance of the two-mode quadrature operator $\hat{X} = \frac{1}{2^{3/2}}(\hat{a}_1 + \hat{a}_1^{\dagger} + \hat{a}_2 + \hat{a}_2^{\dagger})$

Squeezing by Three- and Four-Wave Mixing We conclude this section by noting that since the squeeze operator $\hat{S}(\zeta)$ involves two-photon processes, it resembles the evolution operator $\exp(-i\hat{H}t/\hbar)$ associated with effective two-photon Hamiltonians of the form

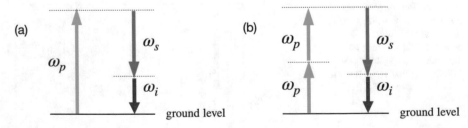

Fig. 2.7 Schematic of (a) three-wave mixing and (b) four-wave mixing, with one, respectively, two pump photons at frequency ω_p being converted into a signal photon at frequency ω_s and an idler photon at frequency ω_i. In many cases, the pump field(s) can be treated classically, with amplitude $E_{\mathrm{pump}}(\omega_p)$. The dotted lines indicate the virtual levels involved in the wave mixing process. For far off-resonant transitions, these levels can be adiabatically eliminated, resulting in effective Hamiltonians of the form (2.157)

$$\hat{H} = \hbar\omega_s \hat{a}_s^\dagger \hat{a}_s + \hbar\omega_i \hat{a}_i^\dagger \hat{a}_i + i\hbar\lambda \hat{a}_s^\dagger \hat{a}_i^\dagger - i\hbar\lambda^* \hat{a}_s \hat{a}_i \, , \tag{2.157}$$

where the coupling strength λ is proportional to the field amplitude $E_{\text{pump}}(\omega_p)$ or to $E_{\text{pump}}^2(\omega_p)$ for three- and four-wave mixing, respectively see Fig. 2.7.[3] In particular, in the degenerate case $\hat{a}_i = \hat{a}_s \equiv \hat{a}$ and $\omega_s = \omega_i \equiv \omega$, the Hamiltonian $\hat{H}$ reduces to

$$\hat{H} = \hbar\omega\hat{a}^\dagger\hat{a} - i\hbar\left[\lambda\hat{a}^{\dagger 2} - \lambda^*\hat{a}^2\right] , \tag{2.158}$$

with an associated interaction picture evolution operator

$$\hat{U}(t) = e^{\zeta\hat{a}^{\dagger 2} - \zeta^*\hat{a}^2} , \tag{2.159}$$

which is nothing but the squeeze operator $\hat{S}(\zeta)$, with $\zeta = -\lambda t$. This suggests that effective two-photon interactions are indeed a good way to generate squeezing. In the non-degenerate case $\omega_1 \neq \omega_2$, the corresponding evolution operator becomes the two-mode squeeze operator

$$\hat{U}(t) = \hat{S}_2(\lambda) = e^{\zeta\hat{a}_1^\dagger\hat{a}_2^\dagger - \zeta^*\hat{a}_1\hat{a}_2} . \tag{2.160}$$

2.4   Photodetection and Correlation Functions

Chapter 6 will discuss quantum measurements in detail, but at this point, we can already consider a practical aspect of this problem that is central to the characterization of optical fields. Specifically, the question that we now address is to determine what property or properties of the field can be inferred from two types of measurements: the detection of the field by absorption, which accesses its intensity, and balanced homodyne detection, which combines the field to be characterized with a local oscillator to access its quadratures. This discussion illustrates the central role of field correlation functions in these measurements.

2.4.1   Detection by Absorption

We consider first a simple detector that operates by absorption [7]. This detector could be a single two-level atom initially in its ground state $|g\rangle$. Its electric dipole interaction with the field can result in the atom undergoing a transition to its excited

---

[3]For a comprehensive discussion of nonlinear optics, see, for example, the excellent text by R. W. Boyd [6].

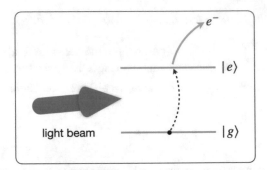

Fig. 2.8 Schematic of a detector operating by absorption. The electric dipole interaction between an incident field and a detector atom induces a transition from its ground state $|g\rangle$ to its excited state $|e\rangle$, and the detection consists in measuring the probability for the atom to be in that state by state-selective ionization, whereby a free electron is emitted only if the atom is in the excited state $|e\rangle$

state $|e\rangle$. The detection consists in measuring the probability for the atom to be in that state, for instance, by state-selective ionization, whereby a free electron is emitted only if the atom is in $|e\rangle$, see Fig. 2.8.

A simple model for the operation of this detector is based on the observation that the probability to absorb a photon at the position $\mathbf{r}$ of the detector and at time t is

$$w_{i\to f} \propto |\langle f|\hat{E}^+(\mathbf{r},t)|i\rangle|^2\,, \qquad (2.161)$$

where $\hat{E}^+(\mathbf{r},t)$ is the positive frequency component of the field and $|i\rangle$ and $|f\rangle$ are the initial and final states of the atom–field system, respectively. Since we are not interested in the final state of the system, just about the counting rate, we can sum over all possible final states and find

$$w_{i\to f} \propto \sum_f |\langle f|\hat{E}^+(\mathbf{r},t)|i\rangle|^2 = \langle i|\hat{E}^-(\mathbf{r},t)\hat{E}^+(\mathbf{r},t)|i\rangle, \qquad (2.162)$$

where we have used the completeness relation $\sum_f |f\rangle\langle f| = 1$. Furthermore, although we do know that the atom starts in the ground state, we typically do not know the initial state $|i\rangle$ of the field precisely. To allow for the corresponding statistical variations, we average the rate (2.162) over $|i\rangle$ using the field density operator $\hat{\rho} = \sum_i p_i |i\rangle\langle i|$, with p_i the probability to be in state $|i\rangle$. Inserting this into Eq. (2.162), we obtain

$$w = \mathrm{Tr}[\hat{\rho}\hat{E}^-(x)\hat{E}^+(x)]\,. \qquad (2.163)$$

This shows that the counting rate at the photodetector is given by the normally ordered, first-order correlation function

$$G^{(1)}(x_1, x_2) \equiv \text{Tr}\{\hat{\rho}\hat{E}^-(x_1)\hat{E}^+(x_2)\} \qquad (2.164)$$

evaluated at $x_1 = x_2$, where $x_i = (\mathbf{r}_i, t_i)$. Here, the qualifier *normally ordered* refers to the fact that the annihilation operators appear at the right of the creation operators, a characteristic of all measurement by absorption processes.

Correlation Functions and Field Coherence Higher order interference experiments require the use of higher order correlation functions like

$$G^{(n)}(x_1 \ldots x_n, y_1 \ldots y_n) = \text{Tr}\{\hat{\rho}\hat{E}^-(x_1) \ldots \hat{E}^-(x_n)\hat{E}^+(y_1) \ldots \hat{E}^+(y_n)\}. \qquad (2.165)$$

Such correlation functions present a close formal analogy to those used in the classical theory of coherence [8]. Pursuing this analogy, a quantum field is said to exhibit nth-order coherence if all of its mth-order correlation functions for $m \leq n$ satisfy

$$G^{(m)}(x_1 \ldots x_m, y_1 \ldots y_m) = \mathcal{E}^*(x_1) \ldots \mathcal{E}^*(x_m)\mathcal{E}(y_1) \ldots \mathcal{E}(y_m), \qquad (2.166)$$

where $\mathcal{E}(x)$ is a complex function. One important concrete illustration is discussed in Problem 2.1, which analyzes the famous Hanbury–Brown experiment where two detectors are used to determine second-order correlation functions of the field.

As an example, consider a single mode of the electromagnetic field in an eigenstate $|n\rangle$, i.e., with density operator $\hat{\rho} = |n\rangle\langle n|$. From Eq. (2.166), a field possessing second-order coherence satisfies

$$G^{(2)}(x_1 x_1, x_1 x_1) = |\mathcal{E}(x_1)|^4 = |G^{(1)}(x_1)|^2. \qquad (2.167)$$

However, directly calculating $G^{(1)}$ and $G^{(2)}$ from Eq. (2.165), we find

$$|G^{(1)}(x_1)|^2 = |\mathcal{E}_\omega \sin Kz|^4 n^2, \qquad (2.168)$$

$$G^{(2)}(x_1 x_1, x_1 x_1) = |\mathcal{E}_\omega \sin Kz|^4 n(n-1); \qquad (2.169)$$

that is, an n-photon state does not possess second-order coherence. This is in sharp contrast with the coherent states $|\alpha\rangle$, which are easily shown to satisfy the general coherence condition (2.166) to all orders, since for all m, we have

$$\langle \alpha | \hat{a}^\dagger(x_1) \hat{a}^\dagger(x_2) \ldots \hat{a}^\dagger(x_m) \hat{a}(y_m) \ldots a(y_1) | \alpha \rangle$$
$$= \mathcal{E}^*(x_1)\mathcal{E}^*(x_2) \ldots \mathcal{E}^*(x_m)\mathcal{E}(y_m) \ldots \mathcal{E}(y_1), \qquad (2.170)$$

where the field amplitudes

$$\mathcal{E}^*(x_i) \equiv \alpha \exp(-i\omega t_i) .$$ (2.171)

The generalization to multimode fields is straightforward.

2.4.2 Balanced Homodyne Detection

So far, we have considered a detector that measures the field intensity. However, there are many circumstances where one needs to gain information on other field observables, most importantly perhaps the field quadratures $\hat{d}_1$ and $\hat{d}_2$ of Eq. (2.122), which are necessary for the characterization of squeezing. Such measurements may appear challenging, as optical fields oscillate at frequencies beyond the response time of electronic detectors. One solution is *balanced homodyne detection*, a technique that exploits interferences between the field to be characterized and a reference field called a local oscillator, which oscillates at the same frequency ω—see e.g. Ref. [9] for a detailed quantum theory of this technique. The interferences are produced by a symmetric beam splitter, whose two output field intensities are then differentially measured by a conventional intensity detector, see Fig. 2.9.

Beam Splitter Hamiltonian To see how this works in some detail, consider first the operation of the beam splitter, see Fig. 2.10. Although in balanced homodyne detection, the local oscillator is typically a classical field, it is useful to consider the more general case of a beam splitter that combines two quantized field modes, a situation that we will encounter again in quantum optomechanics in Chap. 11.

Fig. 2.9 Schematic setup of balanced homodyne detection. A bright coherent beam, called local oscillator (LO), interferes with a signal beam at a beam splitter (BS). The difference current of the two detectors is the output signal of the homodyne detector

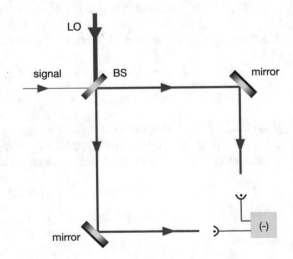

Fig. 2.10 Schematic of a
beam splitter, with input
fields E_a and E_b and output
fields E'_a and E'_b

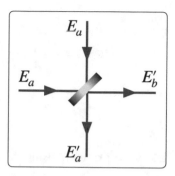

Classically, a beam splitter is a semi-transparent mirror that mixes two modes so
that the outgoing fields E'_a and E'_b are related to the incoming fields E_a and E_b by

$$\begin{pmatrix} E'_a \\ E'_b \end{pmatrix} = \begin{pmatrix} t & r \\ r & t \end{pmatrix} \begin{pmatrix} E_a \\ E_b \end{pmatrix} \equiv U(\theta, \phi) \begin{pmatrix} E_a \\ E_b \end{pmatrix}, \tag{2.172}$$

where the complex, frequency dependent transmission and reflection coefficients t
and r must satisfy the unitarity conditions

$$|t|^2 + |r|^2 = 1,$$

$$tr^* + t^*r = 0. \tag{2.173}$$

These equations show that if t is taken to be real positive, then r is purely imaginary.
They are readily satisfied with $t = \cos(\theta/2)$ and $r = i\exp(i\phi)\sin(\theta/2)$, so that

$$U(\theta, \phi) = \begin{pmatrix} \cos(\theta/2) & ie^{i\phi}\sin(\theta/2) \\ ie^{i\phi}\sin(\theta/2) & \cos(\theta/2) \end{pmatrix}. \tag{2.174}$$

The corresponding quantum description of the beam splitter must be such that
one recovers the classical result in the limit of strong fields. Since it entails a
linear mapping of two input modes onto two output modes, it is described by a
Hamiltonian of the form

$$\hat{V}(t) = -\tfrac{1}{2}\hbar g(t)\left(\hat{a}\hat{b}^\dagger + \text{h.c.}\right), \tag{2.175}$$

see e.g. Ref. [10], where the time-dependent coupling constant $g(t)$ switches the
interaction of the light fields only for the short time interval τ during which the
wave packets of both modes are simultaneously passing through the beam splitter.
The annihilation operators of the outgoing fields $\hat{a}'$ and $\hat{b}'$ are then given by

$$\hat{a}' = \hat{U}^\dagger(\theta, \phi)\hat{a}\hat{U}(\theta, \phi) \quad ; \quad \hat{b}' = \hat{U}^\dagger(\theta, \phi)\hat{b}\hat{U}(\theta, \phi), \tag{2.176}$$

where

$$\hat{U}(\theta, \phi) = \exp\left[-\frac{i}{\hbar} \int dt\, \hat{V}(t)\right] = \exp\left[-i\hat{G}(\phi)\theta/2\right],\qquad(2.177)$$

with

$$\hat{G}(\phi) = -e^{-i\phi}\hat{a}\hat{b}^\dagger + \text{h.c.}\qquad(2.178)$$

and $\theta = \int dt\, g(t)$. Using the Baker–Hausdorff identity (2.128), Eq. (2.176) then gives

$$\hat{a}' = \hat{a} + \frac{i\theta}{2}[\hat{G}, \hat{a}] + \left(\frac{i\theta}{2}\right)^2 [\hat{G}, [\hat{G}, \hat{a}]] + \cdots + \left(\frac{i\theta}{2}\right)^n [\hat{G}, [\hat{G}, [\ldots[\hat{G}, \hat{a}]]]] + \cdots$$

$$(2.179)$$

and similarly for $\hat{b}'$. Since $[\hat{G}, \hat{a}] = e^{i\phi}\hat{b}$ and $[\hat{G}, [\hat{G}, \hat{a}]] = \hat{a}$, it follows that all even order terms in this expansion are proportional to $\hat{a}$ and all odd order terms to $\hat{b}$. Using these commutation relations repeatedly much like in the derivation of Eq. (2.135), we obtain a pair of input–output relationships that are as required formally identical to the classical ones,

$$\hat{a}' = \cos(\theta/2)\hat{a} + ie^{i\phi}\sin(\theta/2)\hat{b},$$
$$\hat{b}' = ie^{-i\phi}\sin(\theta/2)\hat{a} + \cos(\theta/2)\hat{b},\qquad(2.180)$$

with corresponding results for $\hat{a}'^\dagger$ and $\hat{b}'^\dagger$. In particular, $\theta = \pi/2$ corresponds to the case of a symmetric, 50–50 beam splitter.

This result allows us to easily complete the discussion of balanced homodyne detection. We take $\hat{a}$ and $\hat{b}$ to be the annihilation operators of the field to be characterized and of the local oscillator, respectively, and assume that the second field is essentially classical as is usually the case, so that $\hat{b} \to E_{\text{LO}}$. Remembering that the two fields have the same frequency ω, then, we have, for $\theta = \pi/2$ and $\phi = -\pi/2$,

$$\hat{a}' = \tfrac{1}{\sqrt{2}}\left(E_{\text{LO}}e^{i\varphi} + \hat{a}\right),$$
$$\hat{b}' = \tfrac{1}{\sqrt{2}}\left(E_{\text{LO}}e^{i\varphi} - \hat{a}\right).\qquad(2.181)$$

In these expressions, we have also introduced a variable and externally controllable phase φ between the local oscillator and the field to be characterized, as can be achieved, for instance, with a dielectric phase shifter, thereby permitting to perform multiple measurements of the field quadratures. Taking E_{LO} to be real without loss of generality, and assuming that its amplitude is high enough that its relative quantum noise is negligible, the difference in intensities at the two detectors is

$$\hat{I}_-(t) \propto \hat{a}'^\dagger \hat{a}' - \hat{b}'^\dagger \hat{b}' = E_{\mathrm{LO}} \left[\hat{a}(t) e^{-i\varphi} + \hat{a}^\dagger(t) e^{i\varphi} \right] , \tag{2.182}$$

providing a direct measurement of the quadrature $\hat{d}_1 \propto \hat{a}(t) e^{-i\varphi} + \hat{a}^\dagger(t) e^{i\varphi}$. Another choice of the phase φ results likewise in measurements of the quadrature $\hat{d}_2$.

2.5   Quasiprobability Distributions

In many problems, it is useful to describe the state of the field in terms of coherent states, rather than with photon number states. This presents some difficulties, however, since the coherent states are not orthogonal and are overcomplete, as we have seen. On the other hand, this overcompleteness also allows us to obtain a useful diagonal expansion of the density operator in terms of complex matrix elements $P(\alpha)$. This representation can be interpreted as a quasiprobability distribution function, whose dynamics can, under appropriate conditions, be expressed in the form of a Fokker–Planck equation with applications in a number of problems in quantum optics. A number of other quasiprobability distribution descriptions of the electromagnetic field can also be introduced, see e.g. Ref. [11], including the Wigner function $W(\alpha)$ and the Husimi Q-function $Q(\alpha)$.[4]

The $P(\alpha)$ representation is defined in terms of the expansion of the field density operator $\hat{\rho}$ in coherent states as [2, 12]

$$\hat{\rho} = \int d^2\alpha \, P(\alpha) |\alpha\rangle\langle\alpha| , \tag{2.183}$$

where $d^2\alpha = d\mathrm{Re}(\alpha) \, d\mathrm{Im}(\alpha)$. In terms of $P(\alpha)$, the expectation value of an operator $\hat{A}$ is therefore

$$\langle \hat{A} \rangle = \mathrm{Tr}(\hat{\rho}\hat{A}) = \sum_n \langle n| \int d^2\alpha \, P(\alpha) |\alpha\rangle\langle\alpha|\hat{A}|n\rangle$$

$$= \int d^2\alpha \, P(\alpha) \sum_n \langle\alpha|\hat{A}|n\rangle\langle n|\alpha\rangle$$

$$= \int d^2\alpha \, P(\alpha) \langle\alpha|\hat{A}|\alpha\rangle = \int d^2\alpha \, P(\alpha) A(\alpha) , \tag{2.184}$$

---

[4]The reason why these representations are called *quasiprobability* functions is that they are not in general positive definite.

where $A(\alpha) = \langle\alpha|\hat{A}|\alpha\rangle$. Provided that the operator $\hat{A}$ is expressed in normal order, a frequent occurrence in quantum optics, this leads to simple calculations involving only complex numbers. This is easily seen by considering the expectation value of the normally ordered operator $\langle\hat{a}^{\dagger n}\hat{a}^{m}\rangle = \text{Tr}[\hat{a}^{\dagger n}\hat{a}^{m}\hat{\rho}]$. With the definition of (2.183), this gives readily

$$\langle\hat{a}^{\dagger n}\hat{a}^{m}\rangle = \int d^2\alpha\, P(\alpha)\text{Tr}\left[\hat{a}^{\dagger n}\hat{a}^{m}|\alpha\rangle\langle\alpha|\right] = \int d^2\alpha\, P(\alpha)\text{Tr}\left[\hat{a}^{m}|\alpha\rangle\langle\alpha|\hat{a}^{\dagger n}\right]$$

$$= \int d^2\alpha\, P(\alpha)\alpha^{*n}\alpha^{m}\,. \tag{2.185}$$

The $P(\alpha)$ distribution can be expressed as the Fourier transform of the *normally ordered characteristic function*

$$C_N(\lambda) = \text{Tr}(\hat{\rho}\, e^{\lambda\hat{a}^{\dagger}}e^{-\lambda^*\hat{a}})\,, \tag{2.186}$$

where the subscript N stands for "normal order" and λ is a complex number, a formulation that is oftentimes convenient for its evaluation. Substituting Eq. (2.183) into Eq. (2.186) gives readily

$$C_N(\lambda) = \sum_n \langle n|\int d^2\alpha\, P(\alpha)|\alpha\rangle\langle\alpha|e^{\lambda\hat{a}^{\dagger}}e^{-\lambda^*\hat{a}}|n\rangle$$

$$= \sum_n \int d^2\alpha\, P(\alpha)\langle n|e^{-\lambda^*\hat{a}}|\alpha\rangle\langle\alpha|e^{\lambda\hat{a}^{\dagger}}|n\rangle$$

$$= \int d^2\alpha\, P(\alpha)\, e^{\lambda\alpha^*-\lambda^*\alpha}\,, \tag{2.187}$$

which shows that $C_N(\lambda)$ is the Fourier transform of $P(\alpha)$. Hence, $P(\alpha)$ is likewise the Fourier transform of $C_N(\lambda)$,

$$P(\alpha) = \frac{1}{\pi^2}\int d^2\lambda\, e^{\alpha\lambda^*-\alpha^*\lambda}C_N(\lambda)\,. \tag{2.188}$$

Other Characteristic Functions In addition to $C_N(\lambda)$, one can also introduce the antinormally ordered and symmetrically ordered characteristic functions

$$C_A(\lambda) = \text{Tr}(\hat{\rho}\, e^{-\lambda^*\hat{a}}e^{\lambda\hat{a}^{\dagger}}) \tag{2.189}$$

and

$$C_S(\lambda) = \text{Tr}(\hat{\rho}\, e^{\lambda\hat{a}^{\dagger}-\lambda^*\hat{a}})\,. \tag{2.190}$$

From the Baker–Hausdorff relation (2.111), it is easily shown that

$$C_N(\lambda) = C_S(\lambda)e^{|\lambda|^2/2} = C_A(\lambda)e^{|\lambda|^2} .\tag{2.191}$$

The difference in operator ordering of the characteristic functions $C_A(\lambda)$ and $C_S(\lambda)$ as compared to $C_N(\lambda)$ hints at the fact that they are associated with quasidistribution functions adapted to the evaluation of antinormally ordered, respectively symmetrically ordered operators.[5]

Husimi Q-Function Much like $P(\alpha)$ and $C_N(\lambda)$ are Fourier transforms of each other, a similar relation relates the antinormally ordered characteristic function $C_A(\lambda)$ to a probability distribution function $Q(\alpha)$, the Husimi function, via

$$C_A(\lambda) = \frac{1}{\pi} \int d^2\alpha \, \langle\alpha|\hat{\rho}\, e^{-\lambda^*\hat{a}} e^{\lambda\hat{a}^\dagger}|\alpha\rangle = \frac{1}{\pi} \int d^2\alpha \, \langle\alpha|e^{\lambda\hat{a}^\dagger}\hat{\rho}\, e^{-\lambda^*\hat{a}}|\alpha\rangle$$
$$\equiv \int d^2\alpha \, Q(\alpha)e^{\lambda\alpha^* - \lambda^*\alpha} ,\tag{2.192}$$

where

$$Q(\alpha) = \frac{1}{\pi}\langle\alpha|\hat{\rho}|\alpha\rangle .\tag{2.193}$$

and we have used Eq. (2.114). Note that $Q(\alpha)$ is positive, $Q(\alpha) \geq 0$, since $\hat{\rho}$ is a positive operator, hence, a true probability distribution function. The inverse Fourier transform of Eq. (2.192) gives also

$$Q(\alpha) = \frac{1}{\pi^2} \int d^2\lambda \, e^{\alpha\lambda^* - \alpha^*\lambda} C_A(\lambda) .\tag{2.194}$$

The proof that $Q(\alpha)$ is the appropriate distribution to evaluate antinormally ordered correlation functions follows readily from the series of equalities

$$\langle\hat{a}^m\hat{a}^{\dagger n}\rangle = \mathrm{Tr}\left[\hat{\rho}\,\hat{a}^m\hat{a}^{\dagger n}\right] = \frac{1}{\pi}\sum_\ell\langle\ell|\int d^2\alpha\,\hat{\rho}\,\hat{a}^m|\alpha\rangle\langle\alpha|\hat{a}^{\dagger n}|\ell\rangle$$
$$= \frac{1}{\pi}\sum_\ell\int d^2\alpha\langle\alpha|\ell\rangle\langle\ell|\hat{\rho}|\alpha\rangle\alpha^{*n}\alpha^m = \frac{1}{\pi}\int d^2\alpha\langle\alpha|\hat{\rho}|\alpha\rangle\alpha^{*n}\alpha^m$$
$$= \int d^2\alpha\, Q(\alpha)\alpha^{*n}\alpha^m .\tag{2.195}$$

---

[5]Symmetric ordering is an ordering of annihilation and creation operators that is the average of all different products of these operators. For instance, the symmetrically ordered expression for $\hat{a}^\dagger\hat{a}$ is $\{\hat{a}^\dagger\hat{a}\} = \frac{1}{2}(\hat{a}^\dagger\hat{a} + \hat{a}\hat{a}^\dagger)$, and for $\hat{a}^\dagger\hat{a}^2$, it is $\{\hat{a}^\dagger\hat{a}^2\} = \frac{1}{3}(\hat{a}^\dagger\hat{a}^2 + \hat{a}\hat{a}^\dagger\hat{a} + \hat{a}^2\hat{a}^\dagger)$.

Wigner Distribution Finally, the Wigner distribution $W(\alpha)$ is defined as the Fourier transform of the *symmetric* characteristic function $C_S(\lambda)$,

$$C_S(\lambda) = \frac{1}{\pi} \int d^2\alpha \langle \alpha | \hat{\rho}\, e^{\lambda \hat{a}^\dagger - \lambda^* \hat{a}} | \alpha \rangle$$

with

$$W(\alpha) = \frac{1}{\pi^2} \int d^2\lambda\, e^{\alpha\lambda^* - \alpha^*\lambda} C_S(\lambda)$$

$$= \frac{1}{\pi^2} \int d^2\lambda\, \mathrm{Tr}\left[\hat{\rho}\, e^{\lambda(\hat{a}^\dagger - \alpha^*)} e^{-\lambda^*(\hat{a} - \alpha)} \right] e^{-|\lambda|^2/2} . \tag{2.196}$$

The Wigner function, first introduced by E. Wigner in 1932 [13], has a long history and plays a central role in many areas of physics, in particular, in providing a description of quantum mechanics in phase space and in investigating the quantum to classical transition. It is usually defined for mixed states characterized density operator $\hat{\rho}$ as

$$W(q, p) = \frac{1}{2\pi} \int_{-\infty}^{\infty} dy \langle q + \frac{y}{2} | \hat{\rho} | q - \frac{y}{2} \rangle e^{-ipy/\hbar} , \tag{2.197}$$

where q and p are the position and conjugate momentum and $\langle q | \psi \rangle = \psi(q)$. As shown in Problem 2.9, this definition is equivalent to the form (2.196), which is perhaps more frequently used in quantum optics.

An important property of the Wigner function is that its marginals

$$\int dp\, W(q, p) = \langle q | \hat{\rho} | q \rangle \quad \text{and} \quad \int dq\, W(q, p) = \langle p | \hat{\rho} | p \rangle \tag{2.198}$$

give the x and p probability distributions. For a pure state, this yields readily

$$\int dq\, W(q, p) = |\psi(p)|^2 \quad ; \quad \int dp\, W(q, p) = |\psi(q)|^2 . \tag{2.199}$$

With Eqs. (2.18) and (2.19) and the definitions (2.122) of the field quadratures, it is not surprising that the Wigner function can be used to obtain averages of symmetric functions of annihilation and creation operators,

$$\langle \{\hat{a}^{\dagger n} \hat{a}^m\}_S \rangle = \int d^2\alpha\, W(\alpha)\alpha^{*n}\alpha^m \tag{2.200}$$

and, in particular, of correlation functions of the quadratures $\hat{d}_1$ and $\hat{d}_2$. The formal correspondence between the forms (2.196) and (2.197) of the Wigner function is further discussed in detail in Ref. [14].

Summarizing, then, the normally ordered distribution $P(\alpha)$, antinormally ordered Husimi distribution $Q(\alpha)$, and symmetrically ordered Wigner distribution, $W(\alpha)$ distributions are given by

$$P(\alpha) = \frac{1}{\pi^2} \int d^2\lambda\, e^{\alpha\lambda^* - \alpha^*\lambda} C_N(\lambda)\,,$$

$$Q(\alpha) = \frac{1}{\pi^2} \int d^2\lambda\, e^{\alpha\lambda^* - \alpha^*\lambda} C_A(\lambda)\,,$$

$$W(\alpha) = \frac{1}{\pi^2} \int d^2\lambda\, e^{\alpha\lambda^* - \alpha^*\lambda} C_S(\lambda)\,. \tag{2.201}$$

These expressions permit, for example, to express the Wigner distribution and Q-function in terms of the P distribution as

$$W(\alpha) = \frac{2}{\pi} \int d^2\beta\, P(\beta) \exp(-2|\beta - \alpha|^2) \tag{2.202}$$

and

$$Q(\alpha) = \frac{1}{\pi} \int d^2\beta\, P(\beta) \exp(-|\beta - \alpha|^2)\,. \tag{2.203}$$

This shows that both the Husimi Q-function and the Wigner distribution are convolutions of Gaussians with the P-function. Note however that the Q-function is convoluted with a Gaussian of width $\sqrt{2}$ times larger than is the case for the Wigner function. As a consequence, the Q-function is positive definite, as can also be seen directly from its definition, while $P(\alpha)$ and $W(\alpha)$ are not. This non-positivity is illustrated in Fig. 2.11, which shows the Wigner functions of the ground state $|0\rangle$

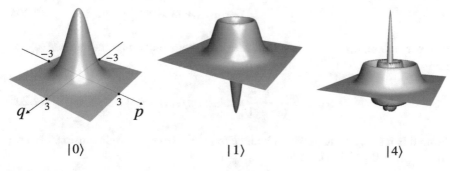

$|0\rangle$ $\qquad$ $|1\rangle$ $\qquad$ $|4\rangle$

Fig. 2.11 Wigner function $W(q, p)$ for the number states $|n = 0\rangle$, $|n = 1\rangle$, and $|n = 4\rangle$, illustrating its non-positivity. Here, q and p are given by $\alpha = \sqrt{\omega/2\hbar}q + i/\sqrt{2\hbar\omega}p$, and the ranges of p and q are $[-3, 3]$, and $[-5, 5]$ for the last three cases. The horizontal plane corresponds in all cases to $W(q, p) = 0$. (See Problem 2.10 for the expression of the Wigner function of the Fock state $|n\rangle$)

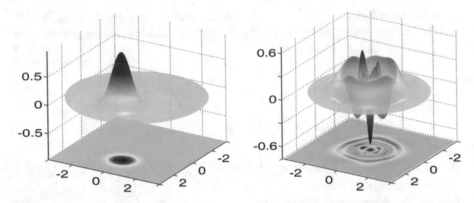

Fig. 2.12 Experimentally measured Wigner functions represented in 3D and 2D: (**a**) coherent state with $\langle n \rangle = 2.5$ and (**b**) $|n = 3\rangle$ Fock state. (From Ref. [15])

and the excited states $|1\rangle$ and $|5\rangle$ of the simple harmonic oscillator, and in Fig. 2.12, which shows the experimental reconstruction of Wigner functions for a coherent state with $\langle n \rangle = 2.5$ photons and the number state $|n = 3\rangle$. Because the coherent state $|\alpha\rangle$ is a displaced vacuum state, its Wigner function is identical to the Wigner function of the state $|n = 0\rangle$, itself a coherent state, simply displaced in the $\{q, p\}$ phase space by α.

Example: $P(\alpha)$ **Function of a Thermal Field** As an illustration of the evaluation of a $P(\alpha)$ distribution, we consider the thermal field described by the density operator (2.85). Its quasidistribution $P(\alpha)$ is best obtained by considering first the $Q(\alpha)$ distribution

$$Q(\alpha) = (1 - e^{-x}) \sum_n e^{-nx} \langle \alpha|n\rangle \langle n|\alpha \rangle$$

$$= (1 - e^{-x}) \exp[-|\alpha|^2 (1 - e^{-x})] \,, \qquad (2.204)$$

where $x = \hbar\omega/k_B T$. Inverting Eq. (2.93) for the average photon number $\langle n \rangle$, we find $1 - e^{-x} = 1/(\langle n \rangle + 1)$, which gives

$$Q(\alpha) = \frac{1}{\langle n \rangle + 1} \exp[-|\alpha|^2/(\langle n \rangle + 1)] \,. \qquad (2.205)$$

From this result, we can readily obtain $P(\alpha)$ since from a two-dimensional Fourier transform of Eq. (2.203), we have

$$\mathcal{F}[Q(\alpha)] = \mathcal{F}[P(\alpha)]\mathcal{F}[\exp{-|\alpha|^2}],$$

and hence

$$P(\beta) = \mathcal{F}^{-1} \frac{\mathcal{F}[Q(\alpha)]}{\mathcal{F}[\exp(-|\alpha|^2)]} .$$

This gives, after carrying out the integrals,

$$P(\alpha) = \frac{1}{\pi \langle n \rangle} \exp[-|\alpha|^2/\langle n \rangle] . \tag{2.206}$$

It is interesting to note that in the classical limit of large mean photon numbers, the expressions for $Q(\alpha)$ and $P(\alpha)$ approach each other. This is because in that limit distinctions depending on the ordering of operators vanish. This point is discussed further in R. J. Glauber's Les Houches Lectures [16], while a more detailed discussion of quasiprobability distributions and their use in quantum optics is presented in the text by D. F. Walls and G. Milburn [17].

The probability of finding n photons in a single-mode thermal field is given by the photon statistics ρ_{nn} of Eq. (2.85). This exponentially decaying distribution contrasts with the Poisson distribution characteristic of a coherent state. The difference between $Q(\alpha)$ and $P(\alpha)$ for the two cases is even more striking, since for thermal light $P(\alpha)$ is given by a Gaussian distribution, as we have seen, while for the coherent state $|\alpha_0\rangle$, it is given by the δ-function $\delta(\alpha - \alpha_0)$. From this and Eqs. (2.202) and (2.203), we immediately find that both the Wigner and the Q-distribution for a coherent state are Gaussian.

While one might be tempted to interpret $P(\alpha)$ as the probability of finding the field in the coherent state $|\alpha\rangle$, this is not correct in general, because $P(\alpha)$ and likewise $W(\alpha)$ are not positive definite and hence cannot be interpreted as probabilities. Sometimes, fields described by a positive $P(\alpha)$ and/or $W(\alpha)$ are referred to as "classical fields." This is however somewhat misleading and does not mean that these fields have vanishing quantum mechanical uncertainties. For example, a coherent state itself is described by a positive definite Dirac delta function P-distribution, but it has minimum, not vanishing, quantum mechanical uncertainties.

The description of electromagnetic fields in terms of quasiprobability distribution functions often permits to replace the quantum mechanical description of the problem by an equivalent description in terms of c-numbers. For instance, as we have seen in Eqs. (2.195) and (2.185), if one is interested in computing antinormally ordered correlation functions, one has

$$\langle \hat{a}^m (\hat{a}^\dagger)^n \rangle = \int d^2\alpha \, Q(\alpha) \alpha^m \alpha^{*n} . \tag{2.207}$$

Similarly, for normally ordered correlation functions,

$$\langle (\hat{a}^\dagger)^m \hat{a}^n \rangle = \int d^2\alpha \, P(\alpha) \alpha^{*m} \alpha^n . \tag{2.208}$$

In particular, correlation functions can readily be computed from the appropriate characteristic function. From the definitions of $C_A(\lambda)$ and $C_N(\lambda)$, one finds readily

$$\frac{\partial^{m+n} C_A(\lambda, \lambda^*)}{(\partial \lambda^*)^m (\partial \lambda)^n} = \mathrm{Tr}[\hat{\rho}\, e^{-\lambda^* \hat{a}} (-\hat{a})^m e^{\lambda \hat{a}^\dagger} (\hat{a}^\dagger)^n] = (-1)^m \langle \hat{a}^m (\hat{a}^\dagger)^n \rangle , \qquad (2.209)$$

where the last equality holds for $\lambda = \lambda^* = 0$. Similarly, under the same conditions,

$$\frac{\partial^{m+n} C_S(\lambda, \lambda^*)}{(\partial \lambda)^m (\partial \lambda^*)^n} = \mathrm{Tr}[\hat{\rho}\, e^{\lambda \hat{a}^\dagger} (\hat{a}^\dagger)^m e^{-\lambda^* \hat{a}} \hat{a}^n] = (-1)^m \langle (\hat{a}^\dagger)^m \hat{a}^n \rangle . \qquad (2.210)$$

Problems

Problem 2.1 (Hanbury Brown and Twiss Experiment)

Recording fields with single-photon detectors as discussed in Sect. 2.4.1 is just one example of photon detection, which can be extended by the use of two or more photodetectors. In the Hanbury Brown and Twiss experiment, a beam of light is split into two beams that are detected by two detectors D_1 and D_2 that work by absorption and perform measurements on the same field, one at time $t = 0$ and the other at time τ, see Fig. 2.13.

(a) For two photodetectors at locations $\mathbf{r}_1$ and $\mathbf{r}_2$, the field matrix element associated with the detection of photon coincidences will be

$$\langle f | \hat{E}^+ (\mathbf{r}_1, t_1) \hat{E}^+ (\mathbf{r}_2, t_2) | i \rangle .$$

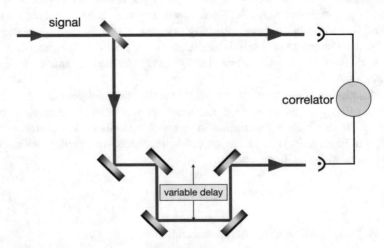

Fig. 2.13 Schematic of the experimental arrangement to measure the normally ordered second-order correlation of the electric field in a Hanbury Brown and Twiss experiment

Following the same approach as the derivation leading to the counting rate of Eq. (2.163) shown that the total rate at which such transitions occur is [7]

$$w_2 \propto \langle \hat{E}^-(\mathbf{r}_1, t_1) \hat{E}^-(\mathbf{r}_1, t_1) \hat{E}^+(\mathbf{r}_2, t_2) \hat{E}^+(\mathbf{r}_1, t_1) \rangle,$$

that is, this arrangement measures the normally ordered second-order correlation of the field

$$G^{(2)}(\mathbf{r}_1, t_1), \mathbf{r}_2, t_2)) = \langle \hat{E}^-(\mathbf{r}_1, t_1) \hat{E}^-(\mathbf{r}_1, t_1) \hat{E}^+(\mathbf{r}_2, t_2) \hat{E}^+(\mathbf{r}_1, t_1) \rangle.$$

For detectors at equivalent positions and stationary fields, this correlation function depends only on $\tau = t_2 - t_1$, that is, $G^{(2)} \to \langle \hat{E}^-(0) \hat{E}^-(\tau) \hat{E}^+(\tau) \hat{E}^+(0) \rangle$, or its normalized form

$$g^{(2)}(\tau) = \frac{G^{(2)}(\tau)}{|G^{(1)}(0)|^2},$$

and fields with $g^{(2)}(0) > 1$ are said to exhibit photon bunching and fields with $g^{(2)}(0) < 1$ exhibit antibunching.

(b) Show that a single-mode thermal field in a Fock state $|n\rangle$ exhibits antibunching, and a thermal field rather than antibunching. What is the result for a coherent state?

Problem 2.2 Calculate the variance of the single-mode electric field operator in the vacuum state.

Problem 2.3 Given that a single-mode field has an average of one photon, what is the probability of having n photons for (a) a Poisson distribution and (b) a thermal distribution? Calculate the variance of a Poisson distribution.

Problem 2.4 Write the operator $\hat{a}^2 \hat{a}^{\dagger 3} \hat{a}$ in normal order and in symmetric order.

Problem 2.5 Express the Wigner function and the Q-distribution of a single-mode field in terms of its $P(\alpha)$ distribution.

Problem 2.6 A photon-added coherent state is the state $|\alpha, 1\rangle = \mathcal{N} \hat{a}^\dagger |\alpha\rangle$. Find the normalization factor $\mathcal{N}$ of this state, and determine both its photon statistics and its Wigner function.

Problem 2.7 More generally, an m-photon-added state of a single-mode field state $|\psi\rangle$ is the state $|\psi, m\rangle = \mathcal{N}_m \hat{a}^\dagger |\psi\rangle$, where $\mathcal{N}_m$ is a normalization constant. Determine the explicit form of state $|\psi, m\rangle$ in the Fock states basis, as well as its photon statistics for $|\psi\rangle = \sum_n c_n |n\rangle$. What is the probability that this state has $p < m$ photons? Find also the photon statistics of the photon-subtracted field $|\psi, -m\rangle \propto \hat{a}^m |\psi\rangle$.

Problem 2.8 A two-mode field is described by the density operator

$$\hat{\rho} = \hat{\rho}_1 \otimes \hat{\rho}_2 ,$$

with $\hat{\rho}_1 = \sum_n p_n |n\rangle \langle n|$ and $\hat{\rho}_2 = \sum_m p_m |m\rangle \langle m|$. What are the density operators of fields "1" an "2"? Consider now a two-mode field described by the density operator

$$\hat{\rho} = \sum_{n,m} p_{nm} |n, m\rangle \langle n, m| .$$

What are the density operators of fields "1" and "2" in that case? Why are they not the same as in the first case? Explain in physical terms the difference between the two situations.

Problem 2.9 Show that the Wigner function

$$W(\alpha) = \frac{1}{\pi^2} \int d^2\lambda \, \mathrm{Tr} \left[\hat{\rho} \, e^{\lambda(\hat{a}^\dagger - \alpha^*)} e^{-\lambda^*(\hat{a} - \alpha)} \right] e^{-|\lambda|^2/2}$$

can be represented as

$$W(q, p) = \frac{1}{2\pi} \int_{-\infty}^{\infty} dy \langle q + \frac{y}{2} |\hat{\rho}| q - \frac{y}{2} \rangle e^{-ipy/\hbar}$$

by introducing the new variables q and p through

$$\alpha = \sqrt{\frac{\omega}{2\hbar}} q + \frac{i}{\sqrt{2\hbar\omega}} p .$$

Hint: Express $\hat{a}$ and $\hat{a}^\dagger$ in terms of the position and momentum operators, and use the facts that $\exp[-ix_0 \hat{p}/\hbar]|x\rangle = |x+x_0\rangle$ and that $(1/2\pi) \int_{-\infty}^{\infty} dx \exp(iax) = \delta(a)$.

Problem 2.10 Calculate the characteristic function $C_N = \mathrm{Tr}(|n\rangle \langle n| e^{\lambda \hat{a}^\dagger} e^{-\lambda^* \hat{a}})$ for a Fock state $|n\rangle$. Use this result to evaluate the Wigner function of that state. Hint:

$$\sum_{\ell=0}^{n} \binom{n}{k} \frac{(-1)^\ell}{\ell!} x^\ell = L_n(x),$$

where $L_n(x)$ is the nth Laguerre polynomial.
Answer:

$$W_{|n\rangle}(q, p) = \frac{(-1)^n}{\pi} e^{-(q^2+p^2)} L_n[2(q^2 + p^2)].$$

Problem 2.11 Using the Baker–Hausdorff relation (2.111) shows that the symmetric, normally ordered and antinormally ordered field characteristic functions are related by

$$C_N(\lambda) = C_S(\lambda)e^{|\lambda|^2/2} = C_A(\lambda)e^{|\lambda|^2}.$$

Problem 2.12 Find the $P(\alpha)$, Husimi, and Wigner distributions for (a) a coherent state $|\alpha\rangle$ and (b) a squeezed state $|\alpha, \zeta\rangle$.

Problem 2.13 Show that the two-mode squeezed vacuum is

$$|TMSV\rangle = \hat{S}_2(\lambda)|0, 0\rangle = \frac{1}{\cosh r} \sum_{n=0}^{\infty} (-e^{-2i\phi} \tanh r)^n |n, n\rangle \qquad (2.211)$$

and that when tracing over one of the modes, say mode 1, the remaining mode is left in the thermal state

$$\mathrm{Tr}_{\text{mode 1}} |TMSV\rangle\langle TMSV| = \frac{1}{\cosh^2 r} \sum_{n=0}^{\infty} \tanh^{2n}(r)|n\rangle\langle n|, \qquad (2.212)$$

with mean photon number $\langle n_2 \rangle = \sinh^2 r$.

References

1. B.J. Shore, P. Meystre, S. Stenholm, Is a quantum standing wave composed of two traveling waves? J. Opt. Soc. Am. **8**, 903 (1991)
2. R.J. Glauber, Coherent and incoherent states of the radiation field. Phys. Rev. **131**, 2766 (1963)
3. D. Stoler, Equivalence classes of minimum uncertainty packets. Phys. Rev. D **1**, 3217 (1970)
4. H.P. Yuen, Two-photon coherent states of the radiation field. Phys. Rev. A **13**, 2226 (1976)
5. D.F. Walls, Squeezed states of light. Nature **306**, 141 (1983)
6. R.W. Boyd, *Nonlinear Optics*, 4th edn. (Academic Press, San Diego, 2020)
7. R.J. Glauber, The quantum theory of optical coherence. Phys. Rev. **130**, 2529 (1963)
8. M. Born, E. Wolf, *Principles of Optics* (Pergamon, Oxford, 1970)
9. M.J. Collett, R. Loudon, C.W. Gardiner, Quantum theory of optical homodyne and heterodyne detection. J. Mod. Optics **81**, 1051 (2009)
10. S. Haroche, J.-M. Raimond, *Exploring the Quantum—Atoms, Cavities and Photons* (Oxford, New York, 2006)
11. C.W. Gardiner, P. Zoller, *Quantum Noise – A Handbook of Markovian and Non-Markovian Quantum Stochastic Methods with Applications to Quantum Optics* (Springer, New York, 2004)
12. E.C.G. Sudarshan, Equivalence of semiclassical and quantum mechanical descriptions of statistical light beams. Phys. Rev. Lett. **10**, 277 (1963)
13. E.P. Wigner, On the quantum correction for thermodynamic equilibrium. Phys. Rev. **40**, 749 (1932)
14. M. Hillery, R.F. O'Connell, M.O. Scully, E.P. Wigner, Distribution functions in physics: Fundamentals. Phys. Rep. **106**, 121 (1984)

15. S. Deléglise, I. Dotsenko, C. Sayrin, J. Bernu, M. Brune, J.-M. Raimond, S. Haroche, Reconstruction of non-classical cavity field states with snapshots of their decoherence. Nature **455**, 510 (2008)
16. R.J. Glauber, in *Quantum Optics and Electronics*, ed. by C. de Witt, A. Blandin, C. Cohen-Tannoudji (Gordon and Breach, New York, 1965), p. 63
17. D.F. Walls, G.J. Milburn, *Quantum Optics* (Springer, Berlin, 2008)

Chapter 3
The Jaynes–Cummings Model

The simplest model of interaction between a quantized electromagnetic field and an atomic system is a single-mode field interacting with a single two-level atom. This is the exactly solvable Jaynes–Cummings model, which describes this interaction in the rotating wave approximation (RWA). After giving its eigenstates and eigenenergies, this chapter discusses quantum Rabi oscillations, Cummings collapse, and quantum revivals and gives an elementary first introduction to spontaneous emission. It then introduces the idea of repeated measurements, a topic that will be revisited several times in later chapters. Removing then the RWA leads us to the quantum Rabi model, for which we outline the main steps of an exact diagonalization.

3.1 The Linchpin of Quantum Optics

The simplest situation that one can think of in the study of the interaction between a quantized electromagnetic field and an atomic system is that of a single-mode field interacting with a two-level atom, see Fig. 3.1. This is the Jaynes–Cummings model, first introduced in the early 1960s by E. T. Jaynes and F. W. Cummings [1] to study some basic aspects of laser theory. What seemed perhaps like an overly simplified model at the time has proven to be of fundamental importance in understanding key aspects of light-matter interaction and has become of added relevance after it was experimentally realized, in particular with Rydberg atoms in cavity quantum electrodynamics (cavity QED), see Chap. 7, or with artificial atoms in circuit quantum electrodynamics (cQED), as will be discussed in Sect. 7.4. The Jaynes–Cummings model is also central to many aspects of quantum information science, where two-level atoms are given the name qubits. In addition, this model presents the considerable advantage of being exactly solvable. As such it is an excellent entry point to the study of quantum optics.

In the Jaynes–Cummings model, the atom and its dipole interaction with the field are still described by the Hamiltonian (1.60), except that the positive and negative frequency components of the field are now the operators of Eq. (2.55), and the free field Hamiltonian must be included. Then, the rotating wave Hamiltonian of the total

© The Author(s), under exclusive license to Springer Nature Switzerland AG 2021
P. Meystre, *Quantum Optics*, Graduate Texts in Physics,
https://doi.org/10.1007/978-3-030-76183-7_3

Fig. 3.1 Schematic of the
Jaynes–Cummings model,
with a single atom at rest
inside a single-mode optical
resonator with perfectly
reflecting mirrors

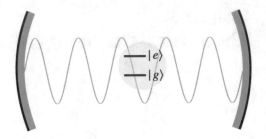

atom–field system becomes

$$\hat{H} = \frac{1}{2}\hbar\omega_0\hat{\sigma}_z + \hbar\omega\left(\hat{a}^\dagger\hat{a} + \frac{1}{2}\right) + \hbar g\left(\hat{\sigma}_+\hat{a} + \hat{a}^\dagger\hat{\sigma}_-\right), \tag{3.1}$$

where the coupling constant

$$g = \frac{d\mathcal{E}_\omega}{\epsilon_0 V}\sin(Kz) \tag{3.2}$$

is the *vacuum Rabi frequency* for the cavity field mode, with V being the field quantization volume and d the electric dipole matrix element of the transition. Here, the $\sin(Kz)$ term accounts for the longitudinal spatial dependence of the cavity field, taken to be a standing wave mode, and we recognize $d\mathcal{E}_\omega/\epsilon_0 V$ as the "electric field per photon". Remember that for a running wave field mode, it is smaller by a factor of $\sqrt{2}$, $\mathcal{E}_\omega \to \mathcal{E}_\omega/\sqrt{2}$, see Eq. (2.74). The Hamiltonian (3.1) defines the Jaynes–Cummings model.

Consider for now an atom located at $z = \pi/2K$ so that $\sin(Kz) = 1$, a simplification that will be removed in later chapters when atomic motion is taken into account. Very much like for the interaction between a classical field and a two-level atom of Chap. 1, the Jaynes–Cummings Hamiltonian can be diagonalized in terms of dressed states, except that we now need an infinite number of them, a result of the infinite dimensionality of the Hilbert space of the field. This is because in the RWA, the dipole interaction only couples pairs of states $|e, n\rangle$ and $|g, n + 1\rangle$ of the atom–field system with the same total excitation number, so that for each of these manifolds, the diagonalization of the Jaynes–Cummings Hamiltonian reduces to that of the semiclassical driven two-level atom. The dressed states introduced in a semiclassical context in Eq. (1.68) can therefore be readily generalized to a two-level atom interacting with a single-mode quantized field where they become[1]

$$|1, n\rangle = \sin\theta_n|e, n\rangle + \cos\theta_n|g, n + 1\rangle$$

---

[1] Note that the ground state $|g, 0\rangle$, with eigenenergy $E_0 = -\hbar\omega_0/2$, is not coupled to any other state, a consequence of the RWA.

$$|2, n\rangle = \cos\theta_n |e, n\rangle - \sin\theta_n |g, n+1\rangle, \tag{3.3}$$

with

$$\tan(2\theta_n) = \frac{-2g\sqrt{n+1}}{\Delta}. \tag{3.4}$$

The corresponding eigenenergies are

$$E_{1n} = \hbar(n + 1/2)\omega + \hbar\Omega_n$$
$$E_{2n} = \hbar(n + 1/2)\omega - \hbar\Omega_n, \tag{3.5}$$

where

$$\Omega_n = \frac{1}{2}\sqrt{\Delta^2 + 4g^2(n+1)} \tag{3.6}$$

is called the *n-photon Rabi frequency*, in analogy with the semiclassical Rabi frequency. As we shall see in Sect. 3.2, this dependence has a profound impact on the dynamics of the atom, leading in particular to quantum collapses and revivals, which are an unambiguous signature of the "granular" nature of the field.

Resonant Interaction At resonance $\omega = \omega_0$, the dressed states (3.3) become

$$|1, n\rangle = \frac{1}{\sqrt{2}}[|g, n+1\rangle + |e, n\rangle]$$

$$|2, n\rangle = \frac{1}{\sqrt{2}}[|g, n+1\rangle - |e, n\rangle], \tag{3.7}$$

with eigenenergies

$$E_{1,n} = \hbar(n + 1/2)\omega + \hbar g\sqrt{n+1}$$
$$E_{2,n} = \hbar(n + 1/2)\omega - \hbar g\sqrt{n+1}, \tag{3.8}$$

resulting in n-dependent energy gaps $\Delta E_n = 2g\sqrt{n+1}$ at the avoided crossings between dressed levels, as illustrated in Fig. 3.2.

Dispersive Limit Another limit of interest is the so-called *dispersive limit*, where the field frequency is far off-resonant from the atomic transition frequency, so that the atom–field detuning $\Delta = \omega_0 - \omega$ is such that

$$|\Delta| \gg g\sqrt{n+1} \tag{3.9}$$

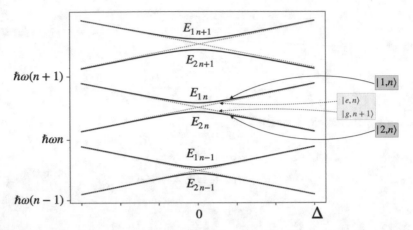

Fig. 3.2 Dressed state energy level diagram of the Jaynes–Cummings model as a function of the detuning $\Delta = \omega_0 - \omega$. The gray dashed lines are the energy eigenvalues of the non-interacting atom–field system, with the bare states $|e, n\rangle$ and $|g, n + 1\rangle$ indicated by dashed arrows. The dressed state energies are the solid black lines, with the states corresponding to the eigenenergies $E_{1,n}$ and E_{2n} indicated by the solid arrows. The manifolds of states with equal total excitation number exhibit avoided crossings with n-dependent energy gaps $\Delta E_n = E_{1n} - E_{2n} = 2g\sqrt{n+1}$ occurring at resonance $\omega = \omega_0$

for all n of interest. In this limit, the eigenenergies (3.5) reduce to

$$E_{1n} = \hbar(n + 1/2)\omega + \frac{1}{2}\hbar\Delta + \frac{\hbar g^2(n+1)}{\Delta} \,,$$

$$E_{2n} = \hbar(n + 1/2)\omega - \frac{1}{2}\hbar\Delta - \frac{\hbar g^2(n+1)}{\Delta} \,, \tag{3.10}$$

with eigenstates approaching the bare states as

$$|1, n\rangle \to |e, n\rangle$$
$$|2, n\rangle \to |g, n + 1\rangle \tag{3.11}$$

for $\Delta > 0$ and

$$|1, n\rangle \to |g, n + 1\rangle$$
$$|2, n\rangle \to |e, n\rangle \tag{3.12}$$

for $\Delta < 0$, see Fig. 3.2. From Eqs. (3.10), (3.11), and (3.12), it follows that the dispersive limit of the Jaynes–Cummings Hamiltonian is

$$H_{\mathrm{JC,eff}} = \frac{1}{2}\hbar\omega_0\hat{\sigma}_z + \hbar\omega\hat{a}^\dagger\hat{a} + \frac{\hbar g^2}{\Delta}\left[(\hat{a}^\dagger\hat{a} + 1)|e\rangle\langle e| - \hat{a}^\dagger\hat{a}|g\rangle\langle g|\right] . \tag{3.13}$$

It is instructive to rewrite this Hamiltonian as

$$H_{\text{JC,eff}} = \frac{1}{2}\hbar\omega_0\hat{\sigma}_z + \hbar\omega\hat{a}^\dagger\hat{a} + \frac{\hbar g^2}{\Delta}\hat{a}^\dagger\hat{a}\,\hat{\sigma}_z + \frac{\hbar g^2}{\Delta}|e\rangle\langle e|$$

$$\approx \frac{1}{2}\hbar\omega_0\hat{\sigma}_z + \hbar\omega\hat{a}^\dagger\hat{a} + \frac{\hbar g^2}{\Delta}\hat{a}^\dagger\hat{a}\,\hat{\sigma}_z, \qquad (3.14)$$

which shows that the term

$$\Delta_{\text{ls}} = \frac{\hbar g^2}{\Delta}\hat{a}^\dagger\hat{a} \qquad (3.15)$$

is an intensity dependent increase, or *light shift*, of the transition frequency ω_0 that results from the elimination of the upper electronic state from the system dynamics.

The additional small term $(g^2/\Delta)|e\rangle\langle e|$, in contrast, is a vacuum induced energy shift of the excited level $|e\rangle$ present also if the field is in the vacuum state $|0\rangle$. It is frequently ignored or effectively incorporated in the frequency ω_e, $\omega_e \to \omega_e + g^2/\Delta \approx \omega_e$. It is sometimes called the *vacuum ac Stark shift* and can also be thought of as a single-mode remnant of the Lamb shift. Importantly, there is no similar shift of the ground state energy. This asymmetry between the states $|e\rangle$ and $|g\rangle$ finds its roots in spontaneous emission, an important point to which we return in Sect. 3.4.

We will revisit the dispersive regime of the Jaynes–Cummings model in Chap. 6 in the context of quantum non-demolition (QND) measurements, in particular when describing continuous weak-field measurements in Sect. 6.3.4, as well as in the discussions of the inverse Stern–Gerlach effect and of the generation of optical Schrödinger cats of Chap. 7.

Entangled States The dressed states (3.3) of the coupled atom–field system are an important example of a class of quantum states called *entangled states*, which are central actors in modern quantum optics and quantum information science. They will be discussed at length in the next chapter, with their role revisited in a number of situations throughout the book, but they are sufficiently important to already deserve a brief introduction in the context of the Jaynes–Cummings model.

The entanglement of two (or more) quantum mechanical objects, in the present case a single mode of the electromagnetic field and a two-level atom, describes situations where the states of these systems exhibit quantum correlations such that they must necessarily be described with reference to each other. One way to describe these peculiar situations is by noting, following Schrödinger, that *the best possible knowledge of a whole system—a pure state—does not necessarily include the best possible knowledge of its parts*. It is important to realize that this lack of knowledge is by no means due to some ignorance on the details of the interaction between the two systems. Rather, it is a key property of quantum systems that differentiates them fundamentally from their classical counterparts.

One can imagine situations where a two-level atom flies through a single-mode cavity, during which time it becomes entangled with the field mode in such a way that the system will be described, say, by the dressed state $|1, n\rangle$ at the end of the interaction. We will see in Chap. 7 how cavity QED environments permit to experimentally realize entangled states called Schrödinger cats (or perhaps more accurately Schrödinger "kitten") in cavity QED by exploiting the Jaynes–Cummings interaction while at the same time reducing the decoherence resulting from dissipation mechanisms to a remarkable level.

3.2 Quantum Rabi Oscillations

Since the dressed states are the eigenstates of the two-level atom interacting with a single mode of the radiation field, we can use them to obtain the state vector of the combined system as a function of time. Writing the Schrödinger equation in integral form as

$$|\psi(t)\rangle = e^{-i\hat{H}t/\hbar}|\psi(0)\rangle, \tag{3.16}$$

where

$$|\psi(0)\rangle = \sum_{n=0}^{\infty}\sum_{j=1}^{2} c_{i,n}|i, n\rangle$$

is the initial state of the system, we insert the identity operator expressed in terms of the dressed states $|j, n\rangle$ to find

$$|\psi(t)\rangle = \sum_{n=0}^{\infty}\sum_{j=1}^{2} \exp\left(-iE_{jn}t/\hbar\right)|j, n\rangle\langle j, n|\psi(0)\rangle, \tag{3.17}$$

where E_{jn} is given by Eq. (3.5). We have seen in the preceding section that the various eigenstate manifolds of the Jaynes–Cummings Hamiltonian are uncoupled. In matrix form and in an interaction picture rotating at the frequency $(n + \frac{1}{2})\Omega$, the dressed state amplitude coefficients inside one such manifold read

$$\begin{bmatrix} c_{2n}(t) \\ c_{1n}(t) \end{bmatrix} = \begin{bmatrix} \exp(\frac{1}{2}i\Omega_n t) & 0 \\ 0 & \exp(-\frac{1}{2}i\Omega_n t) \end{bmatrix}\begin{bmatrix} c_{2n}(0) \\ c_{1n}(0) \end{bmatrix}. \tag{3.18}$$

In particular, for an initially excited atom interacting resonantly with the field, $\Delta = 0$, we have, after returning to the bare states $|e, n\rangle$ and $|g, n\rangle$,

$$|\psi(0)\rangle = \sum_{n}|e, n\rangle = \frac{1}{\sqrt{2}}\sum_{n}(|1, n\rangle - |2, n\rangle) \tag{3.19}$$

and

$$|c_{e,n}(t)|^2 = \cos^2(g\sqrt{n+1}\,t)\,, \tag{3.20}$$

$$|c_{g,n+1}(t)|^2 = \sin^2(g\sqrt{n+1}\,t)\,. \tag{3.21}$$

This shows how the atom "Rabi flops" between the upper and lower levels within each Jaynes–Cummings manifold at the resonant quantum Rabi frequency (3.6). These oscillations have the same form as the semiclassical result (1.67), with the semiclassical Rabi frequency replaced by its n-dependent quantum mechanical value $2g\sqrt{n+1}$.

3.3 Collapse and Revivals

The expressions (3.20) and (3.21) show explicitly that different photon number states undergo periodic oscillations at different Rabi frequencies. This can have important consequences, since quantum fields are not normally in a single number state $|n\rangle$, but rather in a superposition or mixture of such states as we have seen.

Consider for example an initially excited atom interacting with a field initially in a coherent state. Combining the coherent state photon number probability (2.107) with the single-photon state result (3.20), we find that the probability $p_e(t)$ for the atom to be in its excited state at time t is

$$p_e(t) = \sum_n p_n |c_{e,n}(t)|^2 = e^{-|\alpha|^2} \sum_n \frac{|\alpha|^{2n}}{n!} \cos^2(g\sqrt{n+1}\,t)\,. \tag{3.22}$$

For a sufficiently intense field, $|\alpha|^2 \gg 1$, and short enough times $t \ll |\alpha|/g$, Problem 3.1 shows that this sum reduces approximately to

$$p_e(t) \simeq \frac{1}{2} + \frac{1}{2}\cos(2|\alpha|gt)e^{-g^2t^2}\,. \tag{3.23}$$

That this is the case can be intuitively understood by noting that the range of dominant Rabi frequencies in Eq. (3.22) is from $\Omega = g[\langle n\rangle + \sigma_n]^{1/2}$ to $g[\langle n\rangle - \sigma_n]^{1/2}$, where σ_n is the standard deviation of the photon distribution, and the probability (3.22) dephases in a time t_c such that

$$t_c^{-1} = g[\langle n\rangle + \sigma_n]^{1/2} - g[\langle n\rangle - \sigma_n]^{1/2} \simeq g\,.$$

The collapse of the upper state population (3.23) results from the interference of Rabi oscillations at the frequencies of the various number states involved.

Remarkably, though this collapse occurs with a Gaussian envelope that is independent of the mean photon number $\langle n\rangle = |\alpha|^2$. It is called the "Cummings collapse" after the physicist who first predicted it [2]. It can be derived more quantitatively from the following argument. For a large average photon number, $\langle n \rangle = |\alpha|^2 \gg 1$, the atom initially oscillates roughly at a frequency close to $g\sqrt{\langle n \rangle + 1}$. However, the dispersion in Rabi frequencies due to the distribution of photon numbers rapidly changes this behavior. For the Poisson distribution $p(n)$ characteristic of a coherent state, the variance-to-mean ratio is $\sigma_n^2 / \langle n \rangle^2 \simeq 1/\langle n \rangle$, so that for large $\langle n \rangle$, we can expand the square root in Eq. (3.22) as

$$\sqrt{n+1}\, t \approx \sqrt{\langle n \rangle}\left[1 + \frac{(n - \langle n \rangle) + 1}{2\langle n \rangle}\right] t .$$

Carrying the sum over n in the limit $gt \ll \langle n \rangle$ reduces it then to Eq. (3.23). For longer times, the system exhibits a series of "revivals" and "collapses" discussed in detail by J. H. Eberly et al. [3]. Because the photon numbers n are discrete, the n-dependent Rabi oscillations rephase in the revival time

$$t_r \simeq 4\pi\alpha/g = 4\pi \langle n \rangle^{1/2} t_c ,$$

as illustrated in Fig. 3.3.

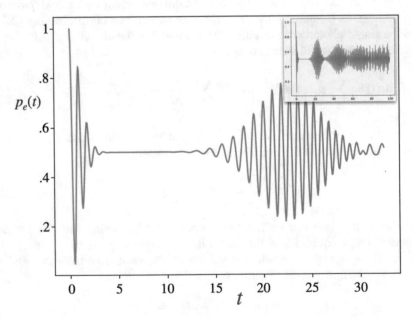

Fig. 3.3 Collapse and revivals in the interaction of a quantized single-mode field initially in a coherent state with $|\alpha|^2 = 12$ with an atom initially in its excited state $|e\rangle$. The small insert shows the long-time dynamics of the upper state population for the same mean photon number $|\alpha|^2 = 12$. Time in units of the inverse vacuum Rabi frequency g

The Cummings collapse is an important illustration of the fact that contrary to a frequently held belief, it is not in general correct to approximate strong quantized fields by their classical counterpart [4]. However, the revival property is a more unambiguous signature of quantum electrodynamics than the collapse: any spread in field strengths will dephase Rabi oscillations, but the revivals are entirely due to the "grainy" nature of the field, so that the atomic evolution is determined by the individual field quanta. The recurring long-time revivals are shown in the small insert of the figure. They are never complete and get broader and broader, eventually overlapping and resulting in a quasi-random time evolution. The Jaynes–Cummings model thus exhibits fascinating nontrivial quantum features, despite its conceptual simplicity. These effects have been observed experimentally in various cavity QED experiments, see e.g. Refs. [5–7]

It may appear rather surprising that while the coherent state is the most classical state allowed by the Heisenberg uncertainty principle, it leads to a result qualitatively different from the classical Rabi flopping formula (1.67). In contrast, the very quantum mechanical number state $|n\rangle$ has a nice, if superficial, semiclassical correspondence. This is because the Jayne-Cummings Hamiltonian eigenstates consist of an infinite number of uncoupled manifolds $\{|e, n\rangle, |g, n + 1\rangle\}$, each of the same form as their classical counterpart (1.68). This property, combined with the fact that the number state and the classical field share the property of a definite intensity, results in the absence of the interferences leading to the Cummings collapse. The indeterminacy in the field phase fundamentally associated with the number state by the Heisenberg principle (but not with the classical field) is not important for Rabi flopping since the atom and field maintain a precise relative phase in the absence of decay processes. In contrast, the coherent state field features a minimum uncertainty phase and associated minimum uncertainty intensity, and it is this latter uncertainty that causes the atom–field relative phase to "diffuse."

3.4 Single-Mode Spontaneous Emission

One intriguing difference between the semiclassical and fully quantum Rabi flopping problems is that in the quantum case (3.20), an initially excited atom Rabi flops even in the absence of any applied field, that is, even for the vacuum state $|n = 0\rangle$. Mathematically, this is because the quantum Rabi-flopping frequency is $\Omega_n = 2g\sqrt{n + 1}$, and hence $\Omega_0 = 2g$ for the vacuum state $|n = 0\rangle$, while the semiclassical expression is $\Omega_r = dE_0/\hbar = 0$ for $E_0 = 0$.

More physically, the reason why a (single-mode) quantized electric field in the vacuum state can drive an excited atom to oscillate between its excited state $|e\rangle$ and ground state $|g\rangle$ is a direct consequence of its form (2.53)

$$\hat{E}(z, t) = \mathcal{E}_\omega(\hat{a} + \hat{a}^\dagger) \sin Kz.$$

With $\hat{a}|0\rangle = 0$ and $\hat{a}^{\dagger}|0\rangle = |1\rangle$, we have that while its expectation value in the vacuum state is

$$\langle 0|\hat{E}(z,t)|0\rangle = 0, \tag{3.24}$$

the expectation value of its intensity does not vanish,

$$\langle 0|\hat{E}^2(z,t)|0\rangle = \mathcal{E}_{\omega}^2\langle 0|(\hat{a}+\hat{a}^{\dagger})^2|0\rangle = \mathcal{E}_{\omega}^2. \tag{3.25}$$

It is these *vacuum fluctuations* of the field that are responsible for driving the atom down from its excited state. These same fluctuations are also responsible for the vacuum frequency shift g^2/Δ of the excited atomic level $|e\rangle$ that we encountered in the discussion of the dispersive Jaynes–Cummings Hamiltonian (3.15).

One may wonder, then, why it is not possible to likewise drive the atom from its ground state $|g\rangle$ to $|e\rangle$ with a vacuum field $|0\rangle$, as is readily seen from Eq. (3.21) for the initial state $|g,0\rangle$. This can be understood by recalling the discussion of photodetection of Sect. 2.4, where we found that the probability (2.162) to excite an atom is given by the normally ordered correlation function of the field (replacing i by g to match the notation to the situation at hand)

$$w_{g\to e} \propto \langle e|\hat{E}^-(\mathbf{r},t)\hat{E}^+(\mathbf{r},t)|e\rangle. \tag{3.26}$$

Conversely, it is easily shown that the probability to emit a photon from the excited state is given by the *antinormally ordered correlation function*

$$w_{e\to g} \propto \langle e|\hat{E}^+(\mathbf{r},t)\hat{E}^-(\mathbf{r},t)|e\rangle, \tag{3.27}$$

that is, absorption is driven by normally ordered field correlations and emission by antinormally ordered ones. The expectation value of the first ones is equal to zero, for a vacuum field $\langle 0|\hat{a}^{\dagger}\hat{a}|0\rangle = 0$, but not the second one, $\langle 0|\hat{a}\hat{a}^{\dagger}|0\rangle = 1$. This is why atoms can undergo spontaneous emission, but not spontaneous absorption.

In the single-mode model considered here, spontaneous emission is followed by the reabsorption of the photon, resulting in periodic oscillations of the atomic population at the vacuum Rabi frequency $2g$. This is not the situation encountered under usual circumstances, in which case once the photon is emitted, it escapes into free space, with no chance of return to the atom to be reabsorbed. This difference results from the fact that free space situations are not adequately described by a single-mode model. They require, instead, a description of the electromagnetic field as a "reservoir" with a continuum of modes, as will be discussed in Chap. 5. Single-mode systems, and more generally systems with tailored densities of field modes, can however be experimentally realized in cavity QED and circuit QED environments, the topic of Chap. 7.

3.5 Repeated Field Measurements

We saw in Sect. 2.4 how a detector operating by absorption measures the normally ordered field correlation function $\langle \hat{E}^-(\mathbf{r}, t)\hat{E}^+(\mathbf{r}, t)\rangle$. In many cases, one is interested in monitoring the evolution of a system as a function of time, and this requires performing a sequence of measurements on that system. A simple extension of the Jaynes–Cummings model provides an example of a model system that can achieve this goal.

We consider an idealized detection scheme where a stream of two-level atoms is used as probes to learn about the dynamics of a single-mode field. The atoms enter the field one at a time in their excited state $|e\rangle$ and interact with it for a time τ. Information on the field is then extracted from the measurement of the final state of the successive atoms after their exit from the interaction region. Despite the fact that it is a caricature of a realistic measurement protocol, this model already teaches us important lessons on the back action of quantum measurements, and illustrates the care that must be exercised in performing them and understanding their impact on the observed system. It may therefore prove useful in preparation for the quantum trajectories method of Sect. 5.4.1, and the issues that it raises will be expanded upon in much detail in Chap. 6.

Assuming for simplicity that the field is diagonal in the energy eigenstates $\{|n\rangle\}$, as the ith atom enters the field at time t_i, the state of the system is

$$\hat{\rho}(t_i) = |e\rangle\langle e| \sum_n p_n(t_i)|n\rangle\langle n|, \tag{3.28}$$

where $p_n(t_i)$ is the field photon statistics at t_i. The atom–field interaction is described by the Jaynes–Cummings Hamiltonian $\hat{H}$ of Eq. (3.1), so that the field density operator at the end of its interaction with the atom is

$$\hat{\rho}_f(t_i + \tau) = \mathrm{Tr}_{\mathrm{atom}}\left[\hat{U}(\tau)\hat{\rho}(t_i)\hat{U}^\dagger(\tau)\right], \tag{3.29}$$

where $\hat{U}(\tau) = \exp(-i\hat{H}\tau/\hbar)$. It follows directly from Eqs. (3.20) and (3.21) that the field photon statistics at that time is therefore

$$p_n(t_i + \tau) = \langle n|\hat{\rho}_f(t_i + \tau)|n\rangle$$
$$= p_{n-1}(t_i)\sin^2(g\sqrt{n+1}\,\tau) + p_n(t_i)\cos^2(g\sqrt{n}\,\tau). \tag{3.30}$$

However, a measurement of the state of the atom changes this result. Specifically, the field density operator will then be projected to

$$\hat{\rho}_f = \mathrm{Tr}_{\mathrm{atom}}\left[|s\rangle\langle s|\hat{U}(\tau)\hat{\rho}(t_i)\hat{U}^\dagger(\tau)\right], \tag{3.31}$$

where $|s\rangle = |e\rangle$ or $|g\rangle$ for an atom measured to be in its excited or ground state, respectively, with corresponding photon statistics

$$p_{n|e}(t_i + \tau) = \mathcal{N}_e \, p_n(t_i) \cos^2(g\sqrt{n+1}\,\tau)\,, \qquad (3.32)$$

$$p_{n|g}(t_i + \tau) = \mathcal{N}_g \, p_{n-1}(t_i) \sin^2(g\sqrt{n}\,\tau)\,, \qquad (3.33)$$

where $\mathcal{N}_e$ and $\mathcal{N}_g$ are normalization constants. Ignoring field dissipation during the intervals between probe atoms, one or the other of these photon statistics, depending on the result of the measurement, will become the field initial condition at the time t_{i+1} when the next atom starts interacting with it.

A numerical simulation of this measurement sequence proceeds by choosing the initial photon statistics $p_n(t_1)$ and interaction time τ. This permits to compute the probability

$$p_e(t_1 + \tau) = \text{Tr}[|e\rangle\langle e|\hat\rho(\tau)] = \sum_{n=0}^{\infty} p_n(t_1) \cos^2(g\sqrt{n+1}\tau) \qquad (3.34)$$

to measure the first atom in the excited state at time $t_1 + \tau$. A random number generator then returns a uniform deviate between 0 and 1, and the atom is said to have been measured in the excited state $|e\rangle$ if $0 \le r < p_e$, and in the ground state otherwise. This results in an updated field density operator $\hat\rho_f(t_1 + \tau)$, with photon statistics given by either Eq. (3.32) or (3.33), that will be the initial condition for the next atom.

Figure 3.4 shows selected results from two typical sequences of 12 measurements each, for a field initially in a thermal state with mean photon number $\langle n \rangle = 20$.

Atom	p_e	Result	$\bar n$	σ^2		Atom	p_e	Result	$\bar n$	σ^2
1	0.467	$|g\rangle$	19.5	21.9		1	0.467	$|e\rangle$	21.7	18.7
2	0.240	$|g\rangle$	20.1	21.6		2	0.728	$|e\rangle$	22.4	17.7
3	0.185	$|g\rangle$	21.2	20.9		3	0.820	$|e\rangle$	22.8	17.2
4	0.179	$|g\rangle$	23.0	20.0		4	0.867	$|e\rangle$	23.1	16.9
5	0.193	$|g\rangle$	25.5	18.7		5	0.896	$|e\rangle$	23.2	16.7
6	0.211	$|g\rangle$	29.0	16.9		6	0.914	$|e\rangle$	23.3	16.6
7	0.223	$|g\rangle$	25.4	14.5		7	0.927	$|e\rangle$	23.3	16.6
8	0.218	$|g\rangle$	38.0	12.0		8	0.937	$|g\rangle$	24.0	16.4
9	0.196	$|e\rangle$	25.4	16.9		9	0.832	$|e\rangle$	24.0	16.4
10	0.450	$|g\rangle$	32.6	14.1		10	0.858	$|e\rangle$	23.8	16.5
11	0.423	$|g\rangle$	40.2	10.2		11	0.875	$|e\rangle$	23.6	16.5
12	0.367	$|e\rangle$	32.3	13.4		12	0.887	$|g\rangle$	26.1	15.4

Fig. 3.4 The left and right tables show extracts from the records of two typical numerical experiments. For each of the successive probe atoms, p_e is the probability to be in the excited state at the end of its interaction with the field, "Result" is the outcome of the measurement of the atomic state, and $\bar n$ and σ^2 are the updated mean photon number $\langle n \rangle$ and variance of the field following that measurement outcome. (From Ref. [8])

This table illustrates important features of repeated measurement sequences. First of all, measuring an exiting atom in its excited state $|e\rangle$ *does not* mean that nothing happened to the field and that the mean photon number $\langle n \rangle$ is conserved. Rather, its photon statistics is reshuffled according to Eq. (3.32). This implies in particular that the mean photon number is changed, since

$$\sum_n n p_n(t_i) \neq \mathcal{N}_e \sum_n n\, p_n(t_i) \cos^2(g\sqrt{n+1}\, t_{\text{int}})\,. \tag{3.35}$$

Likewise, measuring the exiting atom in its ground state does not mean that the mean photon number has increased by one.

These results are not surprising: before the measurement, the mean photon number $\langle n \rangle$ is known only within its variance σ^2, and not conserving it if the state of the atom is the same before and after the interaction does not violate any conservation law. It is only for a number state, $p_n = \delta(n-m)$, at the time of injection of the ith atom, that the conservation of $\langle n \rangle$ is guaranteed if the atom is measured in its excited state at the end of the interaction. The successive measurements also tend to reduce σ^2, although this is not always the case, and consistently with the previous comment, as σ^2 is reduced, one reaches a regime of much better conservation of energy.

The main message of this section is that quantum mechanics permits the simulation of typical realizations of measurement sequences on a single quantum system and that these measurements typically change the state of the system in significant ways. The changes in photon statistics and mean photon number observed in our specific measurement scheme are an example of *measurement back action*, a fundamental feature of quantum measurements to which we will return at length in Chap. 6.

3.6 The Quantum Rabi Model

We mentioned that it is usually inconsistent to describe an atom as a two-state system, thereby ignoring all other levels that may be coupled by the optical field, while not performing the rotating wave approximation at the same time. This is based on the simple fact that "a theory is only as good as its weakest element." However "usually" does not mean "always," and experimental advances, in particular in circuit QED, have resulted in situations where the dipole coupling constant g can be large enough that this statement needs to be reconsidered. There are indeed cases where it is necessary to go beyond the Jaynes–Cummings model and consider its more general version, the *quantum Rabi model*. This is an extension of the model originally developed by E. Rabi for the case of classical fields [9, 10] and characterized by the Hamiltonian

$$\hat{H} = \frac{1}{2}\hbar\omega_0\hat{\sigma}_z + \hbar\omega\left(\hat{a}^\dagger\hat{a} + \tfrac{1}{2}\right) + \hbar g\hat{\sigma}_x\left(\hat{a} + \hat{a}^\dagger\right)\,, \tag{3.36}$$

which is the Jaynes–Cummings Hamiltonian, extended to include the counter-rotating terms in the dipole interaction. It reduces to the Jaynes–Cummings model in the limit $g, \Delta \ll \omega, \omega_0$.

Due to the mathematical difficulties associated with the inclusion of the counter-rotating terms, the Rabi Hamiltonian has been mostly analyzed either numerically or by approximate methods appropriate for the particular situations and sets of parameters at hand. More specifically, in addition to the atom–field interaction, which is characterized by the vacuum Rabi frequency g, experiments always include dissipation mechanisms as well. The most important ones in quantum optics are typically spontaneous emission and cavity losses, which are characterized by decay rates γ and κ, respectively, as will be discussed in Chap. 7. The *strong coupling regime* of the Rabi model is usually defined as that regime where $g \gg \kappa, \gamma$, the *ultrastrong coupling regime* by $g \geq \omega/10, \omega_0/10$, and the *deep strong coupling regime* by $g \gg \omega, \omega_0$.

It is beyond the scope of this section to cover all aspects of the Rabi model and the various situations in which it is being applied, see, for instance, Refs. [11–15] for more details. Indeed, such a review might merit its own monograph. Instead, we concentrate here on just one recent development, the discovery of an exact solution to this model [16]. While its analytical form is not particularly transparent, it is of sufficient importance to merit our attention. In the following, we outline the main steps of its derivation, using the Bogoliubov transformation approach of Ref. [17].

Bogoliubov Diagonalization To reflect the role of the two-level atom in the Rabi model, we proceed by expressing its eigenstates $|\psi\rangle$ as the two-component wave functions

$$|\psi\rangle = \begin{pmatrix} \psi_1 \\ \psi_2 \end{pmatrix}. \tag{3.37}$$

Inserting this form in the time-independent Schrödinger equation $\hat{H}|\psi\rangle = E|\psi\rangle$, then, gives

$$\hbar\left(\frac{\omega_0}{2} + \omega\hat{a}^\dagger\hat{a}\right)\psi_1 + \hbar g\left(\hat{a} + \hat{a}^\dagger\right)\psi_2 = E\psi_1,$$

$$\hbar\left(-\frac{\omega_0}{2} + \omega\hat{a}^\dagger\hat{a}\right)\psi_2 + \hbar g\left(\hat{a} + \hat{a}^\dagger\right)\psi_1 = E\psi_2, \tag{3.38}$$

or, with $\phi_1 = \psi_1 + \psi_2$ and $\phi_2 = \psi_1 - \psi_2$,

$$\hat{H}\begin{pmatrix} \phi_1 \\ \phi_2 \end{pmatrix} \equiv \hbar\begin{pmatrix} \omega\hat{a}^\dagger\hat{a} + g(\hat{a} + \hat{a}^\dagger) & \omega_0/2 \\ \omega_0/2 & \omega\hat{a}^\dagger\hat{a} - g(\hat{a} + \hat{a}^\dagger) \end{pmatrix} = E\begin{pmatrix} \phi_1 \\ \phi_2 \end{pmatrix}. \tag{3.39}$$

We now introduce the Bogoliubov transformation[2]

$$\hat{A} = \hat{a} + g/\omega \quad ; \quad \hat{A}^\dagger = \hat{a}^\dagger + g/\omega \, , \tag{3.40}$$

in terms of which the Hamiltonian $\hat{H}$ becomes

$$\hat{H}' = \hbar \begin{pmatrix} \omega \hat{A}^\dagger \hat{A} - g^2/\omega & \omega_0/2 \\ \omega_0/2 & \hat{A}^\dagger \hat{A} - 2g(\hat{A} + \hat{A}^\dagger) + 3g^2/\omega \end{pmatrix} , \tag{3.41}$$

as well as the trial form of the eigenstates of $\hat{H}$

$$\begin{pmatrix} \phi_1 \\ \phi_2 \end{pmatrix} = \sum_{n=0}^{\infty} \sqrt{n!} \begin{pmatrix} e_n |n_A\rangle \\ f_n |n_A\rangle \end{pmatrix} . \tag{3.42}$$

Here, the states $|n_A\rangle$ are the "extended coherent states"[3]

$$|n_A\rangle = \sum_{n=0}^{\infty} \frac{(A^\dagger)^n}{\sqrt{n!}} |0_A\rangle \, , \tag{3.43}$$

and $|0_A\rangle$ is the displaced vacuum state $|0\rangle$ of the field mode,

$$|0_A\rangle = e^{-(g/\omega)(\hat{a}^\dagger - \hat{a})} |0\rangle \, , \tag{3.44}$$

which we recognize with Eq. (2.112), the coherent state $| - g/\omega\rangle$. The diagonalization of $\hat{H}'$ amounts to determining the expansion coefficients e_n and f_n and the associated eigenenergies E_n. Substituting Eq. (3.42) into Eq. (3.39) with the form (3.41) of $\hat{H}'$ and multiplying by the state $\langle m_A|$ yield readily [17]

$$e_n = -\frac{\omega_0/2}{n\omega - g^2/\omega - E/\hbar} f_n \tag{3.45}$$

and

$$(n\omega + 3g^2/\omega - E/\hbar) f_n - 2g(n+1) f_{n+1} - 2g f_{n-1} = \omega_0 e_n \, . \tag{3.46}$$

---

[2]Bogoliubov transformations are linear, canonical transformations of creation and annihilation operators of the general form

$$\hat{A} = u\hat{a} + v\hat{a}^\dagger + w \quad ; \quad \hat{A}^\dagger = u^\star \hat{a}^\dagger + v^\star \hat{a} + w^\star.$$

For the transformation to be canonical, it must preserve the bosonic commutation relation, $[\hat{A}, \hat{A}^\dagger] = 1$, resulting in the condition $|u|^2 - |v|^2 = 1$ or $u = \exp(i\theta_1) \cosh r$, $v = \exp(i\theta_2) \sinh r$.

[3]Importantly, the states $|n_A\rangle$ should not be confused with the familiar Fock states $|n\rangle$ of the field mode, despite the similar notation.

This expression can be expressed as the recursion relation

$$\eta f_\eta = \Omega(\eta - 1) f_{\eta-1} - f_{\eta-2}, \tag{3.47}$$

with

$$\Omega(n) = \frac{1}{2g} \left(n\omega + 3g^2/\omega - E/\hbar - \frac{\omega_0^2/4}{n\omega - g^2/\omega - E/\hbar} \right), \tag{3.48}$$

and one can choose $f_0 = 1$ and $f_1 = \Omega(0)$, up to the normalization of the state.

However, we are not yet quite done and still need to determine the eigenenergies E. This is achieved by repeating the same procedure with a second Bogoliubov transformation

$$\hat{B} = \hat{a} - g/\omega \quad ; \quad \hat{B}^\dagger = \hat{a}^\dagger - g/\omega \tag{3.49}$$

and requiring that the resulting form of the eigenstates be the same as given by Eq. (3.42). As we now show, this additional requirement will provide us with a final equation that will yield the eigenenergies E.

In terms of the transformation (3.49), the Hamiltonian (3.39) becomes

$$\hat{H}' = \hbar \begin{pmatrix} \omega \hat{B}^\dagger \hat{B} + 2g(\hat{B} + \hat{B}^\dagger) + 3g^2/\omega & \omega_0/2 \\ \omega_0/2 & \hat{B}^\dagger \hat{B} - g^2/\omega \end{pmatrix}, \tag{3.50}$$

and following the same procedure as before, we introduce the eigenstate

$$\begin{pmatrix} \phi_1' \\ \phi_2' \end{pmatrix} = \sum_{n=0}^\infty \sqrt{n!} \begin{pmatrix} f_n'|n_B\rangle \\ e_n'|n_B\rangle \end{pmatrix}. \tag{3.51}$$

This leads now to

$$e_n' = -\frac{\omega_0/2}{n\omega - g^2/\omega - E/\hbar} f_n' \tag{3.52}$$

and

$$\eta f_n' = \Omega'(n - 1) f_{n-1}' - f_{n-2}', \tag{3.53}$$

again with $f_0 = 1$ and $f_1 = \Omega(0)$ and an overall normalization of the state. Requiring that the eigenstates resulting from the two Bogoliubov transformations represent the same physical state, we must have

$$\begin{pmatrix} \phi_1 \\ \phi_2 \end{pmatrix} = r \begin{pmatrix} \phi_1' \\ \phi_2' \end{pmatrix}, \tag{3.54}$$

and multiplying these equations by $\langle 0|$ gives

$$\left(\sum_{n=0}^{\infty} e_n g^n\right)\left(\sum_{n=0}^{\infty} e'_n g^n\right) - \left(\sum_{n=0}^{\infty} f_n g^n\right)\left(\sum_{n=0}^{\infty} f'_n g^n\right) = 0. \tag{3.55}$$

Since f_n and f'_m satisfy the same recurrence relation, it follows that $f_n = f'_n$ and $e_n = e'_n$. With Eq. (3.45), this equation reduces simply to $G_+ G_- = 0$, where

$$G_\pm \equiv \sum_{n=0}^{\infty} f_n \left(1 \mp \frac{\hbar\omega_0/2}{E + \hbar g^2/\omega - n\hbar\omega}\right) g^n. \tag{3.56}$$

Therefore, the full energy spectrum of the Rabi model is given by the zeros of the two functions G_+ and G_-, indicative of the existence of two distinct manifolds of eigenfunctions and eigenenergies.

Energy Spectrum   Figure 3.5 shows $G_\pm(x)$ as functions of $x = (E + g^2/\omega)/\hbar$. Their zeros give the eigenenergy spectrum of the system, which is shown in Fig. 3.6. It consists, as just mentioned, of two manifolds, one of them associated with G_+ and the other with G_-. We observe that G_+ and G_- also exhibit simple poles at $x = 0, \omega, 2\omega, \ldots$ These poles correspond to special values of g, ω_0, and ω such that

$$x = (E + g^2/\omega)/\hbar = 0, 1, 2, \ldots \tag{3.57}$$

Ref. [16] calls the associated energies $E_n^e = \hbar(n\omega - g^2/\omega)$ the *exceptional spectrum*.

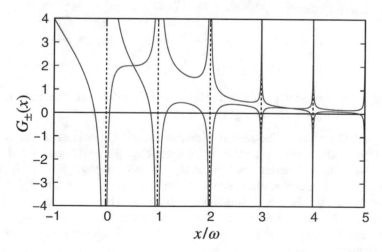

Fig. 3.5 Functions $G_+(x)$ (red lines) and $G_-(x)$ (blue lines) as a function of the variable x/ω, where $x = (E + g^2/\omega)/\hbar$. The zeros of these functions for the regular eigenenergy spectrum of the system. The poles at locations $x = n\omega, n = 0, 1, 2, \ldots$, give the so-called exceptional spectrum. (From Ref. [16])

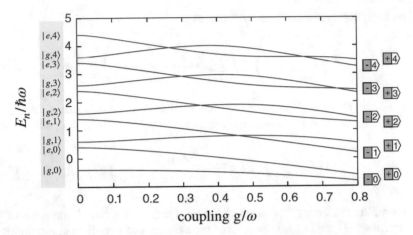

Fig. 3.6 Eigenenergy spectrum of the Rabi model for $\omega_0 = 0.8$ and $0 \leq g \leq 0.8$, in units of ω. The eigenenergy spectrum consists of two intersecting ladders of levels, each corresponding to a different parity subspace, the $G_+ = 0$ (red) and $G_- = 0$ (blue) subspaces. Within each subspace, the states are labeled with ascending numbers, shown on the right-hand side of the figure. This labeling does not change with g because no lines intersect within spaces of fixed parity. On the left side of the graph, the states for $g = 0$ are labeled in the gray box by the bare states $|e, n\rangle$ and $|g, n\rangle$ of the atom–field system. (Figure adapted from Ref. [16])

Figure 3.6 shows the two manifolds of the Rabi spectrum as functions of g. Importantly, there are no level crossings within each subspace of levels, allowing their unique labeling with two quantum numbers $\pm$ and $n = 0, 1, \ldots$ For $g \to 0$, pairs of levels within alternating manifolds converge to pairs of bare states $\{|e, n\rangle, |g, n + 1\rangle\}$, with

$$|-, n\rangle \to |e, n\rangle \quad ; \quad |-, n + 1\rangle \to |g, n + 1\rangle , \quad n \text{ odd} ,$$

$$|+, n\rangle \to |e, n\rangle \quad ; \quad |+, n + 1\rangle \to |g, n + 1\rangle , \quad n \text{ even} . \tag{3.58}$$

For $g \ll 1$, the eigenstates of the Rabi Hamiltonian converge therefore as should be expected to the Jaynes–Cummings dressed states.

Conserved Quantities The reason why the dressed eigenstates of the Jaynes–Cummings model form an infinite set of manifolds $\{|1, n\rangle, |2, n\rangle\}$ is that in the rotating wave approximation the bare levels are only coupled pairwise, $|e, n\rangle \leftrightarrow |g, n + 1\rangle$. Mathematically, this is due to the fact that the operator $\hat{C} \equiv \hat{a}^\dagger \hat{a} + \hat{\sigma}_+ \hat{\sigma}_-$ commutes with the Jaynes–Cummings Hamiltonian. However, this is no longer the case for the quantum Rabi Hamiltonian. As pointed out in Ref. [16], there is however another conserved quantity in that case in addition to energy; this is the *parity operator*

$$\hat{\Pi} \equiv -\hat{\sigma}_z (-1)^{\hat{a}^\dagger \hat{a}} , \tag{3.59}$$

see Problem 3.7. It is this symmetry that leads to a decomposition of the state space of the Rabi Hamiltonian into two subspaces of infinite dimension.

Problems

Problem 3.1 Following the steps outlined after Eq. (3.23), show that for large enough $\langle n \rangle$ and times $gt \ll \langle n \rangle$, the probability for an atom driven by a coherent state $|\alpha\rangle$ to be in its excited state $p_e(t)$ at time t undergoes a Cummings collapse in a time that is independent of $\langle n \rangle$ and with

$$p_e(t) \simeq \tfrac{1}{2} + \tfrac{1}{2}\cos(2|\alpha|gt)\exp(-g^2 t^2).$$

Problem 3.2 Consider a two-photon Jaynes–Cummings model described by the Hamiltonian

$$\hat{H} = \tfrac{1}{2}\hbar\omega_0\hat{\sigma}_z + \hbar\omega\left(\hat{a}^\dagger\hat{a} + \tfrac{1}{2}\right) + \hbar g\left(\hat{\sigma}_+\hat{a}^2 + \hat{a}^{\dagger 2}\hat{\sigma}_-\right). \tag{3.60}$$

(a) Find the eigenstates and eigenenergies of this system.
(b) For a field initially in a coherent state and an initially excited atom, find and plot the probability $p_e(t)$ for the atom still to be in that state at time t. Compare and discuss the differences between that solution and the single-photon transition case of Sect. 3.3.

Problem 3.3 Considering the two-photon Jaynes–Cummings model of Problem 3.2

(a) find and plot the probability $p_e(t)$ for the atom to be in its excited state as a function of time for the initial state $|\psi(0)\rangle = |e, \alpha\rangle$, where $|\alpha\rangle$ is a coherent state with $|\alpha|^2 = 10$.
(b) Find and plot also the resulting photon statistics $p_n(t)$ as a function of time.

Problem 3.4 Consider now the dispersive limit of the Jaynes–Cummings Hamiltonian

$$H_{\mathrm{JC,eff}} = \frac{1}{2}\hbar\omega_0\hat{\sigma}_z + \hbar\omega\hat{a}^\dagger\hat{a} + \frac{\hbar g^2}{\Delta}\left[(\hat{a}^\dagger\hat{a} + 1)|e\rangle\langle e| - \hat{a}^\dagger\hat{a}|g\rangle\langle g|\right].$$

It is possible to approximate it as

$$H_{\mathrm{JC,eff}} \approx \frac{1}{2}\hbar\omega_0\hat{\sigma}_z + \hbar\omega\hat{a}^\dagger\hat{a} + \frac{\hbar g^2}{\Delta}\hat{a}^\dagger\hat{a}\left[|e\rangle\langle e| - |g\rangle\langle g|\right].$$

Carry out this approximation and discuss what is involved in this step in physical terms. When is it justified, and what condition(s) must be satisfied to make this approximation acceptable?

Problem 3.5 In the dispersive Jaynes–Cummings model of Problem 3.4, find and plot the probability $p_e(t)$ for the atom to be in its excited state as a function of time for the initial state $|\psi(0)\rangle = |e, \alpha\rangle$, where $|\alpha\rangle$ is a coherent state with $|\alpha|^2 = 10$, $\hbar\omega_0/\Delta = 0.02\omega_0$. Find and plot also the resulting photon statistics $p_n(t)$ as a function of time. Evaluate also $\langle n \rangle(t)$, and explain this result.

Problem 3.6 Considering the resonant Rabi Hamiltonian

$$\hat{H} = \hbar\omega\left(\frac{\omega_0}{2}\sigma_z + \hat{a}^\dagger\hat{a}\right) + \hbar\sigma_x(\hat{a} + \hat{a}^\dagger),$$

determine under which condition(s), the counterrotating terms of the interaction can be treated as a small perturbation. To lowest order in the corrections due to the no-rotating terms of the interaction, evaluate then the probability $p_e(t)$ for the atom to be in its upper state for an atom initially in the ground state and a field (a) in a coherent state and (b) in its ground state.

Problem 3.7 (a) Show that the operator $\hat{C} \equiv \hat{a}^\dagger\hat{a} + \hat{\sigma}_+\hat{\sigma}_-$ commutes with the Jaynes–Cummings Hamiltonian (3.1) and (b) show also that the parity operator $\hat{\Pi} \equiv -\hat{\sigma}_z(-1)^{\hat{a}^\dagger\hat{a}}$ commutes with the Rabi Hamiltonian (3.36).

Problem 3.8 Consider a field with initial photon statistics p_n and coupled for a time τ with an atom initially in its excited state $|e\rangle$ by the Jaynes–Cummings Hamiltonian.

(a) Find the atom–field density operator at the end of the interaction.
(b) Assuming now that the atom is measured to be in its excited state $|e\rangle$ after the interaction, determine the resulting field density operator, and do the same for a measurement finding the atom in the ground state $|g\rangle$.
(c) Considering the two explicit cases of Poisson photon statistics and of a thermal field with initial mean photon number $\langle n \rangle = 9$, plot the initial and final photon statistics of the field, and find the change in mean photon number following these measurements.
(d) Determine also the change in von Neumann entropy of the field following these measurements.

Problem 3.9 Carry out the derivation of Eqs. (3.55) and (3.56).

References

1. E.T. Jaynes, F.W. Cummings, Comparison of quantum and semiclassical radiation theories with application to the beam maser. Proc. IEEE **51**, 89 (1963)
2. F.W. Cummings, Stimulated emission of radiation in a single mode. Phys. Rev. **140**, A1051 (1965)
3. J.H. Eberly, N.B. Narozhny, J.J. Sanchez-Mondragon, Periodic spontaneous collapse and revival in a simple quantum model. Phys. Rev. Lett. **44**, 1323 (1980)
4. P. Meystre, A. Quattropani, H. Baltes, Quantum-mechanical approach to Rabi-flopping. Phys. Lett. **49A**, 85 (1974)
5. G. Rempe, H. Walther, N. Klein, Observation of quantum collapse and revival in a one-atom maser. Phys. Rev. Lett **58**, 353 (1987)
6. M. Brune, F. Schmidt-Kaler, A. Maali, J. Dreyer, E. Hagley, J.M. Raimond, S. Haroche, Quantum Rabi oscillation: A direct test of field quantization in a cavity. Phys. Rev. Lett. **76**, 1800 (1996)
7. S. Haroche, J.-M. Raimond, *Exploring the Quantum—Atoms, Cavities and Photons* (Oxford, New York, 2006)
8. P. Meystre, Repeated quantum measurements on a single harmonic oscillator. Optics Lett. **13**, 669 (1997)
9. I.I. Rabi, On the process of space quantization. Phys. Rev. **49**, 324 (1936)
10. I.I. Rabi, Space quantization in a gyrating magnetic field. Phys. Rev. **49**, 324 (1936)
11. Q. Xie, H. Zhong, M.T. Batchelor, C. Lee, The quantum Rabi model: solution and dynamics. J. Phys. A Math. Theor. **50**, 11300 (2017)
12. J.S. Pedernales, I. Lizuain, S. Felicetti, G. Romero, E. Solano, Quantum Rabi model with trapped ions. Phys. Rep. **5**, 15472 (2015)
13. J. Braumueller, M. Marthaler, A. Schneider, A. Stehli, H. Rotzinger, M. Weides, A.V. Ustinov, Analog quantum simulation of the Rabi model in the ultra-strong coupling regime. Nature Commun. **8**, 779 (2017)
14. D.Z. Rossatto, C.J. Villas-Boas, M. Sanz, E. Solano, Spectral classification of coupling regimes in the quantum Rabi model. Phys. Rev. A **96**, 013849 (2017)
15. A.F. Kockum, A. Miranowicz, S. De Liberato, S. Savastava, F. Nori, Ultrastrong coupling between light and matter. Nat. Rev. Phys. **1**, 19 (2019)
16. D. Braak, On the integrability of the Rabi model. Phys. Rev. Lett. **107**, 100401 (2011)
17. Q.H. Chen, C. Wang, T. Liu, K.L. Wang, Exact solvability of the quantum Rabi model using Bogoliubov operators. Phys. Rev. A **86**, 023822 (2012)

Chapter 4
Composite Systems and Entanglement

Besides wave–particle duality the most puzzling aspect of quantum systems is arguably quantum entanglement, whereby the state of a subsystem of a composite quantum system cannot be described independently of the state of its other subsystem(s). Following a discussion of the EPR paradox, we introduce formally the concepts of entanglement, maximum entanglement, and monogamy of entanglement. We then turn to Bell's inequalities and summarize their most recent experimental tests. This is followed by a discussion of the no-cloning theorem and of two important applications of quantum entanglement: quantum teleportation and quantum key distribution.

In contrast to their classical counterparts, quantum systems present a number of counter-intuitive features: they can exist in a superposition of several states and can exhibit either wave-like or particle-like behavior, depending on the circumstances. But most puzzling perhaps is *quantum entanglement*, whereby the state of subsystems of a composite quantum system cannot be described independently of the state of the others. Perhaps the most famous consequence of quantum entanglement is the "spooky action at a distance" discussed by Einstein, Podolsky, and Rosen in their famous paper [1] that concluded (erroneously)[1] that quantum mechanics does not give a complete description of reality. This chapter gives an introduction to some of the central aspects of quantum entanglement, concentrating largely on two-particle entanglement.

After a brief summary of the Einstein–Podolsky–Rosen (EPR) paradox, whose extraordinary merit is to point out a fundamental implication of quantum entanglement and forces one to confront the profound difference between the classical and quantum worlds, we introduce some central aspects of entanglement, including entanglement entropy and entanglement monogamy. We then turn to the Bell's

---

[1] Attaching the qualifier "erroneous" to that paper, while technically correct, is profoundly misleading: few papers in physics have had as much impact in advancing our understanding of nature and of the quantum world. I suspect that many of us wish we would be capable of writing such an "erroneous" paper.

inequalities and summarize their most recent experimental tests. This is followed by a discussion of two important applications of quantum entanglement, quantum teleportation and quantum cryptography, a discussion that will also lead us to introduce the no-cloning theorem central to these applications. This chapter deals almost exclusively with two-level systems, or qubits, in pure states, as they are of particular relevance for applications in quantum information. Mixed states will then take center stage in Chaps. 5 and 6.

4.1 The EPR Paradox

In a famous 1935 paper titled "Can Quantum-Mechanical Description of Physical Reality be Considered Complete?" A. Einstein, B. Podolsky, and N. Rosen puzzled on the implications of quantum entanglement and the associated "spooky action at a distance" and argued that the description of physical reality provided by quantum mechanics is incomplete as it failed to account for the existence of "elements of reality." It took several decades to finally put this argument to rest, following a series of increasingly sophisticated experimental tests that quantum mechanics passed with flying colors. But the importance of the EPR paper cannot be overstated: not only does the resolution of the EPR paradox have fundamental implications for the interpretation of quantum mechanics, but the deeper understanding of quantum entanglement resulting from its resolution also paves the way to remarkable new developments, most notably perhaps in quantum information science and quantum metrology.

We begin by briefly reviewing the EPR argument, in the slightly modified form put forward by D. Bohm. Consider a source in which pairs of identical spin-1/2 particles are produced, say, by the photodissociation of a diatomic molecule prepared in the singlet state $S = 0$. Upon emerging from the source, the two particles fly toward two space-like separated Stern–Gerlach magnets acting as analyzers and detectors. Long after the particles are emitted, an observer orients the first magnet so as to measure the spin component $S_a = \mathbf{a} \cdot \mathbf{S_1}$ of particle 1 along $\mathbf{a}$. For a spin-1/2 particle, the result of this measurement is $\pm \hbar/2$. Because the total spin of the system is zero, we then know for sure that the spin of the second particle along that same direction is $\mp \hbar/2$.

At this point, EPR introduce the concept of *reality*: "If, without in any way disturbing a system we can predict with certainty (i.e. with probability equal to unity) the value of a physical quantity, then there is an element of physical reality corresponding to this quantity." EPR further require that "every element of the physical reality must have a counterpart in the physical theory."

According to this criterion, we can attribute an element of physical reality to the spin component S_a. However, the observer could just as well have chosen to set the detector 1 in direction $\mathbf{a}$', thus measuring the spin component $S_{a'}$ of the first particle. In that way, he would have inferred, without in any way disturbing particle "2," its spin component $S_{a'}$. It follows that there is also an element of physical reality attached to $S_{a'}$. But in quantum mechanics the Pauli's uncertainty principle states that one cannot predict precise values for non-commuting observables. Thus, as stated by Einstein in a letter to Max Born, "...one must consider the description given by quantum mechanics as an incomplete and indirect description of reality, destined to be later replaced by an exhaustive and direct description..."

The EPR argument was refuted by many of the founders of quantum mechanics, but for many years, it seemed that no experiment was able to determine which was the correct attitude. The situation has now changed drastically, due largely to the seminal contributions of J. S. Bell. A fascinating collection of his contributions to that topic can be found in Ref. [2].

4.2 Quantum Entanglement

Roy Glauber, the father of Quantum Optics, was fond of saying that "paradoxes are cleverly stated incorrect statements." The EPR paradox results from the fact that it implicitly treats the photodissociation products, in the example of the previous section, as consisting of two separate particles. But this is not correct: the dissociation products still comprise a single quantum system, albeit with the property of possibly being strongly delocalized. That this is the case can be seen by considering its von Neumann entropy S, which we already encountered when discussing thermal states of the electromagnetic field, see Eq. (2.77).

The von Neumann entropy S, like the Gibbs entropy in the classical world, has an important interpretation in terms of information content: basically, it characterizes the missing information about the system. If computed in a log 2 basis, this information is measured in *bits*. Pure states have zero entropy, as follows immediately from the fact that their density operator is idempotent, $\hat{\rho}^2 = \hat{\rho}$, and hence $S = -\text{Tr}[\hat{\rho} \log_2 \hat{\rho}] = 0$.[2] There is no missing information in a pure state. At the other extreme, in a Hilbert space of dimension d the entropy can be at most $S_{\text{max}} = \log d$ bits. This happens for a *maximally mixed state*, where all eigenstates $|m\rangle$ of the system are equally probable so that its density operator is proportional to the identity operator,

$$\hat{\rho}_{mm} = (1/d)\hat{I} . \tag{4.1}$$

For a two-state system, the maximally mixed state is characterized precisely by one bit ($\log_2 2$) of missing information.

---

[2] We now use the expression of S that omits Boltzmann's constant, as is traditionally done in the context of quantum information. For notational simplicity we omit the subscript 2 in the following when no ambiguity is possible. That is, "log" implies a log 2 basis, while 'ln' is reserved for natural logarithms.

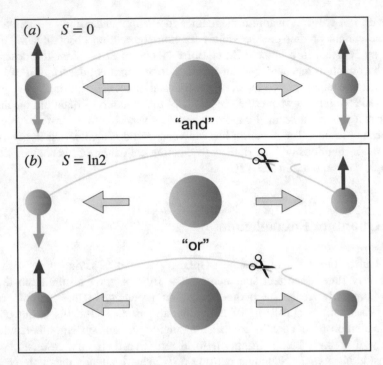

Fig. 4.1 Schematic representation of the dissociation process. (**a**) State of the system after the dissociation process, with the curved line indicating that the two dissociation partners still form a unique quantum system, in a pure state and with zero von Neumann entropy, rather than individual particles. (**b**) Maximally mixed state resulting from "cutting the cord" and considering the particles separately, with a loss of log 2 in information as compared with the pure state of (**a**)

Consider then again the photodissociation experiment. After the molecule dissociates, the state of the system is

$$|\psi\rangle = \frac{1}{\sqrt{2}} [|\uparrow\rangle_1| \downarrow\rangle_2 + | \downarrow\rangle_1| \uparrow\rangle_2] , \tag{4.2}$$

where $|\uparrow\rangle_1$ stands for spin up for the left-propagating component of the dissociation product, with corresponding notations for the other kets. This is clearly a pure state, and hence, its entropy is equal to zero, that is, it contains the maximum possible information about the system, see Fig. 4.1a. But then look at just one of its components, say component 1. This is achieved by tracing over component 2, resulting in the reduced density operator

$$\hat{\rho}_1 = \mathrm{Tr}_2 |\psi\rangle\langle\psi| = (1/2) [|\uparrow\rangle_1\langle\uparrow |_1 + | \downarrow\rangle_1\langle\downarrow |_1] . \tag{4.3}$$

This is no longer a pure state, but rather the maximally mixed state (4.1). Its von Neumann entropy is exactly one bit, $S_1 = \log 2$, indicating that the subsystem is in a state of maximum missing information, as illustrated in Fig. 4.1b. Together, the two subsystems contain all possible information about the system, but separately they contain none. That is, they form an inseparable whole, not individual particles. This property, which is at the heart of quantum entanglement, is a fundamental aspect of quantum mechanics with no equivalent in the classical world. As we shall see, it is the essential ingredient for a number of emerging applications of quantum mechanics, in particular in quantum cryptography and quantum information science more generally, as well as in quantum metrology.

4.2.1 Schmidt Decomposition and Maximum Entanglement

Entangled pure states are profoundly different from separable states, which for bipartite systems composed of two subsystems A and B take the general form

$$|\psi\rangle_{AB} = |\psi\rangle_A \otimes |\psi\rangle_B . \tag{4.4}$$

As such they are a subset of the most general pure states

$$|\psi\rangle_{AB} = \sum_{i,j} c_{ij} |i\rangle_A \otimes |j\rangle_B \tag{4.5}$$

where $\{|i\rangle\}_A$ and $\{|j\rangle\}_B$ are complete sets of basis states for these subsystems. One powerful method to determine whether a bipartite state is separable is the Schmidt decomposition theorem, which states that any pure state of a bipartite system can be expressed as

$$|\psi_{AB}\rangle = \sum_{i=1}^{\min(d_A,d_B)} \lambda_i |u_i\rangle_A \otimes |v_i\rangle_B , \tag{4.6}$$

where λ_i are the so-called Schmidt coefficients, with $\sum_i \lambda_i = 1$, and the Schmidt vectors $\{|u_i\rangle_A\}$, $\{|v_i\rangle\}_B$ form basis sets of the subsystems A and B of dimensions d_A and d_B, respectively. That this is the case can be shown by first expressing the matrix of coefficients $\mathbf{C} \equiv (c_{ij})$ in terms of its singular value decomposition[3] as

$$\mathbf{C} = \sum_k \alpha_k |u_k\rangle_A \langle v_k|_B, \tag{4.7}$$

---

[3]Remember that the singular value decomposition of an $(m \times n)$ real or complex matrix $\mathbf{M}$ is a factorization of the form $\mathbf{M} = \mathbf{U}\boldsymbol{\Sigma}\mathbf{V}^*$, where $\mathbf{U}$ is an $(m \times m)$ real or complex unitary matrix, $\mathbf{V}$ is an $(n \times n)$ real or complex unitary matrix, and $\boldsymbol{\Sigma}$ is an $(m \times n)$ rectangular diagonal matrix with non-negative real numbers on the diagonal. If $\mathbf{M}$ is real, then $\mathbf{U}$ and $\mathbf{V}^T = \mathbf{V}^*$ are real orthogonal matrices. The diagonal entries Σ_{ii} of $\boldsymbol{\Sigma}$ are called the singular values of $\mathbf{M}$, and the number of non-zero singular values is the rank of $\mathbf{M}$. The columns of $\mathbf{U}$ and of $\mathbf{V}$ are called the left-singular (ket) vectors and right-singular (bra) vectors of $\mathbf{M}$, respectively.

where α_k are the singular values of $\mathbf{C}$, and the singular kets $|u_k\rangle_A$ and $|v_k\rangle_B$ are basis sets for the subsystems A and B, respectively.

Substituting then $c_{ij} = \langle i|\mathbf{C}|j\rangle$ into Eq. (4.5) gives

$$|\psi\rangle_{AB} = \sum_{i,j} \sum_k \alpha_k \langle i|u_k\rangle |i\rangle_A \otimes \langle v_k|j\rangle |j\rangle_B , \qquad (4.8)$$

or, with $\sum_i |i\rangle\langle i| = \sum_j |j\rangle\langle j| = 1$,

$$|\psi\rangle_{AB} = \sum_k \alpha_k |u_k\rangle_A \otimes |v_k^*\rangle_B . \qquad (4.9)$$

Since the singular values α_k can in general be complex, the last step consists in introducing their amplitudes $\lambda_k = |\alpha_k|$ and absorbing the remaining phase factors $\exp(-i\phi_k)$ into $|u_k\rangle$, $|u_k\rangle \rightarrow |u_k \exp(-i\phi_k)\rangle$, completing the proof of the Schmidt decomposition (4.6).

The coefficients λ_k are called Schmidt coefficients, and the Schmidt rank of the state $|\psi\rangle_{AB}$ is the number of non-zero coefficients in the Schmidt decomposition (4.6). If only one such coefficient is needed, the state is simply $|\psi\rangle_{AB} = |u_k\rangle_A |v_k^*\rangle_B$ and hence is separable. A pure state is entangled if and only if its Schmidt rank is larger than one. The maximum Schmidt rank of a system is the dimension of the Hilbert space of the smallest of its subsystems, $\min(d_A, d_B)$. A pure state with this Schmidt rank is *maximally entangled*. For spin-$\frac{1}{2}$ subsystems, this rank is 2.

Entanglement Entropy The entropy of one of the subsystems of a bipartite system is called its *entanglement entropy*. It follows directly from the Schmidt decomposition of the state $|\psi\rangle_{AB}$ that the reduced density operators of the two subsystems are

$$\hat{\rho}_A = \sum_{i=1}^{\min(d_A,d_B)} \lambda_i^2 |u_i\rangle\langle u_i| \quad ; \quad \hat{\rho}_B = \sum_{i=1}^{\min(d_A,d_B)} \lambda_i^2 |v_i\rangle\langle v_i| , \qquad (4.10)$$

from which it follows that their entanglement entropies are equal,

$$S_A = S_B = -\sum_i \lambda_i^2 \log\left(\lambda_i^2\right) . \qquad (4.11)$$

Since for a separable state only one Schmidt coefficient is different from zero, $\lambda_1 = 1$, we have that $S_A = S_B = 0$ in that case. An important consequence of this result is that for bipartite pure states the von Neumann entropy of the reduced states is a well-defined measure of entanglement since it is equal to zero if and only if the system is in a product state.

4.2.2 Monogamy of Entanglement

The quantum mechanical violation of Bell's inequalities, the topic of Sect. 4.3, will show quantitatively how bipartite entangled states can exhibit stronger correlations than any correlations that can be generated classically. Importantly, though, these correlations cannot be extended to or be shared with a third party. This property, called *monogamy of entanglement*, is in contrast with the situation in classical physics, where a joint probability distribution that is shared between two members A and B of a bipartite system can in principle also be shared with a third party C, in the sense that the marginal distributions on A and B are the same as those on A and C. It forms the basis for the secure quantum key distribution that will be discussed in Sect. 4.4.

Specifically, consider a system of three qubits A, B, and C and suppose that the subsystems A and B share an entangled state with the most general Schmidt representation[4]

$$|\psi\rangle_{AB} = \lambda_0|0, 0\rangle_{AB} + \lambda_1|1, 1\rangle_{AB}. \tag{4.12}$$

The question that we wish to answer is whether it is possible to find a tripartite state $|\psi\rangle_{ABC}$ such that its reduced density operators on A and B, and on A and C, both yield a state of the form (4.12). A natural choice would be the state

$$|\psi\rangle_{ABC} = \lambda_0|0, 0, 0\rangle_{ABC} + \lambda_1|1, 1, 1\rangle_{ABC}, \tag{4.13}$$

which appears to provide the same correlations between A and B as between A and C. From the corresponding density matrix $\hat{\rho}_{ABC} = |\psi\rangle\langle\psi|_{ABC}$ we find readily

$$\hat{\rho}_{AB} = \text{Tr}_C\,\hat{\rho}_{ABC} = |\lambda_0|^2|0, 0\rangle\langle 0, 0|_{AB} + |\lambda_1|^2|1, 1\rangle\langle 1, 1|_{AB},$$

$$\hat{\rho}_{AC} = \text{Tr}_B\,\hat{\rho}_{ABC} = |\lambda_0|^2|0, 0\rangle\langle 0, 0|_{AC} + |\lambda_1|^2|1, 1\rangle\langle 1, 1|_{AC}, \tag{4.14}$$

that is, both density operators are mixtures of tensor products of states of the two subsystems involved. If the density operator of a system comprised two subsystems A and B can be cast in the form

$$\hat{\rho}_{AB} = \sum_i p_i\,\hat{\rho}_{A,i} \otimes \hat{\rho}_{B,i}, \tag{4.15}$$

with $\sum_i p_i = 1$, then the state of the system is said to be separable, with p_i interpreted as the probability for the system to be in the state $\hat{\rho}_{A,i} \otimes \hat{\rho}_{B,i}$. The

---

[4]Here and in the rest of the book, we omit the tensor product symbol $\otimes$ when no ambiguity is possible, as this lightens the notation significantly. For instance, the state $|1\rangle_A \otimes |1\rangle_B$ will then be written simply $|1\rangle_A|1\rangle_B$, or even more compactly $|1, 1\rangle_{AB}$, with or without comma, whenever possible.

reduced density operators (4.14) describe therefore separable mixed states, with probabilities $|\lambda_0|^2$ and $|\lambda_1|^2$ to be in the uncorrelated product states $|0, 0\rangle\langle0, 0|$ and $|1, 1\rangle\langle1, 1|$ of the subsystems AB and AC.

The fact that partial traces should produce the mixed states (4.14) rather than pure states is due to the fact that they erase the correlations, and hence the information, associated with the entanglement between the party that is traced over and the other two parties. As a result they do produce the same classical correlations between the two subsystems that are kept, whether they are AB, AC, or BC, but no quantum entanglement remains. Entanglement monogamy refers to this impossibility to share the same entanglement between three parties: if a bipartite system is maximally entangled, then it is unentangled from everything else. Importantly, though, entanglement can be *partly* shared between three parties, so that if A is only party entangled with B, then it can also share some entanglement with C, an aspect of entanglement monogamy to which we now turn.

Coffman, Kundu, and Wooters Inequality Distributing entanglement between several parties is a topic of considerable importance in quantum information processing and the object of ongoing research. A detailed discussion of this complex topic is beyond the scope of this book, and we limit our discussion again to the relatively simple but important case of the sharing of entanglement between three qubits in a pure state. V. Coffman and coworkers [3] showed that in this case the trade-off in the amount of entanglement that can be shared between these three parties can be quantified through an entanglement measure called the *concurrence*.[5]

Consider first the parties A and B and the associated pair of qubits, characterized by a density operator $\hat{\rho}_{AB}$ expressed on the basis $\{|00\rangle, |01\rangle, |10\rangle, |11\rangle\}$. We proceed by introducing the spin-flipped density operator

$$\tilde{\rho}_{AB} = (\hat{\sigma}_y \otimes \hat{\sigma}_y)\hat{\rho}_{AB}^*(\hat{\sigma}_y \otimes \hat{\sigma}_y), \tag{4.16}$$

where the asterisk stands for the complex conjugate of the density matrix elements on that basis, and the $\hat{\sigma}_y$ are the Pauli operators

$$\hat{\sigma}_y = \begin{pmatrix} 0 & -i \\ i & 0 \end{pmatrix}$$

acting on the states of the two subsystems. Since density operators are positive operators, $\hat{\rho}_{AB}$ must have real non-negative eigenvalues, and therefore so does $\tilde{\rho}_{AB}$, as well their product $\hat{\rho}_{AB}\,\tilde{\rho}_{AB}$. Calling λ_i the square roots of the four eigenvalues of $\hat{\rho}_{AB}\,\tilde{\rho}_{AB}$, the concurrence of $\hat{\rho}_{AB}$ is defined in terms of these eigenvalues arranged

---

[5]The concept of concurrence was originally introduced by W. Wooters [4] to quantify the resources required to create a given entangled state, the so-called *entanglement of formation*.

in decreasing order as

$$C_{AB} = \max\{\lambda_1 - \lambda_2 - \lambda_3 - \lambda_4, 0\}. \tag{4.17}$$

For a system in a pure state Problem 4.2 shows that C_{AB} reduces to

$$C_{AB} = 2\sqrt{\det(\hat{\rho}_A)} \tag{4.18}$$

with $\hat{\rho}_A = \mathrm{Tr}_B \hat{\rho}_{AB}$. It is also easily verified that we then have $C_{AB} = 1$ for a completely entangled state and $C_{AB} = 0$ for an unentangled state.

With this property of the concurrence of pure states at hand, the question that we wish to address is the following: given a general pure state $|\psi\rangle$ of three qubits A, B, and C, what is the relationship between the concurrences of the subsystems A and B and of the subsystems A and C?

Problem 4.3 demonstrates that the density operator $\hat{\rho}_{AB}$ of a system of 2 qubits in a pure state has at most two non-zero eigenvalues, and hence so does $\tilde{\rho}_{AB}$. Equation (4.17) gives then

$$\begin{aligned}
C_{AB}^2 = (\lambda_1 - \lambda_2)^2 &= \lambda_1^2 + \lambda_2^2 - 2\lambda_1\lambda_2 \\
&= \mathrm{Tr}(\hat{\rho}_{AB}\tilde{\rho}_{AB}) - 2\lambda_1\lambda_2 \leq \mathrm{Tr}(\hat{\rho}_{AB}\tilde{\rho}_{AB}),
\end{aligned} \tag{4.19}$$

so that, after carrying out the same argument for parties A and C we have

$$C_{AB}^2 + C_{AC}^2 \leq \mathrm{Tr}(\hat{\rho}_{AB}\tilde{\rho}_{AB}) + \mathrm{Tr}(\hat{\rho}_{AC}\tilde{\rho}_{AC}). \tag{4.20}$$

To interpret the right-hand side of this inequality we first expand it explicitly for a general three-qubit pure state $|\psi\rangle = \sum_{ijk} c_{ijk}|i, j, k\rangle$ with $i, j, k \in \{0, 1\}$. Problem 4.4 shows that after some algebra this permits to re-express its two terms as

$$\mathrm{Tr}(\hat{\rho}_{AB}\tilde{\rho}_{AB}) = 2[\det(\hat{\rho}_A) + \det(\hat{\rho}_B) - \det(\hat{\rho}_C)],$$

$$\mathrm{Tr}(\hat{\rho}_{AC}\tilde{\rho}_{AC}) = 2[\det(\hat{\rho}_A) + \det(\hat{\rho}_C) - \det(\hat{\rho}_B)], \tag{4.21}$$

so that Eq. (4.20) reduces to

$$C_{AB}^2 + C_{AC}^2 \leq 4\det(\hat{\rho}_A). \tag{4.22}$$

We now turn to the subsystem BC and observe that any general pure state $|\psi\rangle$ of the tripartite system can be expressed as $|\psi\rangle = |0\rangle_A|\alpha\rangle_{BC} + |1\rangle_A|\beta\rangle_{BC}$, with

$$|\alpha\rangle_{BC} = \sum_{jk} c_{0jk}|j, k\rangle_{BC} \quad ; \quad |\beta\rangle_{BC} = \sum_{jk} c_{1jk}|j, k\rangle_{BC}.$$

This means that even though the state space of BC is four-dimensional, only two of those dimensions are necessary to express the state of the full system ABC, a consequence of the facts that A is a single qubit and the state of the whole system is pure. The subsystem BC can therefore be thought of as an *effective qubit*, and the full system ABC can be decomposed into the subsystems A and BC, each composed of one qubit, one of them real and the other effective. With Eq. (4.18), it is then possible to interpret $2\sqrt{\det(\hat{\rho}_A)}$ as the concurrence of the $A(BC)$ system,

$$\mathcal{C}_{A(BC)} = 2\sqrt{\det(\hat{\rho}_A)}\,. \tag{4.23}$$

When substituted into Eq. (4.22), this result yields finally the Coffman, Kundu, and Wooters monogamy inequality

$$\mathcal{C}_{AB}^2 + \mathcal{C}_{AC}^2 \leq \mathcal{C}_{A(BC)}^2\,. \tag{4.24}$$

This shows that if the qubit A possesses a certain amount of entanglement with the subsystem BC, then the entanglement that it shares individually with the subsystems B and C cannot be larger than that entanglement. Whatever entanglement is shared, say, with B is not available to be shared with C.

The extension of this discussion to the case of mixed states is beyond the scope of this brief introduction. It is complicated in particular by the fact that the system BC can no longer be considered as an effective qubit, as discussed in Ref. [3]. Reference [5] also proves the important result that the bipartite entanglement in multipartite states of qubits satisfies a Coffman, Kundu, and Wooters inequality as well. Because of the importance for quantum cryptography and quantum information science of understanding the trade-offs involved in the amount of quantum entanglement that can be shared between multiple parties, this remains a very active topic of research, in particular when considering its generalization to higher dimensions.

4.3 Bell's Inequalities

Armed with our understanding of basic aspects of quantum entanglement we now return to the EPR paradox and discuss its profound and ground-breaking extension, published by John Bell in a 1964 paper titled "On the Einstein Podolsky Rosen Paradox" [6]. The fundamental importance of that work is that it moved the EPR argumentation to a point where it became experimentally and quantitatively testable. How? By allowing for the EPR analyzers 1 and 2 to be set at *different* angles **a** and **b**, rather than at the same angle **a**, and measuring the joint probabilities of obtaining a given outcome, say $+\hbar/2$ for the spin components S_a and S_b of the two particles.[6] Bell showed that correlation experiments of this type permit to distinguish between the predictions of quantum mechanics and those of a class of theories called "local realistic hidden variable theories" of nature. What hides behind this term is what would be broadly considered to be the fundamental ingredients of any "reasonable" theory of nature. Hopefully this will become more clear as we go along.

To see how this works more concretely, consider again an experimental arrangement where a source emits two correlated particles "1" and "2." This could be for instance a diatomic molecule dissociating into two atoms, a pair of photons emitted in a three-level cascade, or a number of other possible situations. Two detectors measure then some property of these particles for settings **a** and **b** of the analyzers.

Let us denote by $p_1(a)$ and $p_2(b)$ the probabilities of detecting particle 1, resp. particle 2, for these settings. If we had a complete theory at hand, these probabilities would depend on all parameters $\{\lambda\}$ describing the emission process in the source. But in the absence of such a theory, we have no way to know, or measure, or even guess what these parameters might be. They are hidden, out of our control—hence the "hidden variables" theory. What we detect in a series of experiments is some average over them,

$$p_1(a) = \int d\lambda \rho(\lambda) p_1(a, \lambda),\tag{4.25}$$

where $d\lambda$ is a (unknown) measure over the space of hidden variables and $\rho(\lambda)$ is some weight function. (For simplicity we write λ instead of $\{\lambda\}$.) Similarly,

$$p_2(b) = \int d\lambda \rho(\lambda) p_2(b, \lambda).\tag{4.26}$$

Suppose now that we could actually control the hidden parameters $\{\lambda\}$ and know their value precisely. We could then ask the joint probability $p_{12}(a, b, \lambda)$ of detecting both particles for detector settings **a** and **b**. If the detectors are space-like separated, and their settings chosen long after the particles have been emitted from the source, the result at one detector should be unaffected by the result at the setting of the other. This is the principle of *locality*: no influence of any kind can travel faster than the speed of light. Thus the counting rates at detectors 1 and 2 must be uncorrelated,

$$p_{12}(a, b, \lambda) = p_1(a, \lambda) p_2(b, \lambda).\tag{4.27}$$

---

[6]We did learn in the previous section that in the quantum context we should avoid using the word "particles" to describe the two subsystems. But since the Bell argument is not or at least not yet about quantum physics at this point we will continue to use it for now.

However, this does not imply that the joint probability actually measured is uncorrelated. Integrating over the hidden variables, we have

$$p_{12}(a, b) = \int d\lambda \rho(\lambda) p_1(a, \lambda) p_2(b, \lambda). \qquad (4.28)$$

The weight function $\rho(\lambda)$, which contains all information about the hidden variables in the source, leads in general to a non-factorizable joint probability distribution $p_{12}(a, b)$—these are correlations through a common cause.

A simple theorem [7] states that for any four numbers $0 \leq x, x', y, y' \leq 1$ we have

$$- 1 \leq xy - xy' + x'y + x'y' - x' - y' \leq 0. \qquad (4.29)$$

Noting that probabilities lie between 0 and 1, and choosing two possible directions **a** and **a'**, respectively **b** and **b'** for the analyzers 1 and 2, we have therefore

$$- 1 \leq p_1(a, \lambda) p_2(b, \lambda) - p_1(a, \lambda) p_2(b', \lambda) + p_1(a', \lambda) p_2(b, \lambda)$$
$$+ p_1(a', \lambda) p_2(b', \lambda) - p_1(a', \lambda) - p_2(b, \lambda) \leq 0, \qquad (4.30)$$

or, with Eq. (4.28),

$$- 1 \leq p_{12}(a, b, \lambda) - p_{12}(a, b', \lambda) + p_{12}(a', b, \lambda)$$
$$+ p_{12}(a', b', \lambda) - p_1(a', \lambda) - p_2(b, \lambda) \leq 0. \qquad (4.31)$$

Integrating this last equation over the hidden variables yields then

$$- \int d\lambda \rho(\lambda) \leq p_{12}(a, b) - p_{12}(a, b') + p_{12}(a', b)$$
$$+ p_{12}(a', b') - p_1(a') - p_2(b) \leq 0. \qquad (4.32)$$

The left-hand side of this double inequality is equal to -1 if $\int d\lambda \rho(\lambda) = 1$, but we actually do not need it. Keeping the right-hand side only yields

$$p_{12}(a, b) - p_{12}(a, b') + p_{12}(a', b) + p_{12}(a', b') - p_1(a') - p_2(b) \leq 0, \qquad (4.33)$$

which is a form of *Bell's inequalities* due to J. F. Clauser and M. A. Horne.

Another form of Bell's inequalities, due to J. F. Clauser, M. A. Horne, A. Shimony, and R. A. Holt [8], results from a similar argument but holds when there are only two possible outcomes for the various measurements, call them $\uparrow$ and $\downarrow$. We proceed by introducing the correlation function

$$E(a, b) = p(\uparrow, \uparrow \,|a, b) - p(\uparrow, \downarrow \,|a, b) - p(\downarrow, \uparrow \,|a, b) + p(\downarrow, \downarrow \,|a, b), \qquad (4.34)$$

where $p(\uparrow, \uparrow \,|a, b)$ is the probability of getting the outcome $(\uparrow, \uparrow)$ for detector settings a and b. With Eq. (4.28) $E(a, b)$ can be expressed, again with the help of Eq. (4.28), as

$$
\begin{aligned}
E(a, b) &= \int d\lambda \rho(\lambda) p_1(\uparrow \,|a, \lambda) p_2(\uparrow \,|b, \lambda) - \int d\lambda \rho(\lambda) p_1(\uparrow \,|a, \lambda) p_2(\downarrow \,|b, \lambda) \\
&\quad - \int d\lambda \rho(\lambda) p_1(\downarrow \,|a, \lambda) p_2(\uparrow \,|b, \lambda) + \int d\lambda \rho(\lambda) p_1(\downarrow \,|a, \lambda) p_2(\downarrow \,|b, \lambda) \\
&= \int d\lambda \rho(\lambda) [\, p_1(\uparrow \,|a, \lambda) - p_1(\downarrow \,|a, \lambda)] [\, p_2(\uparrow \,|b, \lambda) - p_2(\downarrow \,|b, \lambda)] \\
&\equiv \int d\lambda \rho(\lambda) \bar{A}(a, \lambda) \bar{B}(b, \lambda) \,,
\end{aligned}
\tag{4.35}
$$

where $\bar{A}$ and $\bar{B}$ stand for the first and second square brackets. It follows that

$$
E(a, b) \pm E(a, b') = \int d\lambda \rho(\lambda) \bar{A}(a, \lambda) \left[\bar{B}(b, \lambda) \pm \bar{B}(b', \lambda) \right] .
\tag{4.36}
$$

Since the p_i's are probabilities, we have that $0 \leq p_i < 1$, and

$$
|\bar{A}(a, \lambda)| \quad \text{and} \quad |\bar{B}(a, \lambda)| \leq 1
\tag{4.37}
$$

so that

$$
|E(a, b) \pm E(a, b')| \leq \int d\lambda \rho(\lambda) |\bar{B}(b, \lambda) \pm \bar{B}(b', \lambda)| .
\tag{4.38}
$$

Likewise we also have

$$
|E(a', b) \mp E(a', b')| \leq \int d\lambda \rho(\lambda) |\bar{B}(b, \lambda) \mp \bar{B}(b', \lambda)| ,
\tag{4.39}
$$

and, with Eq. (4.37),

$$
|\bar{B}(b, \lambda) \pm \bar{B}(b', \lambda)| + |\bar{B}(b, \lambda) \mp \bar{B}(b', \lambda)| \leq 2 ,
\tag{4.40}
$$

giving finally the Clauser–Horne–Shimony–Holt inequality

$$
|E(a, b) \pm E(a, b')| + |E(a', b) \mp E(a', b')| \leq 2 .
\tag{4.41}
$$

Aspect Experiments Bell's inequalities have now been tested in a number of situations. One momentous early series of experiments was performed by A. Aspect et al. [9]. Instead of spins, as in the Bohm version of the EPR paradox, the system they used consisted of pairs of entangled photons emitted in a radiative atomic

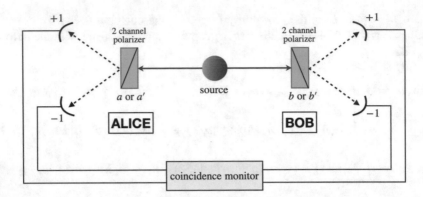

Fig. 4.2 Schematic of an experimental setup used to test Bell's inequalities with photon pairs. A source of pairs of entangled photons sent to space-like separated 2-channel polarizers that are independently set at random by two people, call them Alice and Bob, long after the photons have left the source, and the resulting signals are collected at (**a**) coincidence monitor

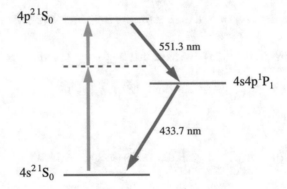

Fig. 4.3 Energy levels of the calcium atom, showing the two-photon pumping scheme (grey arrows) and the two photons at frequencies ν_1 and ν_2 correlated in polarization emitted by the atom during its decay back to the ground state (green and blue arrows)

cascade in Calcium, as sketched in Fig. 4.2, whereby the $4p^2\ {}^1S_0$ level populated by two-photon excitation decays back to the $4s^2\ {}^1S_0$ state over the $4s4p^1P_1$ level, emitting two photons at wavelengths of about $\lambda_1 \approx 551$ nm and $\lambda_2 \approx 423$ nm, see Fig. 4.3. Because the change in angular momentum in the transition is $J = 0 \rightarrow J = 1 \rightarrow J = 0$, no net angular momentum is carried by the pair of photons. For emitted photons counter-propagating in the $\pm z$ directions, the state of polarization of the total system must therefore be of the form

$$|\psi\rangle = \frac{1}{\sqrt{2}}[|\uparrow\rangle_1|\uparrow\rangle_2 + |\rightarrow\rangle_1|\rightarrow\rangle_2] , \tag{4.42}$$

where $|\uparrow\rangle$ represents the polarization of a photon along the x-axis and $|\rightarrow\rangle$ along the y-axis, and the subscript i labels the photon of wavelength λ_i.

Quantum Mechanical Predictions The quantum mechanical prediction for the measurement of photon pair polarizations is easily derived. The single probabilities $P_\pm(a)$ and $P_\pm(b)$ of getting the results $\pm$ for the photon at wavelength λ_i, $i = 1, 2$, are of course equal to 1/2, and Problem 4.5 shows that the joint detection probabilities are

$$P_{++}(a, b) = P_{--}(a, b) = \tfrac{1}{2} \cos^2(a, b)\,,$$
$$P_{+-}(a, b) = P_{-+}(a, b) = \tfrac{1}{2} \sin^2(a, b)\,, \tag{4.43}$$

so that the correlation function (4.34) becomes

$$E(a, b) = P_{++}(a, b) + P_{--}(a, b) - P_{+-}(a, b) = P_{-+}(a, b) = \cos 2(a, b)\,. \tag{4.44}$$

Problem 4.5 also shows that the Clauser–Horne–Shimony–Holt combination of correlation functions (4.41), for instance

$$S \equiv |E(a, b) - E(a, b')| + |E(a', b) + E(a', b')| \tag{4.45}$$

reaches a maximum of $2\sqrt{2}$ for $(a, b) = (b, a') = (a', b') = \pi/8$, thereby strongly violating this form of Bell's inequality.

Eliminating the Loopholes The early experimental results of A. Aspect and coworkers, for angles of $22.5°$ between the polarizers a and b, b and a', and a' and b', and an angle of $67.5°$ between the polarizers a and b' used to analyze the correlations between the emitted photons, demonstrated an excellent agreement with quantum mechanics, with $|E(a, b) \pm E(a, b')| + |E(a', b) \mp E(a', b')| \simeq 2.7$, a definite violation of Bell's inequality (4.41), see Fig. 4.4. As such they confirmed to a high degree of confidence the incompatibility between quantum mechanics and local realistic hidden variable theories. Subsequent experiments further improved very significantly on these early results.

Despite their remarkable success, the original Bell tests suffered however from three loopholes, referred to as the "locality," "fair sampling," and "free will" loopholes. These offered a potential "escape route" to local realistic hidden variable theories.

The "locality" loophole raises the possibility that a local realistic theory might rely on some type of signal sent from one entangled particle to its partner, perhaps in the form of a signal containing information about the specific measurement carried out on the first particle. In the Aspect experiments, that loophole was closed by a fast random setting of the two polarizers, while the pairs of photons, separated by a large distance, were in flight between the source and the detectors. Photons traveling in different directions could then be measured in a space-like separated configuration.

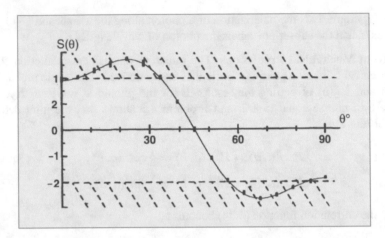

Fig. 4.4 $S(\theta) = \cos(a, b) - \cos(a, b') + \cos(a', b) + \cos(a', b')$ for $\theta \equiv (a, b) = (b, a') = (a', b') = (a, b')/3$ as a function of θ. The indicated experimental errors are ± 2 standard deviations. The solid curve is the quantum mechanical prediction for the actual experiment. For an ideal experiment, the curve would exactly reach $2\sqrt{2}$, but in the original experiment by A. Aspect and coworkers reached 2.697 ± 0.015. (From Ref. [10])

Since according to relativity no influence can travel faster than the speed of light, this approach allowed to remove the locality loophole.

The "fair sampling" loophole took longer to close: The issue here is that detectors do not have a unit efficiency, so that a fraction of the particles emitted by the source will not be detected. It might therefore be argued that a subset of detected particles would violate Bell's inequality, although the entire ensemble can be described by a local realistic theory. A highly efficient experimental setup is therefore necessary to demonstrate a conclusive Bell violation without having to assume that the detected particles represent a "fair" sample. It can be shown that this requires detectors whose probability of detecting a photon when its partner has been detected is higher than 2/3 [11, 12]. This was first achieved in 2013 in two optical experiments [13, 14].

While at that point the locality and fair sampling loopholes had both been individually closed, it took another 2 years to close them simultaneously, in three series of experiments using electron spins in one case [15] and entangled photons with rapid setting generation, together with highly efficient superconducting detectors in the others [16, 17], see also the viewpoint article [18] where A. Aspect comments on these experiments.

The "free will" loophole refers to the fact that if for some reason the random polarization selection is not really random, but correlated to other aspects of the experiment, then the outcome of the Bell test could be affected. To close this loophole, J. Handsteiner and coworkers [19] used the random nature of the color changes in starlight to decide the setting of their polarization detectors. In subsequent work M. H. Li, D. Rauch, and their respective coworkers [20, 21] carried out experiments where the detector settings were based on real time measurements of the wavelength of photons emitted billions of years ago by high redshift quasars. These experiments are also consistent with the nonlocality requirement but do assume fair sampling. Since any influence trying to engineer the outcome of this experiment would have had to act prior to the photons leaving their cosmic source, this pushes back to an impressive 0.8 Gigayears the most recent time by which any local realistic influences could have been exploited to engineer the observed violation of Bell inequalities!

4.4 Quantum Key Distribution

4.4.1 The BB84 Protocol

Entangled states, in addition to their seminal importance in exposing the profound differences between the classical and quantum worlds, also find remarkable applications in quantum metrology and quantum information processing. While practical quantum computers may still be relatively far into the future, other applications of quantum information science and technology are already being implemented. A particularly elegant example is offered by quantum cryptography and quantum teleportation, to which we now turn. This will allow us to also introduce another remarkable aspect of quantum physics, the no-cloning theorem.

The central element of quantum cryptography is the distribution of secret communication keys: the sender and receiver of coded messages (call them Alice and Bob to follow the tradition) must share a "key" that allows Alice to encrypt the message and Bob, but nobody else, to decrypt it. The only completely secure method of encryption is a secret, random key that connects the plain text to the encoded text.

One way to establish such a key would be for Alice and Bob to meet and agree upon it, or for a trusted agent to be sent from one to the other to distribute the key. The difficulty and inefficiency of such an operation are obvious (in particular to anybody who has read a spy novel or watched a spy movie) so more efficient but less secure methods are generally used. Most current key distribution systems derive their security from the use of convoluted algorithms or intractable problems, a simple example being the factoring of two large prime numbers. Yet, as the security of these methods relies on the assumption that the eavesdropper does not possess the advanced techniques capable to defeat such algorithms, there is a need for key distribution techniques that are absolutely secure, especially with the potential on the horizon of quantum computers that open up powerful potential avenues to break the code.

The remarkable advantage of quantum key distribution is that it offers unconditional security, rather than a security guaranteed by the limitations of current technology. This is because it is bound only by the laws of quantum mechanics, more specifically by the fact that quantum measurements are accompanied by an unavoidable back action, a feature that we have already encountered at in Sect. 3.5, and that quantum states cannot be cloned, a property that will be discussed in some detail in Sect. 4.4.2.

The first quantum key distribution protocol was proposed in 1984 by C. H. Bennett and G. Brassard [22], and for that reason it is called the BB84 protocol. In that scheme, which is easily explained in optical terms, Alice sends Bob a string of bits coded into the polarization of individual photons, the essential point being that the successive bits are imprinted into polarization basis states varied *at random* from one bit to the next.

Assume for concreteness that the first basis is a horizontal/vertical basis $\oplus$, and the second one a left-diagonal/right-diagonal basis $\otimes$. Alice decides that in the $\oplus$ basis the bit "1" would be mapped to a vertically polarized photon $\uparrow$ and the bit "0" to a horizontally polarized photon $\rightarrow$. Similarly, in the basis $\otimes$ the bit "1" would correspond to a diagonally polarized photon at $45°$, and the bit "0" to the polarized photon at $135°$. The four qubit states are then

$$|\psi_{0,\oplus}\rangle = |0\rangle \, ,$$

$$|\psi_{1,\oplus}\rangle = |1\rangle \, ,$$

$$|\psi_{0,\otimes}\rangle = \frac{1}{\sqrt{2}} \left[|0\rangle + |1\rangle \right] \, ,$$

$$|\psi_{1,\otimes}\rangle = \frac{1}{\sqrt{2}} \left[|0\rangle - |1\rangle \right] \, .$$

Assuming that Alice wishes to send the string $(1, 0, 0, 1, 1, 1, 0, 1)$ and chooses at random the successive basis sets $(\oplus, \otimes, \otimes, \oplus, \otimes, \oplus, \oplus, \otimes)$, then the polarization of the transmitted photons will be

$$(\uparrow, \nwarrow, \nwarrow, \uparrow, \nearrow, \uparrow, \downarrow, \nearrow) \, ,$$

and it will be transmitted to Bob as the tensor product of n qubits

$$|\psi_{1,\oplus}\rangle \, |\psi_{0,\otimes}\rangle \, |\psi_{0,\otimes}\rangle \, |\psi_{1,\oplus}\rangle \, |\psi_{1,\otimes}\rangle \, |\psi_{1,\oplus}\rangle \, |\psi_{0,\oplus}\rangle \, |\psi_{1,\otimes}\rangle \, .$$

Just like Alice, Bob selects at random the successive basis states of his measuring device. Assume for concreteness that he chooses the sequence

$$(\oplus, \oplus, \otimes, \oplus, \oplus, \oplus, \otimes, \otimes) \, .$$

For those bits where Alice's and Bob's bases are the same (bits 1, 3, 4, 6, and 8 in this example), the measured photon polarization will match exactly that of the emitted photon (assuming noiseless transmission.) In other cases, though, Bob's measurement will give one or the other result at random, and the outcome will be something like

$$(\uparrow, \rightarrow, \nwarrow, \uparrow, \rightarrow, \uparrow, \nearrow, \nearrow).$$

Alice and Bob then communicate over an open channel their choices of polarizations. Keeping only those elements of the bit sequence where their choice coincides, bits number 1, 3, 4, 6, and 8 in our example, results in the generation of the secret key (1,0,1,1,1).

To determine whether eavesdropping has taken place, Alice and Bob generate a significantly longer key than actually needed and exchange the values of the bits for a subset of the values of the key, again on an open channel. The point is that an eavesdropper, Eve, trying to intercept the communication between Alice and Bob has no way of knowing their choice of basis; hence, she will guess the wrong one half the time on average. In those cases, she will therefore randomly modify the polarization of the photon before reinjecting it in the channel, thereby spoiling the perfect correlations that would otherwise observed by Alice and Bob in the verification step. They will then have determined for sure that their communication has been tempered with, and reject the key.

4.4.2 No-cloning Theorem

One might however imagine that Eve would be able to defeat the key distribution scheme if she were in possession of advanced technology that allowed her to clone any ancillary input state $|\psi_i\rangle$. If that were possible, she would then be able to tap the communication channel unnoticed. However, this is forbidden by the "no cloning theorem" [23, 24]. This key ingredient in the development of quantum cryptography states that it is impossible to clone into a system B an arbitrary state of a system A, that is, that there is no "cloning operator" $\hat{U}_c$ that allows to achieve the cloning operation

$$\hat{U}_c |\psi\rangle_A |0\rangle_B = |\psi\rangle_A |\psi\rangle_B \tag{4.46}$$

independently of the state $|\psi\rangle_A$. The proof of the theorem is deceptively simple. Assume first that the cloning operation does work for two orthogonal states $|\psi\rangle_A$ and $|\phi\rangle_A$, that is, that

$$\hat{U}_c |\psi\rangle_A |0\rangle_B = |\psi\rangle_A |\psi\rangle_B \,,$$

$$\hat{U}_c |\phi\rangle_A |0\rangle_B = |\phi\rangle_A |\phi\rangle_B \,, \tag{4.47}$$

and consider then the new state $|\xi\rangle_A = (1/\sqrt{2})[|\psi\rangle_A + |\phi\rangle_A]$. It follows immediately from Eq. (4.47) that

$$\hat{U}_c|\xi\rangle_A|0\rangle_B = \hat{U}_c[|\psi\rangle_A + |\phi\rangle_A] \otimes |0\rangle_B$$

$$= \frac{1}{\sqrt{2}}[|\psi\rangle_A|\psi\rangle_A + |\phi\rangle_A|\phi\rangle_B]\,, \qquad (4.48)$$

rather than the desired state $|\xi\rangle_A|\xi\rangle_B = (1/\sqrt{2})[|\psi\rangle_A + |\phi\rangle_A] \otimes (1/\sqrt{2})[|\psi\rangle_B + |\phi\rangle_B]$. This demonstrates the impossibility to clone an arbitrary quantum state. Hence, Eve cannot clone perfectly the state of the photon sent by Alice. Her spying is detectable in principle at the most fundamental level of quantum mechanics.

4.4.3 Quantum Teleportation

In the discussion of the BB84 quantum key distribution scheme we considered a situation where Alice needed to transfer a known sequence of states under her control to Bob. A more challenging situation is encountered when the state to transfer is unknown. The notion of transferring an object from one to another location by teleporting it in a way that causes it to disappear at the first location and to simultaneously reappear at the second one is familiar to Star Trek aficionados. Teleportation of macroscopic or live objects remains science fiction today, not least because of the massive amount of information that would be required to be transferred. Still, at a much more modest level it is possible to exploit EPR correlations to teleport unknown quantum states of photons and atoms, for atoms over very short distances so far, but for photons over distances of several kilometers. Because the so-called Bell states basis plays a prominent role in this scheme, we proceed by first introducing it before turning to the details of the quantum teleportation scheme.

The Bell States Basis The so-called *Bell states* form a basis of maximally entangled states of a pair of qubits with eigenstates $|0\rangle$ and $|1\rangle$. These could be two-level atoms, two orthogonal states of polarization of a single photon, states of the artificial atoms discussed of Sect. 7.4, or a number of other systems with only two essential states. The Bell states are defined as

$$|\Phi^+\rangle = \frac{1}{\sqrt{2}}(|0,0\rangle + |1,1\rangle)\,, \qquad (4.49)$$

$$|\Phi^-\rangle = \frac{1}{\sqrt{2}}(|0,0\rangle - |1,1\rangle)\,, \qquad (4.50)$$

$$|\Psi^+\rangle = \frac{1}{\sqrt{2}}(|0,1\rangle + |1,0\rangle)\,, \qquad (4.51)$$

$$|\Psi^-\rangle = \frac{1}{\sqrt{2}}(|0,1\rangle - |1,0\rangle)\,, \qquad (4.52)$$

in terms of which

$$|0,0\rangle = \frac{1}{\sqrt{2}}(|\Phi^+\rangle + |\Phi^-\rangle) \quad ; \quad |0,1\rangle = \frac{1}{\sqrt{2}}(|\Psi^+\rangle + |\Psi^-\rangle) ;$$

$$|1,0\rangle = \frac{1}{\sqrt{2}}(|\Psi^+\rangle - |\Psi^-\rangle) \quad ; \quad |1,1\rangle = \frac{1}{\sqrt{2}}(|\Phi^+\rangle - |\Phi^-\rangle) . \qquad (4.53)$$

It is easily shown that these states are orthonormal. They provide a remarkably powerful basis to describe a number of problems in quantum information science involving bipartite systems. We have already seen the role of $|\Phi^+\rangle$ in demonstrations of violations of Bell's inequalities, see e.g. Eq. (4.42) in the summary of Aspect's experiments. We now show how these states are also central to the teleportation of unknown states of a qubit between two distant parties, calling them again Alice and Bob [25].

Teleportation of Unknown States Suppose that Alice wishes to transfer to Bob a state of the qubit

$$|\psi\rangle_C = (\alpha|0\rangle + \beta|1\rangle)_C \qquad (4.54)$$

that is unknown to her. We attach the label C to that unknown qubit for clarity in the following argument, although it belongs to Alice. The protocol goes as follows: first, Alice shares a maximally entangled pair of qubits with Bob, say the Bell state $|\Phi^+\rangle$. It will prove useful to label explicitly which part "belongs" to Alice and which part to Bob, that is,

$$|\Phi^+\rangle_{AB} = \frac{1}{\sqrt{2}}(|0\rangle_A|0\rangle_B + |1\rangle_A|1\rangle_B) . \qquad (4.55)$$

In the presence of the unknown state to be teleported, the total state of the system is

$$|\Psi\rangle = |\psi\rangle_C |\Phi^+\rangle_{AB} = \frac{1}{\sqrt{2}} [\alpha|0\rangle_C + \beta|1\rangle_C] [|0\rangle_A|0\rangle_B + |1\rangle_A|1\rangle_B] , \qquad (4.56)$$

which with Eq. (4.53) can readily be rearranged as

$$|\Psi\rangle = \frac{1}{\sqrt{2}} (\alpha|00\rangle_{CA}|0\rangle_B + \alpha|01\rangle_{CA}|1\rangle_B + \beta|10\rangle_{CA}|1\rangle_B + \beta|11\rangle_{CA}|1\rangle_B)$$

$$= \frac{1}{2} \Big[|\Phi^+\rangle_{CA} \otimes (\alpha|0\rangle + \beta|1\rangle)_B + |\Phi^-\rangle_{CA} \otimes (\alpha|0\rangle - \beta|1\rangle)_B$$

$$+ |\Psi^+\rangle_{CA} \otimes (\alpha|1\rangle + \beta|0\rangle)_B + |\Psi^-\rangle_{CA} \otimes (\alpha|1\rangle - \beta|0\rangle)_B \Big] . \qquad (4.57)$$

While this last step may at first sight seem innocuous, this is not the case: The key point is that when decomposed in that way, the total state $|\Psi\rangle$ is seen to be a superposition of tensor products of four Bell states belonging to Alice—remember, C belongs to Alice—times single qubit states belonging to Bob, with probability amplitudes given by those of the unknown state, albeit not necessarily attached to the proper qubit state $|0\rangle$ or $|1\rangle$, or with the right sign.

With this result in mind, Alice's strategy is straightforward: she performs a Bell state measurement of the local two-qubit system AC, resulting with equal probabilities in one of the four Bell states of her system. The outcome instantly projects Bob's qubit to the corresponding state. If Alice indicates, over a classical channel, that her result is $|\Phi^+\rangle_{CA}$, Bob knows his qubit is already in the desired state. Otherwise, he will apply the unitary transformation $\hat{\sigma}_z$ if her result is $|\Phi^-\rangle_{CA}$, $\hat{\sigma}_x$ if it is $|\Psi^+\rangle_{CA}$, and $i\hat{\sigma}_y$ if it is $|\Psi^+\rangle_{CA}$. This will achieve the Bloch sphere rotations, see Fig. 1.4, required to complete the teleportation of the unknown state, since with Eq. (1.55) we have readily

$$\hat{\sigma}_z \begin{pmatrix} \alpha \\ -\beta \end{pmatrix} = \hat{\sigma}_x \begin{pmatrix} \beta \\ \alpha \end{pmatrix} = i\hat{\sigma}_y \begin{pmatrix} -\beta \\ \alpha \end{pmatrix} = \begin{pmatrix} \alpha \\ \beta \end{pmatrix}. \tag{4.58}$$

Importantly, at the end of the teleportation protocol Alice's original qubit has not remained unchanged, but rather it has become part of an entangled state. Hence, there is no violation of the no-cloning theorem. Also, the quantum teleportation protocol is not instantaneous since it requires the classical communication of the outcome of Alice's measurement, which can proceed no faster than the speed of light. Quantum teleportation of atomic qubits was first demonstrated in two experiments by M. Riebe et al. and M. D. Barrett et al. [26, 27].

Problems

Problem 4.1 Evaluate the spin-flipped density matrix

$$\tilde{\rho}_{AB} = (\hat{\sigma}_y \otimes \hat{\sigma}_y)\hat{\rho}_{AB}^*(\hat{\sigma}_y \otimes \hat{\sigma}_y)$$

for the separable state $|\psi\rangle_{AB} = \frac{1}{\sqrt{2}}|1\rangle_A(|1\rangle + |0\rangle)_B$ and for the entangled state $|\psi\rangle_{AB} = \frac{1}{2}(|11\rangle_{AB} + |00\rangle_{AB})$.

Problem 4.2 Show that for a system in a pure state the concurrence $\mathcal{C}_{AB} = \max\{\lambda_1 - \lambda_2 - \lambda_3 - \lambda_4, 0\}$ reduces to $\mathcal{C}_{AB} = 2\sqrt{\det(\hat{\rho}_A)}$, with $\rho_A = \mathrm{Tr}_B \hat{\rho}_{AB}$. Evaluate it for the two examples of Problem 4.1.

Problem 4.3 Consider a bipartite system A, B consisting of a pair of qubits described by the most general pure state $|\psi\rangle = \sum_{i,j} c_{ij}|i, j\rangle$, where $\{i, j\} = \{0, 1\}$. Show that the associated density operators $\hat{\rho}_{AB}$ and $\tilde{\hat{\rho}}_{AB}$ have at most two eigenvalues that are not equal to zero.

Fig. 4.5 Possible directions
of the pairs of polarizers used
in the optical tests of Bell's
inequality

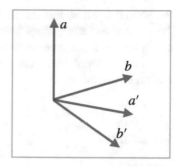

Problem 4.4 For any pure state $|\psi\rangle = \sum_{ijk} c_{ijk} |ijk\rangle$ we have of a tripartite system ABC:

(a) Show that

$$\mathrm{Tr}(\hat{\rho}_{AB}\tilde{\rho}_{AB}) = 2\det(\hat{\rho}_A) - \mathrm{Tr}(\hat{\rho}_B^2) + \mathrm{Tr}(\hat{\rho}_C^2).$$

(b) Using the unicity of the trace, show also that this result implies that

$$\mathrm{Tr}(\hat{\rho}_{AB}\tilde{\rho}_{AB}) = 2\left[\det(\hat{\rho}_A) + \det(\hat{\rho}_B) - \det(\hat{\rho}_C)\right].$$

Problem 4.5 Digging deeper into the Aspect experiments—*after finishing this problem (and even if you do not do it), you should read the account of the loophole-free Bell test experiments by A. Aspect in Ref.* [18].

In the original Aspect experiments, pairs of entangled photons were emitted in a radiative atomic cascade in Calcium, with the $4p^2\,{}^1S_0$ level, populated by two-photon excitation, decaying back to the $4s^2\,{}^1S_0$ state over the $4s4p^1P_1$ level, emitting a photon at $\lambda_1 = 551$ nm and a second photon at $\lambda_2 = 423$ nm. Because the change in angular momentum in the transition is $J = 0 \to J = 1 \to J = 0$, no net angular momentum is carried by the pair of photons, so that for emitted photons counter-propagating in the $\pm z$ direction, the state of polarization of the total system must be of the form

$$|\psi\rangle = \frac{1}{\sqrt{2}}[|\uparrow\rangle_1|\uparrow\rangle_2 + |\rightarrow\rangle_1|\rightarrow\rangle_2].$$

(a) With "+" and "-" labeling the outcome of a measurement where the measured polarization of the photon is parallel or perpendicular to the polarizer's angle, respectively, determines the probabilities $P_{++}(a, b)$, $P_{--}(a, b)$, $P_{+-}(a, b)$, and $P_{-+}(a, b)$ of measuring the pairs of polarization(++), (−−), (+−), and (−+) for the two polarizers at angles (a, b).

(b) Evaluate the Clauser–Horne–Shimony–Holt inequality as a function of the angles (a, b), (a', b), and (a', b') see Fig. 4.5, and find which combination of angles results in a maximum violation of the inequality by quantum mechanics.

Problem 4.6 Show that the Bell states form an orthonormal basis, and determine the unitary transformations in the four-dimensional Hilbert space of the pair of qubits that generate them from the state $|0, 0\rangle$.

Hint: Use tensor products of unitary Pauli matrix transformations acting on the individual qubits.

References

1. A. Einstein, B. Podolsky, N. Rosen, Can quantum-mechanical description of physical reality be considered complete? Phys. Rev. **47**, 777 (1935)
2. J.S. Bell, *Speakable and Unspeakable in Quantum Mechanics*, 2nd edn. (Cambridge University, Cambridge, 2004)
3. V. Coffman, J. Kundu, W.K. Wooters, Distributed entanglement. Phys. Rev. A **61**(5), 052306 (2000)
4. W.K. Wooters, Entanglement of formation of an arbitrary state of two qubits. Phys. Rev. Lett. **80**, 2245 (1998)
5. T.J. Osborne, F. Verstraete, General monogamy inequality for bipartite qubit entanglement. Phys. Rev. Lett. **96**, 220503 (2006)
6. J.S. Bell, On the Einstein-Podolski-Rosen paradox. Physics **1**, 195 (1964)
7. J.F. Clauser, M.A. Horne, Experimental consequences of objective local theories. Phys. Rev. D **10**, 526 (1974)
8. J.F. Clauser, M.A. Horne, A. Shimony, R.A. Holt, Proposed experiment to test local hidden-variable theories. Phys. Rev. Lett. **23**, 880 (1969)
9. A. Aspect, Ph. Grangier, G. Roger, Experimental tests of realistic local theories via Bell's theorem. Phys. Rev. Lett. **47**, 460 (1981)
10. A. Aspect, *Bell's theorem: The Naive View of An Experimentalist, In: Quantum [Un]speakables*, ed. by R.A. Bertlmann, A. Zeilinger (Springer, Berlin, 2002), p. 119
11. N.D. Mermin, The EPR experiment—thoughts about the 'loophole'. Ann. N. Y. Acad. Sci. **480**, 422 (1986)
12. P.H. Eberhard, Background level and counter efficiencies required for a loophole-free Einstein-Podolsky-Rosen experiment. Phys. Rev. A **47**, R747 (1993)
13. M. Giustina, A. Mech, S. Ramelow, B. Wittman, J. Kofler, J. Beyer, A. Lita, B. Calkins, T. Gerrits, S.W. Nam, R. Ursin, A. Zeilinger, Bell violation using entangled photons without the fair-sampling assumption. Nature **497**, 227 (2013)
14. B.G. Christensen, K.T. McCusker, J.B. Altepeter, B. Calkins, T. Gerrits, A.E. Lita, A. Miller, L.K. Shalm, Y. Zhang, S.W. Nam, N. Brunner, C.C.W. Lim, N. Gisin, P.G. Kwiat, Detection-loophole-free test of quantum nonlocality, and applications. Phys. Rev. Lett. **111**, 130406 (2013)
15. B. Hensen, H. Bernien, A.E. Dréau, A. Reiserer, N. Kalb, M.S. Blok, J. Ruitenberg, R.F. Vermeulen, R.N. Schouten, C. Abellàn, W. Amaya, V. Pruneri, M.W. Mitchell, M. Markham, D.J. Twitchen, D. Elkouss, S. Wehner, T.H. Taminiau, R. Hanson, Loophole-free Bell inequality violation using electron spins separated by 1.3 kilometres. Nature **526**, 682 (2015)
16. M. Giustina, M.A. Versteegh, S. Wengerowsky, J. Handsteiner, A. Hochrainer, K. Phelan, F. Steinlechner, J. Kofler, J.-A. Larsson, K. Abellán, W. Amaya, V. Pruneri, M.W. Mitchell, J. Beyer, T. Gerrits, A.E. Lita, L.K. Shalm, S.W. Nam, T. Scheidl, R. Ursin, B. Wittmann, A. Zeilinger, Significant-loophole-free test of Bell's theorem with entangled photons. Phys. Rev. Lett. **115**, 250401 (2015)
17. L.K. Shalm, E Meyer-Scott, B.G. Christensen, P. Bierhorst, M.A. Wayne, M.J. Stevens, T. Gerrits, S. Glancy, D.R. Hamel, M.S. Allman, K.J. Coakley, S.D. Dyer, C. Hodge, A.E. Lita, V.B. Verma, C. Lambrocco, E. Tortoriciand, A.L. Migdall, Y. Zhang, D.R. Kumor, W.H.

Farr, F. Marsili, M.D. Shaw, J.A. Stern, C. Abellán, W. Amaya, V. Pruneri, T. Jennewein, M.W. Mitchell, P.G. Kwiat, J.C. Bienfang, R.P. Mirin, E. Knill, S.W. Nam, Strong loophole-free test of local realism. Phys. Rev. Lett. **115**, 250402 (2015)

18. A. Aspect, Closing the door on Einstein and Bohr's quantum debate. Physics **8**, 123 (2015)

19. J. Handsteiner, A.S. Friedman, D. Rauch, J. Gallicchio, B. Liu, H. Hosp, J. Kofler, D Bricher, M. Fink, C. Leung, A. Mark, H.T. Nguyen, I. Sanders, F. Steinlechner, R. Ursin, S. Wengerowsky, A.H. Guth, D.I. Kaiser, T. Scheidl, A. Zeilinger, Cosmic Bell test: Measurement settings from Milky Way stars. Phys. Rev. Lett. **118**, 060401 (2015)

20. M.-H. Li, C. Wu, Y. Zhang, W.-Z. Liu, B. Bai, Y. Liu, W. Zhang, Q. Zhao, H. Li, Z. Wang, L. You, W.J. Munro, J. Yin, J. Zhang, C.-Z. Peng, X. Ma, Q. Zhang, J. Fan, J.-W. Pan, Test of local realism into the past without detection and locality loopholes. Phys. Rev. Lett. **121**, 080404 (2018)

21. D. Rauch, J. Handsteiner, A. Hochrainer, J. Gallicchio, A.S. Friedman, C. Leung, B. Liu, L. Bulla, S. Ecker, F. Steinlechner, R. Ursin, B. Hu, D. Leon, C. Benn, A. Ghedina, M. Cecconi, A.H. Guth, D.I. Kaiser, T. Scheidl, A. Zeilinger, Cosmic Bell test using random measurement settings from high-redshift quasars. Phys. Rev. Lett. **121**, 080403 (2018)

22. C.H. Bennett, G. Brassard, Quantum cryptography: Public key distribution and coin tossing, in *International Conference in Computers, Systems, and Signal Processing*, Bangalore, India (1984), p. 175

23. W. Wooters, W. Zurek, A single quantum cannot be cloned. Nature **299**, 802 (1982)

24. J.L. Parks, The concept of transition in quantum mechanics. Found. Phys. **1**, 23 (1970)

25. C.H. Bennett, G. Brasssard, C. Crépeau, R. Josza, A. Peres, W.K. Wooters, Teleporting an unknown quantum state via dual and Einstein-Podolski-Rosen channels. Phys. Rev. Lett. **70**, 1895 (1993)

26. M. Riebe, H. Haeffner, C.F. Roos, W. Haensel, J. Benhelm, G.P.T. Lancaster, T.W. Koerber, C. Becher, F. Schmidt-Kaler, D.F.V. James, R. Blatt, Deterministic quantum teleportation with atoms. Nature **429**, 734 (2004)

27. M.D. Barrett, J. Chiaverini, T. Schactz, J. Britton, W.M. Itano, J.D. Jost, E. Knill, C. Langer, D. Leibfried, R. Ozeri, D.J. Wineland, Deterministic quantum teleportation of atomic qubits. Nature **429**, 737 (2004)

Chapter 5
Coupling to Reservoirs

This chapter gives an introduction to the coupling of small systems to reservoirs. After setting the stage with a discussion of spontaneous emission in free space, we introduce three major theoretical methods used in the analysis of system–reservoir interactions in the Markovian limit: the master equation, Langevin equations, and Monte Carlo wave function approaches. We conclude with a brief discussion of the input-output formalism.

The bipartite systems that we have encountered so far consisted largely of small subsystems A and B of one or few particles prepared in pure quantum states. The reason we focused on those situations is that it is for such states that the distinctive quantum mechanical features associated with quantum entanglement, for instance, the violation of Bell inequalities, are most readily apparent. However, a couple of important points need to be kept in mind: first, the separation of a quantum system into separate subsystems is largely arbitrary, dictated either by convenience and/or by experimental considerations. That was, for instance, the case in the discussion of quantum state teleportation, where the system was separated into two subsystems A and B based on the location of its various components. While this was certainly the appropriate decomposition in that case, one can well imagine that another question might be more easily handled by separating the full system into a different set of subsystems. One such situation appeared in the analysis of entanglement distribution, where we separated a system ABC into a subsystem consisting of the qubit A and a second subsystem consisting of the effective qubit (BC). In addition, the simple systems that we have considered so far can never be completely isolated from their environment, even by the most astute and careful experimentalist, and we have largely ignored this key issue so far.

In the simplest cases, it is again possible to think of a system coupled to the environment as a bipartite system, where one of the systems is now "small," and the other, the environment (or reservoir), is "large," in the sense that it contains a large number of degrees of freedom and is characterized by some density of states and some spectral density function $\mathcal{D}(\omega)$. Depending on the specific properties of $\mathcal{D}(\omega)$, the large system may exhibit memory effects or not, that is, be either non-

© The Author(s), under exclusive license to Springer Nature Switzerland AG 2021
P. Meystre, *Quantum Optics*, Graduate Texts in Physics,
https://doi.org/10.1007/978-3-030-76183-7_5

Markovian or Markovian. In the Markovian case, the small system will wind up being characterized by the *effective* irreversible decay of energy and coherence, despite the fact that at the fundamental level, it is described by a Schrödinger (or Dirac in the relativistic case) equation, which is a time-reversible equation. For non-Markovian systems, though, memory effects can result in the small system regaining some of its energy and coherence in a time-dependent fashion.

In general, one has relatively little control over the environment and may only know its average energy, or temperature, and perhaps its mean number of particles. Its state is then described by a density operator maximizing its entropy under the known constraints, following an approach along the lines already discussed in Sect. 2.3.1, but generalized to multimode and/or multiparticle systems. In most cases, it can also be assumed that whatever happens to the small system does not change the state of the reservoir in any significant way—think of a drop of water added to the ocean.

This chapter gives an introduction to the coupling of small systems to reservoirs. After setting the stage with a discussion of spontaneous emission in free space, we introduce three major theoretical approaches used in the analysis of system–reservoir interactions in the Markovian limit: the master equation, Langevin equations, and Monte Carlo wave function approaches. We conclude with a brief discussion of the input–output formalism, which becomes, for example, important in quantum optics situations where the reservoir, or parts of it, serves as a measuring apparatus as well and it becomes necessary to treat it more explicitly than is done in the simplest reservoir formalism approach. Chapter 6 will then go one step further and introduce the idea of a reservoir serving as a "pointer basis" in the context of quantum measurements. Chapter 7 on cavity quantum electrodynamics (cavity QED) will introduce a number of ways to tailor the electromagnetic environment in highly controlled fashion, with applications in basic physics, quantum metrology, and information processing.

5.1   Spontaneous Emission in Free Space

We have seen in Sect. 3.4 that an excited atom interacting with a single-mode electromagnetic field can spontaneously undergo a transition to the ground state even in the absence of cavity photons, $|n = 0\rangle$. It can then reabsorb this photon, resulting in vacuum Rabi oscillations of the atomic populations $p_e(t)$ and $p_g(t)$. The situation is however quite different in free space: the atom interacts not with a single field mode, but rather with an electromagnetic environment that consists of a broad continuum of modes. From the blackbody energy density formula

$$u(\omega, T)\mathrm{d}\omega = \left(\frac{\hbar\omega^3}{\pi^2 c^3}\right) \frac{1}{e^{\hbar\omega/k_B T} - 1}\mathrm{d}\omega, \tag{5.1}$$

it follows that in the visible region and at room temperature, these modes can be taken to an excellent approximation to be at zero temperature, $T \approx 0$, and are for all practical purposes in the vacuum state $\{|0\rangle\}$, as is apparent from Fig. 5.1. The excited

Fig. 5.1 Blackbody power density spectrum (in arbitrary units) as a function of wavelength for temperatures $T = 3000K$, $T = 4000K$, $T = 5000K$, and $T = 6000K$

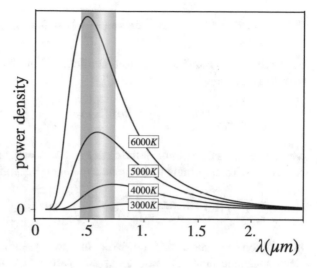

atom can therefore spontaneously emit a photon into any of this infinite number of vacuum modes, and once this happens, spontaneously emitted photons escape for good, with a vanishing probability of being reabsorbed.[1]

5.1.1 Free Space Density of Modes

A quantitative description of spontaneous emission requires a precise formulation of the density of electromagnetic modes $\mathcal{D}(\omega)$ that is specific to the geometry at hand. It can be very broad, as is the case in free space, but could also be narrow or have a complex structure, perhaps with band gaps. These various situations can lead to qualitatively and quantitatively different atomic behaviors, as will be seen in some detail in Chap. 7.

The free space geometry that we consider can be handled by considering field quantization in a three-dimensional cubic cavity of side length L with periodic boundary conditions, with L eventually taken to infinity. Along the $\hat{x}$ direction, the cavity can sustain running modes with wave numbers $K_x = 2\pi n_x/L$, $n_x = 1, 2, \ldots$ Taking differentials of this expression, we find the number of modes between K_x and $K_x + dK_x$ to be $dn_x = dK_x L/2\pi$. Performing the same calculation for the $\hat{y}$ and $\hat{z}$ directions, the number of modes in the volume element $dK_x dK_y dK_z$ is therefore

---

[1]This would however not be the case for transitions in the microwave regime, where the mean number of thermal photons is small, but different from zero at room temperature. The reabsorption of spontaneously emitted photons is also an important consideration in many cavity QED experiments, which can create situations where the photon is forced to return to the atom, maybe by an appropriate arrangement of mirrors.

$$dn = \left(\frac{L}{2\pi}\right)^3 d^3K. \tag{5.2}$$

For large L, it is possible to go to a continuous limit by replacing the summation over K by the integral

$$\frac{1}{V}\sum_K f(\mathbf{K}) \rightarrow \frac{1}{V}\int dnf(\mathbf{K}) = \frac{1}{(2\pi)^3}\int d^3K f(\mathbf{K}), \tag{5.3}$$

where $V = L^3$ is the volume of the cavity and $d^3K = K^2\sin\theta dK d\theta d\phi$ in spherical coordinates. Transforming to frequencies with $\omega = Kc$ gives then

$$\frac{1}{V}\sum_K f(\mathbf{K}) \rightarrow \frac{1}{(2\pi)^3}\int_0^\infty d\omega \frac{\omega^2}{c^3}\int_0^\pi d\theta\sin\theta\int_0^{2\pi} d\phi f(\mathbf{K}). \tag{5.4}$$

In addition, we have to sum over the two polarizations of the transverse electromagnetic field. In performing any particular sum over states, we should insert the desired function $f(\mathbf{K})$ into Eq. (5.4) and carry out the three integrals, taking into account the two possible field polarizations.

5.1.2 Weisskopf–Wigner Theory of Spontaneous Emission

Armed with the three-dimensional free space density of states, we can now proceed with the dynamics of a two-state atom interacting with a field initially in the vacuum state $\{|0_s\rangle\}$, where s labels the field modes. In the electric dipole interaction and rotating wave approximation, the atom–field Hamiltonian is

$$\hat{H} = \hbar\sum_s \omega_s\hat{a}_s^\dagger\hat{a}_s + \hbar\omega_e|e\rangle\langle e| + \hbar\omega_g|g\rangle\langle g| + \hbar\sum_s(g_s\hat{a}_s\hat{\sigma}_+ + \text{h.c.}), \tag{5.5}$$

where the electric dipole interaction Hamiltonian

$$\hat{V} = \hbar\sum_s(g_s\hat{a}_s\hat{\sigma}_+ + \text{h.c.}) \tag{5.6}$$

connects the state $|e\{0\}\rangle$ only to the set of states $|g\{1_s\}\rangle$, which describe an atom in its lower state with one photon in the sth mode and no photons in any other mode. This reduces the most general state vector describing the system to

$$|\psi(t)\rangle = C_{e0}(t)e^{-i\omega_e t}|e\{0\}\rangle + \sum_s C_{g\{1_s\}}(t)e^{-i(\omega_g+\omega_s)t}|g\{1_s\}\rangle. \tag{5.7}$$

Substituting this state into the Schrödinger equation with the Hamiltonian (5.5) and projecting onto the states $|e\{0\}\rangle$ and $|g\{1_s\}\rangle$ give the probability amplitude equations of motion

$$\frac{dC_{e0}(t)}{dt} = -i \sum_s g_s e^{-i(\omega_s - \omega_0)t} C_{g\{1_s\}}(t) , \tag{5.8}$$

$$\frac{dC_{g\{1_s\}}(t)}{dt} = -i g_s^* e^{i(\omega_s - \omega_0)t} C_{e0}(t) , \tag{5.9}$$

with $\omega_0 = \omega_e - \omega_g$. Inserting the formal time integral of the second equation into the first one yields an integro-differential equation for C_{e0} alone,

$$\frac{dC_{e0}(t)}{dt} = -\sum_s |g_s|^2 \int_{t_0}^{t} dt' e^{-i(\omega_s - \omega_0)(t-t')} C_{e0}(t') . \tag{5.10}$$

We now use Eq. (5.4) to convert the sum over states to a three-dimensional integral, which gives

$$\frac{dC_{e0}(t)}{dt} = -\frac{V}{(2\pi c)^3} \int d\omega \omega^2 \int_0^{\pi} d\theta \sin\theta \int_0^{2\pi} d\phi$$

$$\times |g(\omega, \theta)|^2 \int_{t_0}^{t} dt' e^{-i(\omega - \omega_0)(t-t')} C_{e0}(t') . \tag{5.11}$$

Figure 5.2 shows the coordinate system for one running wave with two possible polarizations $\mathbf{e}_1$ and $\mathbf{e}_2$ interacting with the atomic dipole. To calculate $|g(\omega, \theta)|^2$, we evaluate

$$|g(\omega, \theta)|^2 = |\mathcal{E}_\omega / \hbar|^2 \sum_{\sigma=1}^{2} |\langle \alpha | e\mathbf{r} \cdot \mathbf{e}_\sigma | b \rangle|^2$$

$$= |\mathcal{E}_\omega d / \hbar|^2 \sin^2 \theta (\cos^2 \phi + \sin^2 \phi)$$

$$= |\mathcal{E}_\omega d / \hbar|^2 \sin^2 \theta , \tag{5.12}$$

Fig. 5.2 Coordinate system for a plane running wave with wave vector $\mathbf{K}$ and two transverse polarizations along the directions $\mathbf{e}_1$ and $\mathbf{e}_2$. The atomic dipole points in a direction at angle θ with respect to the propagation direction $\mathbf{K}$ and ϕ with respect to $\mathbf{e}_1$

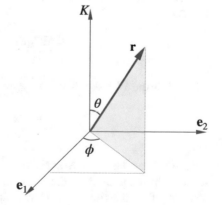

which is independent of the azimuthal coordinate ϕ. Remembering that for running waves, the square of the "electric field per photon" differs from its value for standing waves by a factor of $1/\sqrt{2}$ and is given by

$$\mathcal{E}_\omega = \sqrt{\hbar\omega/2\epsilon_0 V} , \tag{5.13}$$

see Eq. (2.74), and substituting Eq. (5.12) into Eq. (5.11), we encounter the integral

$$\int_0^\pi d\theta \sin^3\theta = \int_{-1}^1 d(\cos\theta)(1 - \cos^2\theta) = \frac{4}{3} . \tag{5.14}$$

With this result, Eq. (5.11) becomes

$$\frac{dC_{e0}(t)}{dt} = -\frac{1}{6\epsilon_0 \pi^2 \hbar c^3} \int d\omega\, \omega^3 |d|^2 \int_{t_0}^t dt'\, e^{-i(\omega-\omega_0)(t-t')} C_{e0}(t') . \tag{5.15}$$

Markov Approximation This equation is nonlocal in time since $C_{e0}(t)$ is a function of the earlier amplitudes $C_{e0}(t')$ and depends therefore on all of its past history. To solve it approximately, we observe that $\omega^3 |d|^2$ varies little in the frequency interval over which the integral over t' has an appreciable value and perform in addition a coarse-grained integration, that is, assume that $C_{e0}(t')$ varies sufficiently slowly that it can be evaluated at $t' = t$ and factored outside the integrals. This is in essence the *Markov approximation*, where all memory effects resulting from the interaction between the atom and the reservoir are being ignored. This results in a remaining time integral that has a highly peaked value at $\omega = \omega_0$, since

$$\lim_{t\to\infty} \int_{t_0}^t dt'\, e^{-i(\omega-\omega_0)(t-t')} = \pi\delta(\omega - \omega_0) - \mathcal{P}\left[\frac{i}{\omega - \omega_o}\right] , \tag{5.16}$$

and allows us therefore to evaluate the product $\omega^3 |d|^2$ at ω_0. The principal part $\mathcal{P}[i/(\omega-\omega_0)]$ leads to a frequency shift related to the Lamb shift, and the $\pi\delta(\omega-\omega_0)$ term gives

$$\frac{dC_{e0}(t)}{dt} = -\frac{\Gamma}{2}C_{e0}(t) , \tag{5.17}$$

where Γ is the Weisskopf–Wigner *spontaneous emission decay rate*

$$\Gamma = \frac{|d|^2 \omega_0^3}{3\pi\epsilon_0 \hbar c^3} = \frac{1}{4\pi\epsilon_0}\frac{4|d|^2\omega_0^3}{3\hbar c^3} . \tag{5.18}$$

As a result of the Markov approximation, the Weisskopf–Wigner theory of spontaneous emission predicts therefore an *irreversible* exponential decay of the upper state population, with no revivals. Although under the action of each individual mode, the atom would have a finite probability to return to the upper state, as in the Jaynes–Cummings vacuum Rabi oscillations, the probability amplitudes for such events effectively interfere destructively when summed over the continuum of free space modes. This is quite a remarkable result, since we started from a system described by the completely reversible dynamics of the Schrödinger equation and wound up with an effective irreversible decay of the upper state atomic population. This irreversible dissipation of the small system energy into the reservoir is of course only approximate and results from ignoring all memory effects in the system dynamics. It is a frequent signature of the interaction of small systems with Markovian reservoirs as will be discussed in a more general context in the following sections.

While the Markov approximation is oftentimes excellent, it is certainly not always justified, and it is not uncommon to face situations where the coupling of small systems to reservoirs is characterized by memory effects such as recurrences or other remnants of reversible behavior. We have already encountered the most extreme such situation in the Jaynes–Cummings model and will see in Chap. 7 that electromagnetic environments can be tailored in ways that result in situations where the spontaneous emission rate can be either enhanced or inhibited from its free space value (5.18).

5.1.3   Superradiance and Subradiance

In addition to the use of tailored electromagnetic environments, radiative decay can also be modified by the presence of other radiators. While based on everyday experience, one might intuitively expect that with more than one atom in an atomic vapor each of them will still decay at the same rate Γ, this need not be the case. Remarkably, under appropriate conditions, even the presence of *just one* additional atom can drastically modify the decay of an atom, a result of constructive or destructive quantum interferences between the one-photon radiation fields emitted by the two atoms. This is perhaps best illustrated by considering the simple example of two co-located identical atoms prepared in the entangled state

$$|\psi\rangle_{\text{atoms}}(0) = \frac{1}{\sqrt{2}} \left[|e, g\rangle \pm |g, e\rangle \right], \tag{5.19}$$

with the electromagnetic field initially in the free space vacuum. This system is again described by the Hamiltonian (5.5), with the only difference that we now have two atoms instead of one so that

$$\hat{H} = \hbar \sum_s \omega_s \hat{a}_s^\dagger \hat{a}_s + \sum_{m=1}^{2} \left[\hbar \omega_e |e\rangle\langle e|_m + \hbar \omega_g |g\rangle\langle g|_m + \hbar \sum_s \left(g_s \hat{a}_s \hat{\sigma}_{m,+} + \text{h.c.} \right) \right]. \tag{5.20}$$

The state of the atom–field system at time t is then

$$|\psi(t)\rangle = C_{eg0}(t)e^{-i(\omega_e+\omega_g)t}|e, g, \{0\}\rangle + C_{ge0}(t)e^{-i(\omega_e+\omega_g)t}|e, g, \{0\}\rangle$$
$$+ \sum_s C_{gg\{1_s\}}(t)e^{-i(2\omega_g+\omega_s)t}|g, g, \{1_s\}\rangle. \tag{5.21}$$

Following the same procedure as in the previous section, we replace the sum over modes by an integral, formally integrate the equation of motion for $C_{gg\{1_s\}}(t)$, and carry out the Weisskopf–Wigner approximation. This yields readily the coupled equations of motion for the excited state probability amplitudes of the two atoms as

$$\frac{d}{dt}C_{eg0}(t) = -\frac{\Gamma}{2}\left[C_{eg0}(t) + C_{ge0}(t)\right],$$
$$\frac{d}{dt}C_{ge0}(t) = -\frac{\Gamma}{2}\left[C_{eg0}(t) + C_{ge0}(t)\right]. \tag{5.22}$$

Contrary to what might have been intuitively expected, but not surprising perhaps following our discussion of quantum entanglement, this result indicates that the decay of one of the atoms is not independent of the presence of a second atom. Importantly, its decay rate can be either increased, decreased, or left unchanged by the presence of the second atom. This is easily seen by adding and subtracting the two equations (5.22) to find

$$\frac{d}{dt}\left[C_{eg0}(t) + C_{ge0}(t)\right] = -\Gamma\left[C_{eg0}(t) + C_{ge0}(t)\right],$$
$$\frac{d}{dt}\left[C_{eg0}(t) - C_{ge0}(t)\right] = 0. \tag{5.23}$$

The second of these equations indicates that if $C_{ge0}(0) = C_{eg0}(0) = 1/\sqrt{2}$, then they remain equal at all times, so that

$$\frac{d}{dt}C_{eg0}(t) = \frac{d}{dt}C_{ge0}(t) = -\Gamma C_{eg0}(t); \tag{5.24}$$

that is, the atoms decay at twice the rate of the isolated atoms. This effect, called *superradiance*, was first predicted by R. H. Dicke in 1954 [1], see also the review [2] for a more detailed presentation. In contrast, for $C_{ge0}(0) = -C_{eg0}(0) = 1/\sqrt{2}$, we have

$$\frac{d}{dt}C_{eg0}(t) = \frac{d}{dt}C_{ge0}(t) = 0, \tag{5.25}$$

and the atoms do not decay at all. Not surprisingly, this is called *subradiance*. Importantly, it is easily shown, for instance, by imposing a random phase between

$C_{eg0}(0)$ and $C_{ge0}(0)$, that if the atoms are in the incoherent mixture

$$\rho_{\text{atoms}}(0) = \frac{1}{2}\big[|e, g\rangle\langle e, g| + |g, e\rangle\langle g, e|\big] \qquad (5.26)$$

rather than an entangled state such as Eq. (5.19), then the two atoms decay at the individual spontaneous decay rate Γ, with equal excited state population probabilities $p_e(t)$ given by

$$\frac{dp_e(t)}{dt} = -\Gamma p_e(t). \qquad (5.27)$$

Superradiance and subradiance originate from quantum interferences in the fields emitted and reabsorbed by the two atoms. When two initially excited atoms radiate into the vacuum field, they act as sources for the evolution of the two-atom ground state probability amplitudes $C_{gg\{1_s\}}(t)$, rather than of the individual atom probability amplitudes $C_{g\{1_s\}}(t)$, see Eq. (5.9). The fields from these two sources can interfere either destructively or constructively, and as a result, they drive the dynamics of the two-atom system differently when the field is repeatedly reabsorbed and reemitted. For a mixed state, the relative phases of the fields originating from the two atoms add at random. Hence, the interferences are washed out and each atom acts independently of the other. This is the situation encountered in everyday situations, but that this is not necessarily the case must be kept in mind, in particular in quantum information applications where optically driven qubits might or might not be entangled.

5.2   Master Equation

The Weisskopf–Wigner theory of spontaneous emission is an example of a general class of problems involving the coupling of a small system to its environment, in this case the continuum of modes of the electromagnetic field. When computing the atomic decay rate, we were not interested in the field itself, but only in its effect on the atomic dynamics. In stark contrast with the situations encountered in Chap. 4, where both subsystems were treated on an equal footing, we never explicitly computed the field dynamics. Instead, we carried out a pair of approximations that resulted in the effectively irreversible decay of the atomic upper state population and the associated transfer of energy to the field. They were (a) the assumption that the probability amplitude $C_{e0}(t)$ varies little during the time interval defined by the inverse bandwidth of the continuum of modes of the electromagnetic field and (b) the replacement of the remaining time integration in Eq. (5.15) by a δ-function.

This chapter discusses two primary methods to describe system–reservoir interactions. The first one is based on the Schrödinger picture and leads to the so-called *master equation*. The alternative Heisenberg approach leads to the introduction of *quantum noise operators* giving the description of the problem a flavor reminiscent of the Langevin approach to stochastic problems in classical physics. In addition to these two approaches, we will also consider the method of Monte Carlo wave functions, which permits to unravel the master equation into "quantum trajectories" of considerable intuitive appeal and numerical convenience.

We consider a generic system consisting of a small system S coupled to a large reservoir R by some interaction $\hat{V}$. It is described by the Hamiltonian

$$\hat{H} = \hat{H}_s + \hat{H}_r + \hat{V} \equiv \hat{H}_0 + \hat{V}, \tag{5.28}$$

which, together with initial conditions, completely specifies the problem at hand. We assume that at the initial time t_0, the small system is described by a density operator $\hat{\rho}_s(t_0)$ with $\mathrm{Tr}_s \hat{\rho}_s(t_0) = 1$, where Tr_s stands for the partial trace over the system. In contrast, we take the reservoir to be a very large system with an immense number of degrees of freedom. In most cases (but not always), it is taken to be in thermal equilibrium at some temperature T and is therefore described by the time-independent density operator $\hat{\rho}_r$ of Eq. (2.83),

$$\hat{\rho}_r(\hat{H}_r) = \frac{e^{-\beta \hat{H}_r}}{\mathrm{Tr}_r\{e^{-\beta \hat{H}_r}\}}, \tag{5.29}$$

with $\beta = 1/k_B T$ and $\mathrm{Tr}_r \hat{\rho}_r(\hat{H}_r) = 1$, with Tr_r the trace over the reservoir. Assuming that the system and reservoir are brought into contact at time $t = t_0$, they initially do not exhibit any correlations, and thus the initial state of the system is described by the factorized density operator

$$\hat{\rho}_{sr}(t_0) = \hat{\rho}_s(t_0) \otimes \hat{\rho}_r(\hat{H}_r). \tag{5.30}$$

Its Schrödinger picture evolution is given by

$$\frac{\partial \hat{\rho}_{sr}}{\partial t} = -\frac{\mathrm{i}}{\hbar}[\hat{H}, \hat{\rho}_{sr}]. \tag{5.31}$$

When solved for all times, it provides a full characterization of the states of both the system and the reservoir.

The goal of the master equation is less ambitious: it is to provide, in the Schrödinger picture, a partial and simplified description that allows us to achieve the limited goal to determine the expectation value of the *system* operators $\hat{O}$ only,

$$\langle \hat{O}(t) \rangle = \mathrm{Tr}_{sr}\{\hat{O}\hat{\rho}_{sr}(t)\}, \tag{5.32}$$

where the trace is over both the system and the reservoir. Since $\hat{O}$ is a system operator, we may rewrite this expression as

$$\langle \hat{O}(t) \rangle = \mathrm{Tr}_s\{\hat{O}\,\mathrm{Tr}_r \hat{\rho}_{st}(t)\} \equiv \mathrm{Tr}_s\{\hat{O}\hat{\rho}_s(t)\}. \tag{5.33}$$

The operator

$$\hat{\rho}_s(t) \equiv \mathrm{Tr}_r \hat{\rho}_{st}(t) \tag{5.34}$$

is called the *reduced density operator* of the system, and all we need to determine the expectation value of system operators is to know $\hat{\rho}_s(t)$ at all times. The equation of motion for $\hat{\rho}_s(t)$ is called a *master equation.*

Our strategy to derive the master equation is quite simple: we solve the problem to second order in perturbation theory, trace over the reservoir, take into account if appropriate that it has a very broad bandwidth to perform the Markov approximation, and obtain directly an equation for $\hat{\rho}_s(t)$ that is valid for times long compared to the inverse bandwidth τ_c of the reservoir.

In doing so, it is important not to confuse τ_c with the characteristic time scale of the free evolution of the system, which can possibly be fast as well. To make sure that things do not get mixed up, we therefore first go into an interaction picture where all free evolutions are eliminated. Assuming that the Hamiltonian $\hat{H}_0$ has no explicit time dependence, the interaction picture density operator $\hat{P}_{sr}(t)$ is related to the Schrödinger picture density operator by the unitary transformation

$$\hat{P}_{sr}(t) = e^{i\hat{H}_0(t-t_0)/\hbar} \hat{\rho}_{sr}(t) e^{-i\hat{H}_0(t-t_0)/\hbar} , \tag{5.35}$$

and the corresponding interaction picture reduced density operator $\hat{\rho}(t)$ for the system[2]

$$\hat{\rho}(t) \equiv \mathrm{Tr}_r\{\hat{P}_{sr}(t)\} \tag{5.36}$$

is related to the Schrödinger picture reduced density operator $\hat{\rho}_s(t)$ by

$$\hat{\rho}_s(t) = e^{-i\hat{H}_s(t-t_0)/\hbar} \hat{\rho}(t) e^{i\hat{H}_s(t-t_0)/\hbar} . \tag{5.37}$$

Differentiating Eq. (5.35) with respect to time, and using Eq. (5.31), yields the equation of motion for $P_{sr}(t)$

$$\frac{\partial \hat{P}_{sr}(t)}{\partial t} = -\frac{i}{\hbar}[\hat{V}_I(t-t_0), \hat{P}_{sr}(t)], \tag{5.38}$$

where

$$\hat{V}_I(t-t_0) = e^{i\hat{H}_0(t-t_0)/\hbar} \hat{V} e^{-i\hat{H}_0(t-t_0)/\hbar} \tag{5.39}$$

---

[2]The notation for the various density operators appearing here, while not necessarily completely satisfying, is motivated by the desire to keep the master equation as free of indices as possible in the interaction picture, where it is usually applied.

is the interaction Hamiltonian in the interaction picture. Finally, differentiating
Eq. (5.37) with respect to time gives

$$\frac{\partial \hat{\rho}_s(t)}{\partial t} = -\frac{i}{\hbar} e^{-i\hat{H}_s(t-t_0)/\hbar} \left([\hat{H}_s, \hat{\rho}(t)] + i\hbar \frac{\partial \hat{\rho}(t)}{\partial t} \right) e^{i\hat{H}_s(t-t_0)/\hbar}, \qquad (5.40)$$

an equation that relates the equations of motion for the Schrödinger and interaction
picture reduced density operators $\hat{\rho}_s$ and $\hat{\rho}$.

With this relationship established, we now proceed to determine the evolution of
$\hat{\rho}(t)$, using as our starting point the evolution of $\hat{P}_{sr}(t)$. In general, system–reservoir
coupling problems are not amenable to an exact solution, and here we solve the
problem to second order in perturbation theory. Specifically, we integrate Eq. (5.38)
from t_0 to t, taking $\hat{P}_{sr}(t) \simeq \hat{P}_{sr}(t_0)$ in the commutator to obtain a first-order
solution for $\hat{P}_{sr}(t)$. We then use this improved value in the commutator, integrating
again to obtain a value of $P_{sr}(t)$ accurate to second order. We find

$$\hat{P}_{sr}(t) = \hat{P}_{sr}(t_0) - \frac{i}{\hbar} \int_{t_0}^{t} dt' [\hat{V}_I(t'-t_0), \hat{P}_{sr}(t_0)]$$

$$-\frac{1}{\hbar^2} \int_{t_0}^{t} dt' \int_{t_0}^{t'} dt'' [\hat{V}_I(t'-t_0), [\hat{V}_I(t''-t_0), \hat{P}_{sr}(t_0)]] + \dots, \quad (5.41)$$

and tracing over the reservoir yields the evolution of the reduced density operator
$\hat{\rho}(t)$. Performing this trace, we can define *a coarse-grained equation of motion* for
$\hat{\rho}(t)$ by

$$\dot{\hat{\rho}}(t) \simeq \frac{\hat{\rho}(t) - \hat{\rho}(t-\tau)}{\tau}, \qquad (5.42)$$

where the time interval $\tau = t - t_0$ is taken to be long compared to the reservoir
memory time τ_c, but short compared to times yielding significant changes in the
system variables. In explicit calculations, it is convenient to shift the time origin by
τ, i.e., to write

$$\dot{\hat{\rho}}(t+\tau) \simeq \frac{\hat{\rho}(t+\tau) - \hat{\rho}(t)}{\tau}. \qquad (5.43)$$

Since we assume that $\hat{\rho}(t)$ does not vary significantly in the time τ, we suppose
that $\partial \hat{\rho}(t)/\partial t$ itself is given by this expression. We further note that the double
commutator in Eq. (5.41) simplifies somewhat since

$$[\hat{V}', [\hat{V}'', \hat{P}_{sr}]] = \hat{V}'\hat{V}''\hat{P}_{sr} - \hat{V}'\hat{P}_{sr}\hat{V}'' + \text{adj.}$$

This is easily shown since $(\hat{A}\hat{B}\hat{C})^\dagger = \hat{C}^\dagger \hat{B}^\dagger \hat{A}^\dagger$ and all operators appearing in this
equation are Hermitian. Combining these observations yields the interaction picture

coarse-grained equation of motion for the system density operator

$$\frac{\partial \hat{\rho}(t)}{\partial t} \simeq -\frac{i}{\hbar \tau} \int_0^\tau d\tau' \mathrm{Tr}_r\{\hat{V}_I(\tau')\hat{P}_{sr}(t)\} - \frac{1}{\hbar^2 \tau} \int_0^\tau d\tau' \int_0^{\tau'} d\tau''$$
$$\times \mathrm{Tr}_r \left[\hat{V}_I(\tau')\hat{V}_I(\tau'')\hat{P}_{sr}(t) - \hat{V}_I(\tau')\hat{P}_{sr}(t)\hat{V}_I(\tau'') \right] + \mathrm{adj}. \quad (5.44)$$

Importantly, we observe that the reduced density operator actually has two time dependencies, t and τ. As we shall see, τ is associated with the dynamics of reservoir operators, and this dependence disappears in case the reservoir is stationary and with infinitely short memory.

To proceed further, it is pedagogically helpful to temporarily sacrifice generality and use explicit forms of the interaction Hamiltonian $\hat{V}_I$ that make the underlying physics of the system–reservoir interaction more transparent. With this goal in mind, we concentrate first on the important case of a simple harmonic oscillator coupled to a bath of harmonic oscillators, before returning to more general considerations in Sect. 5.2.2.

5.2.1   Damped Harmonic Oscillator

The system and reservoir Hamiltonians of a simple harmonic oscillator of frequency Ω coupled to a bath of harmonic oscillators of frequencies ω_j are

$$\hat{H}_s = \hbar \Omega \hat{a}^\dagger \hat{a} \quad (5.45)$$

and

$$\hat{H}_r = \sum_j \hbar \omega_j \hat{b}_j^\dagger \hat{b}_j . \quad (5.46)$$

This is a broadly used model of a reservoir that finds applications in many situations, well past the relatively narrow confines of quantum optics where its most familiar application is in the description of the radiative coupling of an atom to the continuum of modes of the electromagnetic field. More generally, it also accounts for electron–phonon interactions in conductors, mechanical damping by a phonon bath in optomechanics, and many other examples. This is because while different problems require of course different system Hamiltonians $\hat{H}_s$, the general impact of the bath on the system dynamics is usually not very sensitive to the explicit form of $\hat{H}_r$, in particular if it is a Markovian reservoir characterized by a broadband spectrum and hence a very short correlation time compared to the characteristic time(s) of the small system. In addition to this main reason, another motivation for the broad use of this model is that harmonic oscillators are some of the simplest quantum systems, and they make our lives particularly easy.

To complete our model, we assume that the elementary exchange of energy between system and bath consists of the simultaneous creation of a quantum of excitation of the system with annihilation of a quantum in the jth mode of the bath, or the reverse process,

$$\hat{V} = \hbar \sum_j (g_j \hat{a}^\dagger \hat{b}_j + g_j^* \hat{b}_j^\dagger \hat{a}) , \tag{5.47}$$

or, in the interaction picture,

$$\hat{V}_I(t) = \hbar \sum_j g_j \, \hat{a}^\dagger \hat{b}_j \, e^{i(\Omega - \omega_j)(t - t_0)} + \text{adj.} , \tag{5.48}$$

where we made use of the free evolution of the annihilation and creation operators and remembered that system and reservoir operators commute at equal times. We can simplify the form of Eq. (5.48) by writing it in the more compact form

$$\hat{V}_I(\tau) = \hbar \hat{a}^\dagger \hat{F}(\tau) + \hbar \hat{a} \hat{F}^\dagger(\tau), \tag{5.49}$$

where the operator

$$\hat{F}(\tau) \equiv -i \sum_j g_j \hat{b}_j e^{i(\Omega - \omega_j)\tau} \tag{5.50}$$

acts only on states of the reservoir Hilbert space. We will see explicitly in Sect. 5.3 how $\hat{F}(\tau)$ can be interpreted as a *quantum noise operator*.

Reservoir Correlation Functions When tracing over a reservoir in thermal equilibrium, we encounter terms of the type

$$\text{Tr}_r \{ \hat{a}^\dagger \hat{F}(\tau) \hat{P}_{sr}(t) \} = \hat{a}^\dagger \hat{\rho}(t) \text{Tr}_r \{ \hat{F}(\tau) \hat{\rho}_r (\hat{H}_r) \} ,$$

where we have used Eqs. (5.29) and (5.30) and the fact that at time t_0 the interaction picture and Schrödinger density operators are identical. The trace on the right-hand side of this equation is readily identified as the expectation value $\langle \hat{F}(\tau) \rangle$ of the reservoir operator $\hat{F}(\tau)$. It vanishes provided that the reservoir density operator $\hat{\rho}_r$ is diagonal in the Fock states basis, as is the case for the thermal density operator (5.29). We can then cyclically permute reservoir operators under the reservoir trace in the remaining terms of Eq. (5.44) to find

$$\dot{\hat{\rho}}(t) = -\frac{1}{\hbar^2 \tau} \int_0^\tau d\tau' \int_0^{\tau'} d\tau'' \left[\hat{a}^\dagger \hat{a} \hat{\rho}(t) \langle \hat{F}(\tau') \hat{F}^\dagger(\tau'') \rangle_r - \hat{a} \hat{\rho}(t) \hat{a}^\dagger \langle \hat{F}(\tau'') \hat{F}^\dagger(\tau') \rangle_r \right.$$

$$\left. + \hat{a} \hat{a}^\dagger \hat{\rho}(t) \langle \hat{F}^\dagger(\tau') \hat{F}(\tau'') \rangle_r - \hat{a}^\dagger \hat{\rho}(t) \hat{a} \langle \hat{F}^\dagger(\tau'') \hat{F}(\tau') \rangle_r \right.$$

$$+\hat{a}\hat{a}\hat{\rho}(t)\langle\hat{F}^{\dagger}(\tau')\hat{F}^{\dagger}(\tau'')\rangle_{r} - \hat{a}\hat{\rho}(t)\hat{a}\langle\hat{F}^{\dagger}(\tau'')\hat{F}^{\dagger}(\tau')\rangle_{r}$$

$$+ \hat{a}^{\dagger}\hat{a}^{\dagger}\hat{\rho}(t)\langle\hat{F}(\tau')\hat{F}(\tau'')\rangle_{r} - \hat{a}^{\dagger}\hat{\rho}(t)\hat{a}^{\dagger}\langle\hat{F}(\tau'')\hat{F}(\tau')\rangle_{r}\Big] + \text{adj.} \tag{5.51}$$

Averages such as $\langle\hat{F}(\tau')\hat{F}^{\dagger}(\tau'')\rangle_{r}$ are readily identified as first-order correlation functions of the bath. If the bath is stationary, as is the case in thermal equilibrium, see Eq. (5.29), they depend only on the time difference $\mathcal{T} = \tau' - \tau''$, so that

$$\langle\hat{F}(\tau')\hat{F}^{\dagger}(\tau'')\rangle_{r} = \langle\hat{F}(\tau'')\hat{F}^{\dagger}(\tau')\rangle_{r}^{*}.$$

Using Eq. (5.50), we find that these expressions have explicit forms like

$$\langle\hat{F}(\tau')\hat{F}^{\dagger}(\tau'')\rangle_{r} = \sum_{i,j} g_{i}g_{j}^{*}\langle\hat{b}_{i}\hat{b}_{j}^{\dagger}\rangle_{r}e^{i\Omega(\tau'-\tau'')}e^{i(\omega_{j}\tau''-\omega_{i}\tau')}$$

$$= \sum_{i}|g_{i}|^{2}\langle\hat{b}_{i}\hat{b}_{i}^{\dagger}\rangle_{r}e^{i(\Omega-\omega_{i})(\tau'-\tau'')}, \tag{5.52}$$

where the reservoir density operator is assumed to be diagonal in the energy representation—that is, on the Fock states basis—to obtain the second equality. It is important to keep in mind that while this condition is satisfied in thermal equilibrium, it must be relaxed when considering, for example, "squeezed reservoirs."

Markov Approximation Equation (5.52) tells us how fast the bath correlations decay away. We now perform the *Markov approximation,* which we already encountered in Sect. 5.1, and which assumes that this correlation time is infinitely short compared to all times of interest for the system. For example, with the change of variable $\mathcal{T} = \tau' - \tau''$, Eq. (5.52) becomes

$$\int_{0}^{\tau} d\tau' \int_{0}^{\tau'} d\tau'' \langle\hat{F}(\tau')\hat{F}^{\dagger}(\tau'')\rangle_{r} = \int_{0}^{\tau} d\tau' \sum_{i}|g_{i}|^{2}\langle\hat{b}_{i}\hat{b}_{i}^{\dagger}\rangle_{r} \int_{0}^{\tau'} d\mathcal{T} e^{i(\Omega-\omega_{i})\mathcal{T}}.$$
$$\tag{5.53}$$

In the Weisskopf–Wigner theory of spontaneous emission, we replaced the sum over modes with an integral and interpreted the integral over the exponential as a δ-function. Similarly, calling $\mathcal{D}(\omega)$ the density of modes of the reservoir, and assuming that the reservoir has sufficient bandwidth to justify the δ-function approximation of the integral, we find that (5.53) becomes

$$\int_{0}^{\tau} d\tau' \int_{0}^{\tau'} d\tau'' \langle\hat{F}(\tau')\hat{F}^{\dagger}(\tau'')\rangle_{r} = \frac{\gamma\tau}{2}\langle\hat{b}(\Omega)\hat{b}^{\dagger}(\Omega)\rangle_{r}, \tag{5.54}$$

where in analogy with the Weisskopf–Wigner theory, we introduced a decay rate

$$\gamma = 2\pi\mathcal{D}(\Omega)|g(\Omega)|^{2}, \tag{5.55}$$

and we neglected the simple harmonic oscillator analog of the Lamb shift. The factor $|g(\Omega)|^2 \langle \hat{b}(\Omega)\hat{b}^\dagger(\Omega)\rangle_r$ is a measure of the strength of the coupling of the simple harmonic oscillator with the mode of the reservoir that oscillates at its frequency Ω.

When effectively extending the upper limit of the time integration to infinity in the second integral of Eq. (5.53) to approximate it as a δ-function, we implicitly assumed that the reservoir correlation time τ_c is so small that the integrand vanishes after times short enough for second-order perturbation theory to remains valid. Thus, the approximate solution (5.54) is valid for times *short* compared to the decay of the system, but *long* compared to the correlation time of the reservoir. This is the essence of the Markov approximation. We will revisit this approximation in the Heisenberg picture in the next section and confirm that it amounts to assuming that the correlation functions of the bath are effectively δ-correlated, that is, the reservoir loses its memory instantaneously.

From Eq. (5.29), the average $\bar{n}$ of the number operator $\hat{b}^\dagger(\Omega)\hat{b}(\Omega)$ over a thermal distribution is given by

$$\bar{n} \equiv \langle \hat{b}^\dagger(\Omega)\hat{b}(\Omega)\rangle_r = \frac{1}{e^{\beta\hbar\Omega} - 1}, \tag{5.56}$$

and from the bosonic commutation relation $[\hat{b}(\Omega), \hat{b}^\dagger(\Omega)] = 1$, $\langle \hat{b}(\Omega)\hat{b}^\dagger(\Omega)\rangle = \bar{n} + 1$.

Substituting Eqs. (5.54) and (5.56) along with corresponding expressions for the terms with $\langle \hat{F}^\dagger(\tau')\hat{F}(\tau'')\rangle_r$ gives finally the master equation that governs the dynamics of the system reduced density operator in the interaction picture

$$\frac{d\hat{\rho}(t)}{dt} = -\frac{\gamma}{2}(\bar{n} + 1)[\hat{a}^\dagger \hat{a}\hat{\rho}(t) - \hat{a}\hat{\rho}(t)\hat{a}^\dagger]$$

$$-\frac{\gamma}{2}\bar{n}[\hat{\rho}(t)\hat{a}\hat{a}^\dagger - \hat{a}^\dagger\hat{\rho}(t)\hat{a}] + \text{adj.} \equiv \hat{\mathcal{L}}\hat{\rho}, \tag{5.57}$$

where the superoperator $\hat{\mathcal{L}}$ is called the *Liouvillian*. Here, we have neglected terms containing correlation functions like $\langle \hat{F}(\tau')\hat{F}(\tau'')\rangle$ and $\langle \hat{F}^\dagger(\tau')\hat{F}^\dagger(\tau'')\rangle$, an approximation valid if the density operator is diagonal in the energy representation. If the reservoir is at zero temperature, $\bar{n} = 0$, then the remaining terms in the master equation (5.57) result from reservoir vacuum fluctuations only.

Detailed Balance As advertised at the beginning of this section, we can use the master equation (5.57) to determine the time dependence of the expectations value of system operators without needing to worry any longer about the reservoir. For example, the average excitation number $\langle \hat{a}^\dagger\hat{a}\rangle_s = \text{Tr}_s[\hat{a}^\dagger\hat{a}\hat{\rho}(t)]$ in the oscillator is governed by the equation of motion

$$\frac{d\langle \hat{a}^\dagger\hat{a}\rangle_s}{dt} = -\gamma\langle \hat{a}^\dagger\hat{a}\rangle_s + \gamma\bar{n}, \tag{5.58}$$

where we have used the cyclic property of the trace and the boson commutation relation $[\hat{a}, \hat{a}^\dagger] = 1$. This shows that the oscillator is damped by its coupling to the reservoir and that it will reach an equilibrium point with

$$\langle \hat{a}^\dagger \hat{a} \rangle_s = \bar{n} \, ; \tag{5.59}$$

that is, the mean excitation number of the oscillator will equal the excitation number of the thermal reservoir at its frequency Ω.

A useful way to interpret this result is to reexpress Eq. (5.58) as

$$\frac{d\langle \hat{a}^\dagger \hat{a} \rangle_s}{dt} = -\gamma \langle \hat{a}^\dagger \hat{a} \rangle_s (\bar{n} + 1) + \gamma \bar{n} (\langle \hat{a}^\dagger \hat{a} \rangle_s + 1) \, .$$

When written in this form, the rate of change of the mean number $\langle \hat{a}^\dagger \hat{a} \rangle_s$ of system excitations is seen to result from the balance between emission from the system into the bath and from the bath into the system. In both terms, the "+1" is the contribution from spontaneous emission and would vanish if the boson creation and annihilation operators commuted. The other term is the result of stimulated emission. For a reservoir at zero temperature, $\bar{n} = 0$, all that is left is spontaneous decay from the system into the reservoir.

Equation (5.58) is readily solved to give

$$\langle \hat{a}^\dagger \hat{a} \rangle_s (t) = \langle \hat{a}^\dagger \hat{a} \rangle_s (0) e^{-\gamma t} + \bar{n} [1 - e^{-\gamma t}] \, . \tag{5.60}$$

For large times, the average number of excitations in the simple harmonic oscillator (the average number of photons if it describes an electromagnetic field mode) equilibrates to that of the bath oscillator at the same frequency Ω, as we have seen. Similarly, we find that the expectation value of $\langle \hat{a} \rangle$ (proportional to the complex electric field operator for optical fields) obeys the interaction picture equation of motion

$$\frac{d\langle \hat{a} \rangle_s}{dt} = -\frac{\gamma}{2} \langle \hat{a} \rangle_s \tag{5.61}$$

and equilibrates at $\langle \hat{a} \rangle_s = 0$.

It is also instructive to calculate the equation of motion for the diagonal matrix elements $p_n \equiv \langle n | \hat{\rho} | n \rangle$ in the number states representation. We find readily

$$\dot{p}_n = -\gamma (\bar{n} + 1) [n p_n - (n + 1) p_{n+1}] - \gamma \bar{n} [(n + 1) p_n - n p_{n-1}] \, . \tag{5.62}$$

The four terms in this equation represent flows of number probability up and down the harmonic oscillator ladder of energy levels. These are illustrated in Fig. 5.3, which shows the absorptive and emissive roles of the first and second bracketed expressions, respectively. A steady state occurs when *a detailed balance*

Fig. 5.3 Detail of the energy level diagram of the damped harmonic oscillator with the number probability flows of Eq. (5.62) into (green arrows) and out of (red arrows) state $|n\rangle$

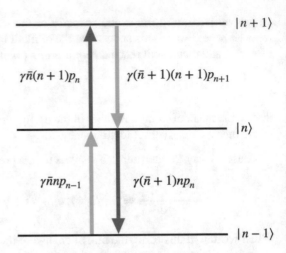

is established between the emissive and absorptive processes, that is, when

$$\gamma(\bar{n} + 1)np_n = \gamma\bar{n}np_{n-1}. \tag{5.63}$$

Note that steady state implies this detailed balance: Eq. (5.63) must be true for the lowest probability p_0 to be constant, which implies that it must be true for p_1 to be constant, and so on up the ladder. This gives the thermal number distribution

$$p_n = \frac{\bar{n}}{\bar{n}+1}p_{n-1} = \left(\frac{\bar{n}}{\bar{n}+1}\right)^n p_0 = e^{-n\hbar\Omega/k_BT}[1 - e^{-\hbar\Omega/k_BT}], \tag{5.64}$$

which, when taking the harmonic oscillator to describe a single-mode optical field, is precisely the thermal photon statistics of Eq. (2.94). As would be intuitively expected, the contact of the oscillator to a thermal reservoir at temperature T thermalizes it at that temperature, resulting in a state of thermal equilibrium between the system and the reservoir.

Damped Two-level Atom It is straightforward to modify the derivation of the master equation (5.57) to describe a two-level atom with upper to lower level decay and damped by a thermal reservoir of simple harmonic oscillators. In this case, the atom–bath interaction Hamiltonian is

$$\hat{V}_I(\tau) = \hbar\hat{\sigma}_+\hat{F}(\tau) + \hbar\hat{\sigma}_-\hat{F}^\dagger(\tau), \tag{5.65}$$

where $\hat{F}(\tau)$ is still given by Eq. (5.50). The derivation of the master equation follows along exactly the same lines as before, with the replacement of $\hat{a}$ by $\hat{\sigma}_-$ and $\hat{a}^\dagger$ by

$\hat{a}_+$. Hence, we find

$$\frac{d\hat{\rho}(t)}{dt} = -\frac{\Gamma}{2}(\bar{n}+1)[\hat{\sigma}_+\hat{\sigma}_-\hat{\rho}(t) - \hat{\sigma}_-\hat{\rho}(t)\hat{\sigma}_+]$$

$$-\frac{\Gamma}{2}\bar{n}[\hat{\rho}(t)\hat{\sigma}_-\hat{\sigma}_+ - \hat{\sigma}_+\hat{\rho}(t)\hat{\sigma}_-] + \text{adj.} \tag{5.66}$$

5.2.2   Lindblad Form

The two examples of master equations (5.57) and (5.66) that we have explicitly considered can be cast in the general form

$$\frac{d\hat{\rho}}{dt} = -\frac{i}{\hbar}[\hat{H}_s, \hat{\rho}] + \hat{\mathcal{L}}[\hat{\rho}], \tag{5.67}$$

where the Liouvillian $\mathcal{L}[\hat{\rho}]$ describes the non-Hermitian evolution of the system due to its coupling to the reservoir and is responsible for irreversible dissipation. It can be shown that in order to preserve the trace of the density operator, $\text{Tr}\hat{\rho} = 1$, for any Markovian system, this term must be of the so-called *Lindblad form*

$$\hat{\mathcal{L}}[\hat{\rho}] = -\frac{1}{2}\sum_i(\hat{C}_i^\dagger\hat{C}_i\hat{\rho} + \hat{\rho}\hat{C}_i^\dagger\hat{C}_i) + \sum_i\hat{C}_i\hat{\rho}\hat{C}_i^\dagger, \tag{5.68}$$

where the $\hat{C}_i$'s are system operators. See e.g. Ref. [3] for the mathematical proof of this result.

That this is the case for the examples that we have explicitly considered is easily seen by substitution. For instance, for the master equation (5.57) describing a damped harmonic oscillator, we have

$$\hat{C}_1 = \sqrt{\gamma(\bar{n}+1)}\hat{a},$$

$$\hat{C}_2 = \sqrt{\gamma\bar{n}}\hat{a}. \tag{5.69}$$

The same expressions hold for the damped two-level atom master equation (5.66), but with $\hat{a}$ replaced by $\hat{\sigma}_-$.

5.2.3   Fokker–Planck Equation

Starting from the master equation (5.57), we now derive a more classical looking equation that provides complementary insights by using quasiprobability distributions introduced in Sect. 2.5. As a concrete example, if the system is a harmonic oscillator, we can use the $P(\alpha)$-distribution (2.183) to expand the system density operator on a coherent states basis as

$$\hat{\rho} = \int d^2\alpha\, P(\alpha)|\alpha\rangle\langle\alpha|\,. \tag{5.70}$$

Substituting this form into the master equation (5.57) gives

$$\int d^2\alpha\, \dot{P}(\alpha, t)|\alpha\rangle\langle\alpha| = -\frac{\gamma}{2}(\bar{n}+1)\int d^2\alpha\, P(\alpha, t)[\hat{a}^\dagger\hat{a}|\alpha\rangle\langle\alpha| - \hat{a}|\alpha\rangle\langle\alpha|\hat{a}^\dagger]$$

$$-\frac{\gamma}{2}\bar{n}\int d^2\alpha\, P(\alpha, t)[|\alpha\rangle\langle\alpha|\hat{a}\hat{a}^\dagger - \hat{a}^\dagger|\alpha\rangle\langle\alpha|\hat{a}] + \text{c.c.} \tag{5.71}$$

With the representation of a coherent state (2.110), $|\alpha\rangle = e^{|\alpha|^2/2}e^{\alpha\hat{a}^\dagger}|0\rangle$, we have that

$$|\alpha\rangle\langle\alpha| = e^{-|\alpha|^2}e^{\alpha\hat{a}^\dagger}|0\rangle\langle0|e^{\alpha^*\hat{a}}\,, \tag{5.72}$$

so that $|\alpha\rangle\langle\alpha|\hat{a}$ may be written as

$$|\alpha\rangle\langle\alpha|\hat{a} = e^{-|\alpha|^2}\frac{\partial}{\partial\alpha^*}[e^{\alpha\hat{a}^\dagger}|0\rangle\langle0|e^{\alpha^*\hat{a}}] = \left(\frac{\partial}{\partial\alpha^*} + \alpha\right)|\alpha\rangle\langle\alpha|\,, \tag{5.73}$$

and similarly

$$\hat{a}^\dagger|\alpha\rangle\langle\alpha| = \left(\frac{\partial}{\partial\alpha} + \alpha^*\right)|\alpha\rangle\langle\alpha|\,. \tag{5.74}$$

Substituting these expressions into Eq. (5.71) and with the definition $\hat{a}|\alpha\rangle = \alpha|\alpha\rangle$ of the coherent state, we find

$$\int d^2\alpha\, \dot{P}(\alpha, t)|\alpha\rangle\langle\alpha| = -\frac{\gamma}{2}(\bar{n}+1)\int d^2\alpha\, P(\alpha, t)\alpha\frac{\partial}{\partial\alpha}|\alpha\rangle\langle\alpha|$$

$$+\frac{\gamma}{2}\bar{n}\int d^2\alpha\, P(\alpha, t)\left(\alpha\frac{\partial}{\partial\alpha} + \frac{\partial^2}{\partial\alpha\partial\alpha^*}\right)|\alpha\rangle\langle\alpha| + \text{adj.} \tag{5.75}$$

We can integrate the right-hand side of this equation by parts and drop the constants of integration since $P(\alpha, t)$ must vanish for $|\alpha| \to \infty$. Thus, for example,

$$\int d^2\alpha\, P(\alpha, t)\alpha\frac{\partial}{\partial\alpha}|\alpha\rangle\langle\alpha| = -\int d^2\alpha\left[\frac{\partial}{\partial\alpha}\big(\alpha P(\alpha, t)\big)\right]|\alpha\rangle\langle\alpha|\,. \tag{5.76}$$

Equation (5.75) becomes, after equating the coefficients of $|\alpha\rangle\langle\alpha|$ in the integrand,

$$\dot{P}(\alpha, t) = \frac{\gamma}{2}\left[\frac{\partial}{\partial\alpha}\Big(\alpha P(\alpha, t)\Big) + \text{c.c.}\right] + \gamma\bar{n}\,\frac{\partial^2}{\partial\alpha\,\partial\alpha^*}P(\alpha, t)\,. \tag{5.77}$$

This expression is in the form of a *Fokker–Planck equation* for the quasiprobability $P(\alpha, t)$ of finding the harmonic oscillator in the coherent state $|\alpha\rangle$ at time t. For reasons explained shortly, the coefficients of the first derivatives on the RHS of Eq. (5.77) are the elements of the *drift matrix*, and the coefficients of the second derivatives compose the *diffusion matrix*. The steady-state solution of this equation is easily verified to be

$$P(\alpha) = \frac{1}{\pi\bar{n}}e^{-|\alpha|^2/\bar{n}}\,, \tag{5.78}$$

which is a thermal distribution with average excitation value $\bar{n}$.

Although it is hard in general to find the time-dependent solution of $P(\alpha, t)$, Eq. (5.77) can readily be used to obtain the rate of change of the expectation value of observables of interest; for instance, for $\bar{n} = 0$,

$$\frac{\mathrm{d}}{\mathrm{d}t}\langle\hat{a}\rangle = \int \mathrm{d}^2\alpha\,\, \alpha\dot{P}(\alpha, t) = -\frac{\gamma}{2}\langle\hat{a}\rangle\,, \tag{5.79}$$

in agreement with Eq. (5.61).

We can gain an intuitive understanding of how a Fokker–Planck equation works by considering the general one-dimensional form

$$\frac{\partial}{\partial t}p(x, t) = -\frac{\partial}{\partial x}(M_1\,p(x, t)) + \frac{\partial^2}{\partial x^2}(M_2\,p(x, t))\,, \tag{5.80}$$

where $M_1(x)$ and $M_2(x)$ are called the first- and second-order moments of the distribution $p(x)$ or alternatively the *drift* and *diffusion* coefficients of the Fokker–Planck equation. In the more general multidimensional case, one speaks of drift and diffusion matrices. As shown in Fig. 5.4, for x values to the left of the peak of $M_1 p$, the slope of $M_1 p$ is positive, which causes a decrease of $p(x)$ in that region, while for x values to the right of the peak of $M_1 p$, the slope is negative, causing an increase of $p(x)$. This results in a movement of the peak toward larger values of x provided M_1 is itself positive. The second derivative at the peak of $M_2 p(x)$ is negative, which according to Eq. (5.80) causes a decrease in $p(x)$, while on either side of the peak of $M_2 p$, the second derivative is positive causing an increase of $p(x)$. This results in a diffusion of $p(x)$. For these reasons, the M_1 term is called the drift term and the M_2 term is called the diffusion term.

It is important to realize that the derivation of a Fokker–Planck equation via quasiprobability distributions does not always lead to well-behaved results. We have seen that $P(\alpha)$ need not be positive and does not lend itself to a simple interpretation as a probability distribution. In situations where $P(\alpha)$ becomes negative or singular, which are typical if truly nonclassical effects are important, we often find that the resulting Fokker–Planck equation has a nonpositive diffusion matrix and hence

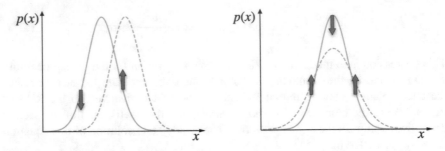

Fig. 5.4 The left diagram shows how the $M_1 p(x, t)$ term in Eq. (5.75) causes a distribution $p(x)$ (solid line) to "drift" along the x-axis and evolve at a later time toward the distribution shown as a dotted line. The right diagram shows the effect of the $M_2 p(x, t)$ term on that same initial distribution, which causes $p(x)$ to diffuse to the dotted line distribution

is not mathematically well behaved. In such situations, one can take advantage of the overcompleteness of the coherent states to introduce generalizations of the $P(\alpha)$ distribution that eliminate this difficulty, typically at the expense of doubling the phase space dimensions.

5.3 Langevin Equations

We now turn to a Heisenberg picture analysis of the same system–reservoir model and show that in that approach the reservoir operators $\hat{F}(t)$ play a role analogous to the Langevin forces familiar from classical statistical mechanics. These *quantum noise operators* are the source of both fluctuations and of the irreversible dissipation of energy from the system to the reservoir.

Focusing again on the case of a damped harmonic oscillator, we readily obtain from the Hamiltonians (5.45), (5.46), and (5.47) the Heisenberg equations of motion for the annihilation operators $\hat{a}(t)$ and $\hat{b}_j(t)$ as

$$\frac{d\hat{a}(t)}{dt} = -i\Omega\hat{a}(t) - i\sum_j g_j \hat{b}_j(t) , \qquad (5.81)$$

$$\frac{d\hat{b}_j(t)}{dt} = -i\omega_j \hat{b}_j(t) - ig_j^* \hat{a}(t) . \qquad (5.82)$$

Integrating the second of these equations formally gives

$$\hat{b}_j(t) = \hat{b}_j(t_0)e^{-i\omega_j(t-t_0)} - ig_j^* \int_{t_0}^{t} dt' \hat{a}(t')e^{-i\omega_j(t-t')}$$

$$\equiv \hat{b}_{\text{free}}(t) + \hat{b}_{\text{radiated}}(t) . \qquad (5.83)$$

The first term $\hat{b}_{\text{free}}$ on the right-hand side of this equation accounts for the free evolution of the reservoir operators $\hat{b}_j$, while the second term $\hat{b}_{\text{radiated}}$ gives the modification of that evolution due to their coupling with the system. It shows that $\hat{a}(t)$ is the source of $\hat{b}_j(t)$. If the small system were, say, a two-level system instead of a harmonic oscillator and the reservoir consisted of a continuum of electromagnetic field modes, then the appropriately modified Eq. (5.83) would show that the atomic polarization is the source of the field.

Inserting Eq. (5.83) into Eq. (5.81), we find then

$$\frac{d\hat{a}(t)}{dt} = -i\Omega\hat{a}(t) - i\sum_j g_j\hat{b}_j(t_0)e^{-i\omega_j(t-t_0)} - \sum_j |g_j|^2 \int_{t_0}^{t} dt'\hat{a}(t')e^{-i\omega_j(t-t')},$$

(5.84)

where the first summation describes the effect of fluctuations and the second one those of radiation reaction from the reservoir on the oscillator dynamics.

As in the Schrödinger picture approach, we move to an interaction picture in order to separate the free evolution of $\hat{a}(t)$ at frequency Ω from the fast evolution with characteristic time τ_c due to the large bandwidth of the bath. This is now done by introducing the slowly varying operator

$$\hat{A}(t) = \hat{a}(t)e^{i\Omega t},$$

(5.85)

with

$$[\hat{A}(t), \hat{A}^\dagger(t)] = 1.$$

(5.86)

From Eq. (5.84), the evolution of $\hat{A}(t)$ is given by

$$\frac{d\hat{A}(t)}{dt} = -\sum_j |g_j|^2 \int_{t_0}^{t} dt'\hat{A}(t')e^{-i(\omega_j-\Omega)(t-t')} + \hat{F}(t),$$

(5.87)

where we have introduced the quantum noise operator $\hat{F}(t)$

$$\hat{F}(t) = -i\sum_j g_j\hat{b}_j(t_0)e^{i(\Omega-\omega_j)(t-t_0)}.$$

(5.88)

This is the same operator that we already encountered in Eq. (5.50), aside from a shift in time origin. Note that this operator varies rapidly in time due to the presence of all the reservoir frequencies. Furthermore, as previously mentioned in the master equation analysis, the expectation value $\langle\hat{F}(t)\rangle_r$ vanishes if the reservoir is described by a density operator diagonal in the energy representation.

We have encountered integrals that resemble the first term on the right-hand side of Eq. (5.87) earlier in this chapter. We handle it in the same fashion here by replacing the sum over modes of the reservoir by an integral and invoke the Markov approximation by claiming that $\hat{A}(t)$ varies little over the inverse reservoir bandwidth. This allows us to extend the limit of integration to infinity. Using the representation (5.16) of the delta function, then, we obtain the *quantum Langevin equation*

$$\frac{d\hat{A}(t)}{dt} = -\frac{\gamma}{2}\hat{A}(t) + \hat{F}(t) \, . \tag{5.89}$$

Its expectation value

$$\frac{d\langle\hat{A}(t)\rangle}{dt} = -\frac{\gamma}{2}\langle\hat{A}(t)\rangle \tag{5.90}$$

is the same as Eq. (5.61) for the expectation value of $\langle\hat{a}(t)\rangle_s$ obtained from the master equation, as it should be since expectation values may not depend on whether they are evaluated in the Heisenberg or the Schrödinger picture.

We should keep in mind however that equations for expectation values are profoundly different in nature from Eq. (5.89), which gives the evolution of the operator itself. Importantly, one cannot expect an operator to have an evolution as simple as that given by Eq. (5.90). If that were the case, its value at time t would be $\hat{A}(t) = \hat{A}(0)\exp(-\gamma t/2)$, and for times long compared to γ, we would have $[\hat{A}(t), \hat{A}^\dagger(t)] \to 0$, in violation of the laws of quantum mechanics, since commutation relations must be valid at all times. It is the rapidly fluctuating operator $\hat{F}(t)$ in (15.69) that guarantees *by construction* that this is the case, even though its expectation value is $\langle\hat{F}(t)\rangle = 0$.

As advertised, the noise operator $\hat{F}(t)$ plays a role similar to that of the Langevin forces in the theory of Brownian motion. In both cases, the associated random force of zero average value leads to dissipation. The only difference here is that this force has now the character of an operator. It is because of this analogy that Eq. (5.89) is sometimes called a quantum Langevin equation and the operator $\hat{F}(t)$ a quantum noise operator. In principle, one can write down a quantum Langevin equation for any system operator, but of course each equation will have a different noise operator; for instance, the adjoint of Eq. (5.89) is

$$\frac{d\hat{A}^\dagger(t)}{dt} = -\frac{\gamma}{2}\hat{A}^\dagger(t) + \hat{F}^\dagger(t) \, . \tag{5.91}$$

Operator Ordering The Heisenberg picture has the appealing feature that the operator equations of motion resemble the corresponding classical equations of motion. As such it can provide useful intuition, but it does present a few pitfalls as well. The most important one has to do with the ordering of operators. System operators commute with reservoir operators at equal times, but not usually at different times. This is, for example, illustrated in Eq. (5.83), which shows that as time evolves, $\hat{b}_j(t)$ acquires some of the character of $\hat{a}(t)$. Importantly, the homogeneous (free field) part of $\hat{b}_j(t)$ alone does not commute with $\hat{a}(t)$ even at equal times, although $[\hat{a}(t), \hat{b}_j(t)] = 0$. For this reason, after separating $\hat{b}_j(t)$ into $\hat{b}_{\text{free}}$ and $\hat{b}_{\text{radiated}}$, we can no longer interchange the order of system and reservoir operators without taking the chance of committing serious errors. Therefore, once we chose the order in which to write system and reservoir operators in the initial Hamiltonian, we must *stick to it*. Here, we always put all operators with a "†" to the left, which is called "normal ordering." Any other ordering will do, provided that it is used consistently throughout the calculation. As we have seen in Sect. 2.5, different quasiprobability distributions are associated with these different orderings.

Although final answers do not depend on the choice of ordering, the physical interpretation of the results is typically different in different orderings. For instance, the normal ordering attributes spontaneous emission to radiation reaction, since $\langle \hat{F}(t) \rangle_r = 0$, while vacuum fluctuations give the Langevin force $\hat{F}(t)$. In contrast, J. Dalibard and coworkers [4] advocate the use of symmetric ordering, which presents the advantage of making the contributions of the free and radiated fields to the system evolution separately Hermitian. With this choice of ordering, radiation reaction and vacuum fluctuations give contributions of equal magnitude to spontaneous emission. These contributions have equal phase and add for the upper level of a two-level atom but have opposite phase and cancel exactly for the lower level.

Correlation Functions In addition to the evolution of simple system operators such as $\hat{A}(t)$ and $\hat{A}^\dagger(t)$, one typically needs to consider the evolution of additional observables as well, for instance, to evaluate the intensity, the spectrum, or higher order correlation functions of a field. In many problems, this can rapidly become rather complex, as the resulting set of quantum Langevin equations does not close in general. Particular care needs to be taken to properly describe the resulting noise correlation functions, which account for important aspects of the system dynamics as we have seen. For example, the equation of motion for the number operator $\hat{a}^\dagger \hat{a}(t) = \hat{A}^\dagger \hat{A}(t)$ is readily found to be

$$\frac{d}{dt}(\hat{A}^\dagger \hat{A}) = -i \sum_j g_j \hat{A}^\dagger \hat{b}_j e^{i\Omega t} + \text{adj.}, \tag{5.92}$$

or, by substitution of Eq. (5.89) and its adjoint

$$\frac{d}{dt}(\hat{A}^\dagger \hat{A}) = -\sum_j |g_j|^2 \hat{A}^\dagger(t) \int_{t_0}^t dt' A(t') e^{i(\Omega - \omega_j)(t-t')}$$

$$-i \sum_j g_j \hat{A}^\dagger(t) \hat{b}_j(t_0) e^{i(\Omega - \omega_j)(t - t_0)} + \text{adj.} \tag{5.93}$$

Performing the Markov approximation as before results then in the "Langevin" equation for the number operator

$$\frac{d}{dt}(\hat{A}^\dagger \hat{A}) = -\gamma \hat{A}^\dagger \hat{A} + \hat{G}_{A^\dagger A}(t), \tag{5.94}$$

where

$$\hat{G}_{A^\dagger A}(t) = \mathrm{i} \sum_j g_j^* \hat{b}_j^\dagger(t_0) \hat{A}(t) e^{-\mathrm{i}(\Omega - \omega_j)(t - t_0)} + \text{adj} \,. \tag{5.95}$$

Here, we put "Langevin" in quotation marks, because we like to think that in such equations the fluctuating force should have zero average. This is not the case for $\langle \hat{G}_{A^\dagger A} \rangle_r$, which can be shown by substitution of a formal integral for $\hat{A}(t)$ to be equal to $\gamma \bar{n}$, see Problem 5.8, and of course $\bar{n} \neq 0$ for $T \neq 0$. To obtain a proper Langevin equation, we introduce a new quantum noise operator that subtracts this expectation value,

$$\hat{\mathcal{G}}_{A^\dagger A}(t) = \hat{G}_{A^\dagger A}(t) - \langle \hat{G}_{A^\dagger A}(t) \rangle_r = \hat{G}_{A^\dagger A}(t) - \gamma \bar{n} \,, \tag{5.96}$$

in terms of which Eq. (5.94) becomes

$$\frac{\mathrm{d}}{\mathrm{d}t}(\hat{A}^\dagger \hat{A}) = -\gamma \hat{A}^\dagger \hat{A} + \gamma \bar{n} + \hat{\mathcal{G}}_{A^\dagger A}(t) \,. \tag{5.97}$$

This equation gives the same evolution as Eq. (5.94) for the mean excitation number, as it should.

Noise Spectral Density In the analysis of the damped harmonic oscillator, we encountered in Eq. (5.54) the correlation function $|g(\Omega)|^2 \langle \hat{b}(\Omega) \hat{b}^\dagger(\Omega) \rangle$, a measure of the intensity of the noise at a given frequency. Such spectral noise densities, which are essentially the Fourier transforms of two-time correlation functions of the reservoir noise operators, provide an important characterization of the interaction between the system and the reservoir, as they determine the rate of noise-induced transitions that eventually lead to the thermalization of the system.

To show that this is the case, consider the generic system–reservoir interaction

$$\hat{V} = -\hat{A} \hat{F}(t) \,, \tag{5.98}$$

where $\hat{A}$ is a system operator and $\hat{F}(t)$ is a (Hermitian) noise operator.[3] We assume that the system is initially in the state $|\psi_i\rangle$ of energy $\hbar \omega_i$, so that in the interaction picture

$$|\psi(t)\rangle = |\psi_i\rangle - \frac{\mathrm{i}}{\hbar} \int_0^t \mathrm{d}\tau \, \hat{V}(\tau) |\psi_i\rangle \,. \tag{5.99}$$

---

[3]Note that the form (5.98) of $\hat{V}$ implies a Hermitian noise operator, so that $\hat{F}^\dagger = \hat{F}$. In general, the individual contributions to an interaction Hamiltonian need not be Hermitian, as we have seen for instance, in the Jaynes–Cummings interaction $\hbar g(\hat{a}\hat{\sigma}_+ + \text{h.c.})$, although of course the full interaction potential must be. Since it is often sufficient to keep track explicitly of just part of the full interaction when carrying out calculations, in the following we allow for the fact that in such situations $\hat{F}$ may not be Hermitian.

The probability $p_f(t)$ for the system to be at time t in some state $|\psi_f\rangle$ of energy $\hbar\omega_f$ and orthogonal to $|\psi_i\rangle$ is therefore

$$p_f(t) = \left| \frac{-i}{\hbar} \int_0^t d\tau \langle \psi_f | \hat{A} | \psi_i \rangle e^{-i\omega_{fi}\tau} \hat{F}(\tau) \right|^2$$

$$= \frac{|d|^2}{\hbar^2} \int_0^t d\tau \int_0^t d\tau' e^{-i\omega_{fi}(\tau-\tau')} \hat{F}(\tau) \hat{F}^\dagger(\tau'), \qquad (5.100)$$

where $d = \langle \psi_f | \hat{A} | \psi_i \rangle$, $\omega_{fi} = \omega_f - \omega_i$, and the second line in Eq. (5.100) accounts for the fact that $\hat{F}$ may not be Hermitian in a specific calculation, see footnote 2. Accounting also for the fact that $\hat{F}$ is a quantum noise operator, the average probability $\langle p_f(t) \rangle$ to be in the final state is therefore

$$\langle p_f(t) \rangle = \frac{|d|^2}{\hbar^2} \int_0^t d\tau \int_0^t d\tau' e^{-i\omega_{fi}(\tau-\tau')} \langle \hat{F}(\tau) \hat{F}^\dagger(\tau') \rangle. \qquad (5.101)$$

For times t larger than the noise correlation time τ_c of the reservoir, the limits of the integral over τ can be extended to infinity. Using time translational invariance, we then have

$$\langle p_f(t) \rangle \approx \frac{|d|^2}{\hbar^2} \int_0^t d\tau' \int_{-\infty}^{\infty} d\tau \, e^{-i\omega_{fi}\tau} \langle \hat{F}(\tau) \hat{F}^\dagger(0) \rangle \qquad (5.102)$$

or, introducing the *noise spectral density*

$$S_{FF}(\omega) \equiv \int_{-\infty}^{\infty} d\tau e^{i\omega\tau} \langle \hat{F}(\tau) \hat{F}^\dagger(0) \rangle, \qquad (5.103)$$

$$\langle p_f \rangle(t) \approx \frac{|d|^2}{\hbar^2} S_{FF}(-\omega_{fi}) t, \qquad (5.104)$$

compare with Eq. (5.54). Hence, the transition rate between the initial state $|\psi_i\rangle$ and a final state $|\psi_f\rangle$ separated in frequency by ω_{fi} is

$$A_{i \to f} = \frac{|d|^2}{\hbar^2} S_{FF}(-\omega_{fi}). \qquad (5.105)$$

As an example, consider a system consisting of a simple harmonic oscillator of frequency Ω, and take the quadrature $\hat{X} = \hat{a} + \hat{a}^\dagger$ as the system operator $\hat{A}$. For transitions between neighboring levels, the spectral density spectrum needs only be evaluated at $\pm\Omega$, and

$$A_{n \to n-1} = n S_{FF}(\Omega) \quad ; \quad A_{n-1 \to n} = n S_{FF}(-\Omega). \qquad (5.106)$$

For a reservoir in thermal equilibrium, we must have, invoking detailed balance,

$$A_{n-1 \to n} = e^{-\hbar\Omega/k_B T} A_{n \to n-1}, \tag{5.107}$$

so that

$$S_{FF}(\Omega) = e^{\hbar\Omega/k_B T} S_{FF}(-\Omega). \tag{5.108}$$

This is an important relationship that we will encounter again in the context of quantum optomechanics of Chap. 11.

Fluctuation–Dissipation Theorem   We conclude this section by deriving, again for the simple case of a damped harmonic oscillator of frequency Ω, an important relationship between the correlation functions of the noise operators and the damping coefficient γ.

We proceed once more by considering the correlation function of the noise operators $\hat{F}(t)$ and $\hat{F}^\dagger(t)$ and converting the sum over modes to an integral. Introducing the reservoir average number of quanta at frequency ω via $\langle \hat{b}(\omega)\hat{b}^\dagger(\omega)\rangle = \bar{n}(\omega) + 1$ gives then, in the continuous limit of Eq. (5.88),

$$\langle \hat{F}(t')\hat{F}^\dagger(t'')\rangle_r = \int d\omega\, \mathcal{D}(\omega)|g(\omega)|^2[\bar{n}(\omega) + 1]e^{i(\Omega-\omega)(t'-t'')}. \tag{5.109}$$

Assuming as before that the correlation time of the reservoir is short compared to all times of interest for the system, we can evaluate $\mathcal{D}(\omega)|g(\omega)|^2\bar{n}(\omega)$ at Ω and remove it from the integral, giving

$$\langle \hat{F}(t')\hat{F}^\dagger(t'')\rangle_r = \mathcal{D}(\Omega)|g(\Omega)|^2[\bar{n}(\Omega) + 1] \int d\omega e^{i(\Omega-\omega)(t'-t'')}. \tag{5.110}$$

For a broadband reservoir, we can extend the limits of integration to infinity, thereby performing once again the Markov approximation. Using Eq. (5.55) for γ and the integral representation of the δ-function

$$\int_{-\infty}^{\infty} d\omega e^{i(\Omega-\omega)(t'-t'')} = 2\pi\delta(t'-t''), \tag{5.111}$$

we find

$$\langle \hat{F}(t')\hat{F}^\dagger(t'')\rangle_r = \gamma(\bar{n}(\Omega) + 1)\delta(t'-t''). \tag{5.112}$$

Similarly, we have for the normally ordered correlation function

$$\langle \hat{F}^\dagger(t')\hat{F}(t'')\rangle_r = \gamma\bar{n}(\Omega)\delta(t'-t''). \tag{5.113}$$

This shows that the Markov approximation amounts to assuming that the correlation functions of the noise operators are δ-correlated in time, a mathematical statement of the fact that it assumes that the reservoir has no memory. Under that approximation, the noise operator correlation functions depend on the operator ordering, but not on time ordering.

Integrating both sides of Eq. (5.113) over $\tau = t' - t''$ yields a simple example of the *fluctuation–dissipation* theorem that relates the first-order correlation function of the quantum noise operator $\hat{F}(t)$ to the dissipation rate γ,

$$\gamma = \bar{n}(\Omega)^{-1} \int_{-\infty}^{\infty} \mathrm{d}\tau \, \langle \hat{F}^\dagger(\tau) \hat{F}(0) \rangle_r \,. \tag{5.114}$$

5.4 Monte Carlo Wave Functions

We mentioned in Sect. 5.2.2 that the master equation of a small system coupled to a Markovian bath must be of the general Lindblad form

$$\frac{\mathrm{d}\hat{\rho}}{\mathrm{d}t} = -\frac{\mathrm{i}}{\hbar}[\hat{H}_s, \hat{\rho}] + \mathcal{L}[\hat{\rho}] \,,$$

where $\hat{H}_s$ is the system Hamiltonian, $\mathcal{L}[\hat{\rho}]$ is the Liouvillian

$$\mathcal{L}[\hat{\rho}] = -\frac{1}{2} \sum_i (\hat{C}_i^\dagger \hat{C}_i \hat{\rho} \mid \hat{\rho} \hat{C}_i^\dagger \hat{C}_i) + \sum_i \hat{C}_i \hat{\rho} \hat{C}_i^\dagger \,,$$

and the $\hat{C}_i$'s are system operators. This equation can be recast in the form

$$\frac{\mathrm{d}\hat{\rho}}{\mathrm{d}t} = -\frac{\mathrm{i}}{\hbar}(\hat{H}_{\mathrm{eff}}\hat{\rho} - \hat{\rho}\hat{H}_{\mathrm{eff}}^\dagger) + \mathcal{L}_{\mathrm{jump}}[\hat{\rho}] \,, \tag{5.115}$$

where we have introduced the *non-Hermitian* effective Hamiltonian

$$\hat{H}_{\mathrm{eff}} \equiv \hat{H}_s - \frac{\mathrm{i}\hbar}{2} \sum_i \hat{C}_i^\dagger \hat{C}_i \tag{5.116}$$

and the "quantum jump" Liouvillian

$$\mathcal{L}_{\mathrm{jump}}[\hat{\rho}] \equiv \sum_i \hat{C}_i \hat{\rho} \hat{C}_i^\dagger \,. \tag{5.117}$$

For example, for the case of a damped harmonic oscillator, the effective Hamiltonian is

$$\hat{H}_{\mathrm{eff}} = \hbar\Omega \hat{a}^\dagger \hat{a} - \mathrm{i}\hbar\gamma \, (\bar{n} + 1/2) \, \hat{a}^\dagger \hat{a} \,, \tag{5.118}$$

with

$$\mathcal{L}_{\text{jump}}[\hat{\rho}] = \gamma(2\bar{n} + 1)\hat{a}\hat{\rho}\hat{a}^{\dagger} + \gamma\bar{n}\hat{a}\hat{\rho}\hat{a}^{\dagger}, \tag{5.119}$$

while for a two-level atom from upper to lower level decay, we have

$$\hat{H}_{\text{eff}} = \tfrac{1}{2}\hbar\omega_0\hat{\sigma}_z - i\hbar\Gamma\left(\bar{n} + \tfrac{1}{2}\right)\hat{\sigma}_+\hat{\sigma}_- \tag{5.120}$$

and

$$\mathcal{L}_{\text{jump}}[\hat{\rho}] = \Gamma(\bar{n} + 1)\hat{\sigma}_-\hat{\rho}\hat{\sigma}_+ + \gamma\bar{n}\hat{\sigma}_+\hat{\rho}\hat{\sigma}_-. \tag{5.121}$$

The evolution of the system density operator can therefore be thought of as resulting from two contributions: a Schrödinger-like part governed by the non-Hermitian effective Hamiltonian $\hat{H}_{\text{eff}}$ and a "quantum jump" part resulting from $\mathcal{L}_{\text{jump}}[\hat{\rho}]$, with the "quantum jump" qualifier becoming clear shortly. This decomposition is the basis of the quantum trajectories approach to the solution of the master equation, to which we now turn.

5.4.1   Quantum Trajectories

The quantum trajectory method starts by considering the evolution of pure states of the small system and carries out a statistical average over an ensemble of them in the end. But in contrast to the situation for closed systems, where this is straightforwardly achieved by solving the Schrödinger equation, the system evolution is now intrinsically stochastic, as it results from the combination of a Schrödinger-like, but non-Hermitian evolution and random "quantum jumps."

We proceed by expressing the density operator as a statistical mixture of state vectors

$$\hat{\rho} = \sum_{\psi} P_{\psi}|\psi\rangle\langle\psi|, \tag{5.122}$$

the summation over ψ resulting from a classical average over the various states that the system can occupy with probabilities P_{ψ}. Introducing that expression into the master equation (5.115), we have

$$\sum_{\psi} P_{\psi}\left[|\dot{\psi}\rangle\langle\psi| + |\psi\rangle\langle\dot{\psi}| = -\frac{i}{\hbar}\left(\hat{H}_{\text{eff}}|\psi\rangle\langle\psi| - |\psi\rangle\langle\psi|\hat{H}_{\text{eff}}^{\dagger}\right) + \sum_i \hat{C}_i|\psi\rangle\langle\psi|\hat{C}_i^{\dagger}\right]. \tag{5.123}$$

If we restrict ourselves for now to a single representative state vector $|\psi\rangle$ in the mixture, we recognize that the first term on the right-hand side of this equation can be simply interpreted as resulting from the non-Hermitian, but Schrödinger-like evolution of $|\psi\rangle$ under the influence of $\hat{H}_{\text{eff}}$,

$$i\hbar|\dot{\psi}\rangle = \hat{H}_{\text{eff}}|\psi\rangle . \tag{5.124}$$

Things are more tricky for the second term, which is clearly not a Schrödinger-like term. Rather, it seems to result from a discontinuous evolution whereby the state $|\psi\rangle$ is projected—or "jumps"—onto one of the possible states

$$|\psi\rangle \rightarrow |\psi\rangle_i = \hat{C}_i|\psi\rangle . \tag{5.125}$$

This is precisely what motivates calling $\mathcal{L}_{\text{jump}}[\hat{\rho}]$ a "quantum jump" Liouvillian.

The decomposition of the evolution of the representative state vector $|\psi\rangle$ into a Schrödinger-like part and a quantum jump contribution suggests therefore an elegant way to solve master equations by carrying out an ensemble average over the evolution of a large number of such state vectors. It proceeds by first selecting an arbitrary state vector $|\psi\rangle$ out of the initial ensemble and evolving it for a short time δt under the influence of $\hat{H}_{\text{eff}}$ only. For sufficiently small time intervals, this gives

$$|\tilde{\psi}(t+\delta t)\rangle = \left(1 - \frac{i\hat{H}_{\text{eff}}\delta t}{\hbar}\right)|\psi(t)\rangle . \tag{5.126}$$

An important consequence of the non-Hermitian nature of $\hat{H}_{\text{eff}}$ is that $|\tilde{\psi}(t+\delta t)\rangle$ is not normalized. Rather, the square of its norm is

$$\langle\tilde{\psi}(t+\delta t)|\tilde{\psi}(t+\delta t)\rangle = \langle\psi(t)|\left(1 + \frac{i\hat{H}_{\text{eff}}^{\dagger}\delta t}{\hbar}\right)\left(1 - \frac{i\hat{H}_{\text{eff}}\delta t}{\hbar}\right)|\psi(t)\rangle \equiv 1 - \delta p , \tag{5.127}$$

where to the lowest order in δt

$$\delta p = \frac{i\delta t}{\hbar}\langle\psi(t)|\hat{H}_{\text{eff}} - \hat{H}_{\text{eff}}^{\dagger}|\psi(t)\rangle = \delta t \sum_i \langle\psi(t)|\hat{C}_i^{\dagger}\hat{C}_i|\psi(t)\rangle \equiv \sum_i \delta p_i . \tag{5.128}$$

This lack of norm preservation results from the fact that we have so far ignored the effects of $\mathcal{L}_{\text{jump}}[\hat{\rho}]$. The "missing norm" δp must therefore be accounted for by the states $|\psi\rangle_i$ resulting from the jumps part of the evolution. It is consistent with Eq. (5.128) to interpret this observation as a result of the fact that $\mathcal{L}_{\text{jump}}[\hat{\rho}]$ projects the system into the state $|\psi\rangle_i = \hat{C}_i|\psi\rangle$ with a probability δp_i such that $\sum_i \delta p_i = \delta p$. Hence, the next step of a Monte Carlo simulation consists in deciding whether a jump occurred or not. Numerically, this is achieved by choosing a uniform random variate $0 \le r \le 1$. If its value is larger than δp, no jump is said to have occurred, and the next integration step proceeds from the *normalized* state vector

$$|\psi(t + \delta t)\rangle = \frac{|\tilde{\psi}(t + \delta t)\rangle}{\||\tilde{\psi}(t + \delta t)\rangle\|}. \tag{5.129}$$

If, on the other hand, $r \le \delta p$, then a jump is said to have occurred. The state vector $|\tilde{\psi}(t + \delta t)\rangle$ is then projected to the normalized new state

$$|\psi(t + \delta t)\rangle = \frac{\hat{C}_i|\psi(t)\rangle}{||\hat{C}_i|\psi(t)\rangle||} = \sqrt{\frac{\delta t}{\delta p_i}} \hat{C}_i|\psi(t)\rangle \tag{5.130}$$

with probability $\delta p_i / \delta p$, with δp_i given by Eq. (5.128). This state is then taken as the initial condition for the next integration step. The procedure is repeated for as many iterations as desired and yields a possible time evolution of the initial state vector $|\psi\rangle$, sometimes called a "quantum trajectory." Clearly, the random nature of the jumps implies that different trajectories will be obtained in successive simulations of the system evolution from the same initial state.

In some cases, it is possible to interpret the reservoir to which the small system is coupled as a "measurement apparatus," a point to which we return briefly in the next section and in more detail in Chap. 6, in particular in the "pointer basis" discussion of Sect. 6.4. In such situations, the individual quantum trajectories may be interpreted as "typical" of a single sequence of measurements on the system. It is not normally possible to say for sure whether a given numerical realization will be achieved in practice or not, though. Nonetheless, the individual Monte Carlo wave function trajectories can often provide one with useful intuition about the way a given system behaves in the laboratory.

We still need to prove that in an ensemble average sense, the predictions of the Monte Carlo wave function simulations are identical to those of the corresponding master equation. This is easily done by considering the quantity

$$\hat{\varrho}(t) \equiv \sum_{\psi} P_{\psi} \langle |\psi(t)\rangle \langle \psi(t)| \rangle_{\text{traj}},$$

which is a double average over both a large number of Monte Carlo wave function trajectories resulting from a given initial state and a representative set of initial states necessary to reproduce the initial density operator (5.122). Consider first the average over trajectories for a fixed initial state: by construction, the Monte Carlo wave function algorithm implies that

$$\hat{\varrho}(t + \delta t) = (1 - \delta p) \frac{|\tilde{\psi}(t + \delta t)\rangle}{\sqrt{1 - \delta p}} \frac{\langle \tilde{\psi}(t + \delta t)|}{\sqrt{1 - \delta p}}$$

$$+\delta p \sum_i \frac{\delta p_i}{\delta p} \left(\sqrt{\frac{\delta t}{\delta p_i}} \hat{C}_i |\psi(t)\rangle \right) \left(\sqrt{\frac{\delta t}{\delta p_i}} \langle \psi(t)| \hat{C}_i^\dagger \right), \quad (5.131)$$

where the average over trajectories is accounted for by the probabilities δp and δp_i. With Eq. (5.126), one has therefore to the lowest order in δt

$$\hat{\varrho}(t + \delta t) = \hat{\varrho}(t) + \frac{i\delta t}{\hbar} [\hat{\varrho}(t), \hat{H}_s] + \delta t \mathcal{L}_{\text{jump}}[\varrho(t)]. \quad (5.132)$$

In the case of a mixed initial state, one needs in addition to perform also an average over the distribution P_ψ of initial states, as already indicated. But this step is trivial, since Eq. (5.131) is linear in $\hat{\varrho}$. Hence, the result of the double averaging yields for $\hat{\varrho}$ an evolution identical to that given by the master equation (5.115). The two approaches are therefore equivalent, provided that the initial conditions for $\hat{\rho}$ and $\hat{\varrho}$ are the same.

An important practical advantage of the Monte Carlo wave function approach occurs in situations where the number of states N that need to be considered is large. Since $\hat{\rho}$ scales as N^2, such problems can easily stretch the capabilities of even the largest computers. In contrast, the Monte Carlo simulations deal with state vectors only, whose dimensions scale as N. Hence, the memory requirements are significantly reduced, the trade-off being the additional CPU time normally required in order to achieve good statistical accuracy.

In addition to these practical considerations, the Monte Carlo wave function method also provides one with additional physical insight into the way a physical system behaves in a single experiment. Consider, for example, the problem of spontaneous decay by a two-level atom at zero temperature. In that case,

$$\hat{H}_{\text{eff}} = \tfrac{1}{2}\hbar\omega\hat{\sigma}_z - \tfrac{1}{2}i\hbar\Gamma\hat{\sigma}_+\hat{\sigma}_- \quad (5.133)$$

and

$$\mathcal{L}_{\text{jump}}[\hat{\rho}] = \Gamma\hat{\sigma}_-\hat{\rho}\hat{\sigma}_+. \quad (5.134)$$

In this example, single quantum trajectories will illustrate the distribution of times after which the system undergoes a transition from the upper to the lower state.

Note also that when taking the expectation value of the right-hand side of the master equation between $\langle e|$ and $|e\rangle$, the contribution of $\mathcal{L}_{\text{jump}}[\hat{\rho}]$ vanishes, and we obtain an effective Hamiltonian that describes the upper state population decay averaged over a large number of experiments, see Problem 5.11. Hence, if all we are interested in is the evolution of the upper state $|e\rangle$, it is sufficient to consider the evolution of the system under the influence of the effective Hamiltonian (5.120) only. This justifies *a posteriori* the phenomenological treatment of atomic decay introduced in Sect. 1.3. For finite reservoir temperatures, single trajectories will also illustrate the successive quantum jumps between the atomic states driven by both spontaneous and stimulated emission.

5.5 Input–Output Formalism

While the system–reservoir approach that we considered so far in this chapter can
be extremely powerful, it does have its limitations, as there are many situations
where treating the environment as a thermal bath whose state remains essentially
unchanged is inappropriate. This is, for example, the case in cavity QED, the topic
of Chap. 7. In these systems, atoms or artificial atoms are confined inside a resonator
where they interact with an intracavity field. This small atom–field system is in turn
coupled, typically through mirrors, to external fields that drive it and/or serve as
external probes of its dynamics. Clearly, in such circumstances, the state of the
probe cannot be assumed to remain unchanged. M. Collett and C. Gardiner [5] have
developed an input–output formalism that permits to describe such open quantum
systems. That approach does not involve tracing over a quantum bath, that is,
it makes no assumption on its quantum state. Rather, it determines its dynamics
assuming that the system dynamics is known.

We consider again the Hamiltonian (5.28)

$$\hat{H} = \hat{H}_s + \hat{H}_r + \hat{V},$$

where for concreteness, the system is taken to be a single intracavity field mode of
frequency Ω with annihilation and creation operators $\hat{a}$ and $\hat{a}^\dagger$,

$$\hat{H}_s = \hbar\Omega\hat{a}^\dagger\hat{a} \tag{5.135}$$

coupled to a continuum of external field modes with Hamiltonian

$$\hat{H}_r = \int d\omega\, \hbar\omega\, \hat{b}^\dagger(\omega)\hat{b}(\omega) \tag{5.136}$$

by the interaction Hamiltonian

$$\hat{V} = \hbar \int d\omega g(\omega) \left[\hat{b}(\omega)\hat{a}^\dagger + \hat{a}\hat{b}^\dagger(\omega) \right] \tag{5.137}$$

with

$$[\hat{b}(\omega), \hat{b}^\dagger(\omega')] = \delta(\omega - \omega'). \tag{5.138}$$

We have not specified explicitly the limits of integration over frequencies, but it is
a good approximation to extend them to $\pm\infty$ in the following, much like in the
Weisskopf–Wigner theory of spontaneous emission of Sect. 5.1.

At first sight, this might appear to be precisely the problem that we already
considered when studying the damped harmonic oscillator, except that as advertised
the continuum of modes will no longer be treated as a reservoir whose state remains
effectively unchanged by its coupling to the intracavity field.

The Heisenberg equations of motion for the operators $\hat{a}(t)$ and $\hat{b}(\omega, t)$ are

$$\frac{d\hat{a}(t)}{dt} = \frac{i}{\hbar}[\hat{H}_s, \hat{a}] - i \int d\omega g(\omega)\hat{b}(\omega, t), \qquad (5.139a)$$

$$\frac{d\hat{b}(\omega, t)}{dt} = -i\omega\hat{b}(\omega, t) - ig^*(\omega)\hat{a}(t). \qquad (5.139b)$$

Formally integrating the second of these equations gives readily

$$\hat{b}(\omega, t) = \hat{b}(\omega, t_0)e^{-i\omega(t-t_0)} - ig^*(\omega) \int_{t_0}^{t} dt'\hat{a}(t')e^{-i\omega(t-t')}, \qquad (5.140)$$

where $t_0 < t$.

Not surprisingly, this is essentially the same equation as Eq. (5.83) which we encountered in the Langevin approach to system–reservoir interactions of Sect. 5.3. There we recognized the first term on the RHS of that equation as describing the free evolution of the field, while the second term accounts for the coupling of mode "ω" to the intracavity field. Alternatively, since the time-dependent Schrödinger evolution is reversible, we may also express $\hat{b}(\omega, t)$ in terms of fields at later times as

$$\hat{b}(\omega, t) = \hat{b}(\omega, t_1)e^{-i\omega(t-t_1)} + ig^*(\omega) \int_{t}^{t_1} dt'\hat{a}(t')e^{-i\omega(t-t')}, \qquad (5.141)$$

with $t_1 > t$. Physically, the two forms (5.140) and (5.141) of $\hat{b}(t)$ can be thought of as corresponding to solving the Heisenberg equations of motion for $t > t_0$ in terms of boundary conditions that describe "input fields" or for $t < t_1$ in terms of "output fields." Inserting the solution (5.140) into Eq. (5.139a) gives

$$\frac{d\hat{a}(t)}{dt} = \frac{i}{\hbar}[\hat{H}_s, \hat{a}(t)] \qquad (5.142)$$

$$-i \int d\omega g(\omega)\hat{b}(\omega, t_0)e^{-i\omega(t-t_0)} - \int d\omega|g(\omega)|^2 \int_{t_0}^{t} dt'\hat{a}(t')e^{-i\omega(t-t')}$$

As in the Weisskopf–Wigner theory of spontaneous emission, we assume that $g(\omega)$ is approximately independent of ω over the frequency range of interest and set

$$|g(\omega)|^2 = \kappa/2\pi. \qquad (5.143)$$

With the relation

$$\int_{-\infty}^{\infty} d\omega e^{-i\omega(t-t')} = \delta(t - t'), \qquad (5.144)$$

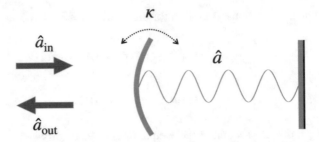

Fig. 5.5 Schematic of the coupling of the intracavity field $\hat{a}$ to the input and output fields $\hat{a}_{\text{in}}$ and $\hat{a}_{\text{out}}$. Here, the mirror on the right side of the resonator is assumed to be perfectly reflecting. The generalization of the input–output description to the case where both mirrors are semi-transparent in the subject of Problem 5.4

the third term on the right-hand side of Eq. (5.143) can then be interpreted like in system–reservoir theory as describing the decay of the intracavity field at rate $\kappa/2$.

The second term is more interesting. In Sect. 5.3, we interpreted it as a noise operator, but in the present context, it describes more generally the effect on the intracavity field of the external field at some initial time $t_0 < t$, that is, of the *input field*, see Fig. 5.5,

$$\hat{a}_{\text{in}}(t) \equiv -\frac{1}{\sqrt{2\pi}} \int d\omega\, \hat{b}(\omega, t_0) e^{-i\omega(t-t_0)} . \tag{5.145}$$

The factor of $1/\sqrt{2\pi}$ in this expression guarantees that the input field satisfies bosonic commutation relations,

$$[\hat{a}_{\text{in}}(t), \hat{a}_{\text{in}}^{\dagger}(t')] = \delta(t - t') , \tag{5.146}$$

as can easily be verified with the help of Eq. (5.144). We then obtain the evolution of the intracavity field $\hat{a}(t)$ as

$$\frac{d\hat{a}(t)}{dt} = \frac{i}{\hbar}[\hat{H}_s, \hat{a}(t)] - \frac{\kappa}{2}\hat{a}(t) + \sqrt{\kappa}\hat{a}_{\text{in}}(t) , \tag{5.147}$$

where the term that was interpreted as a noise operator in Sect. 5.3 appears now explicitly as the input field.

Similarly, we may relate $\hat{a}(t)$ to the *output field*

$$\hat{a}_{\text{out}}(t) \equiv \frac{1}{\sqrt{2\pi}} \int d\omega\, \hat{b}(\omega, t_1) e^{-i\omega(t-t_1)} \tag{5.148}$$

as

$$\frac{d\hat{a}(t)}{dt} = \frac{i}{\hbar}[\hat{H}_s, \hat{a}(t)] - \frac{\kappa}{2}\hat{a}(t) - \sqrt{\kappa}\hat{a}_{\text{out}}(t) . \tag{5.149}$$

Subtracting Eq. (5.149) from Eq. (5.147) relates the input and output fields to the intracavity field as

$$\hat{a}_{out}(t) + \hat{a}_{in}(t) = \sqrt{\kappa}\hat{a}(t). \tag{5.150}$$

Note importantly that $\hat{a}_{in}(t)$ and $\hat{a}_{out}(t)$ have units of $1/\sqrt{time}$ and represent therefore photon fluxes in and out of the cavity.

It is possible to gain additional insight into the relationship between the intracavity and output fields in case the dynamics of $\hat{a}(t)$ is linear. The corresponding Heisenberg equations of motion can then be cast in the simple form

$$\frac{d}{dt}\hat{\mathbf{a}}(t) = \hat{\mathbf{A}}\hat{\mathbf{a}}(t) - \frac{\kappa}{2}\hat{\mathbf{a}}(t) + \sqrt{\kappa}\hat{\mathbf{a}}_{in}(t), \tag{5.151}$$

where we have introduced the short-hand notation

$$\hat{\mathbf{a}} = \begin{bmatrix} \hat{a} \\ \hat{a}^{\dagger} \end{bmatrix} \quad ; \quad \hat{\mathbf{a}}_{in} = \begin{bmatrix} \hat{a}_{in} \\ \hat{a}_{in}^{\dagger} \end{bmatrix}, \tag{5.152}$$

and $\hat{\mathbf{A}}$ is a 2×2 matrix that accounts for the intracavity Hamiltonian dynamics of the field. Taking the Fourier transform of this equation, we get

$$\left[\hat{\mathbf{A}} + (i\omega - \kappa/2)\,\hat{\mathbf{I}}\right]\hat{\mathbf{a}}(\omega) = -\sqrt{\kappa}\,\hat{\mathbf{a}}_{in}(\omega), \tag{5.153}$$

with $\hat{\mathbf{I}}$ the 2×2 identity operator. Combining this result with a similar equation relating $\mathbf{a}(\omega)$ to $\mathbf{a}_{out}(\omega)$ yields then

$$\hat{\mathbf{a}}_{out}(\omega) = -\left[\hat{\mathbf{A}} + (i\omega + \kappa/2)\,\hat{\mathbf{I}}\right]\left[\hat{\mathbf{A}} + (i\omega - \kappa/2)\,\hat{\mathbf{I}}\right]^{-1}\hat{\mathbf{a}}_{in}(\omega). \tag{5.154}$$

For the case of a simple harmonic oscillator $\hat{H}_s = \hbar\Omega\hat{a}^{\dagger}\hat{a}$, this gives finally

$$\hat{\mathbf{a}}(\omega) = \frac{\sqrt{\kappa}}{\kappa/2 + i(\Omega - \omega)}\hat{\mathbf{a}}_{in}(\omega), \tag{5.155}$$

and

$$\hat{\mathbf{a}}_{out}(\omega) = \frac{\kappa/2 - i(\Omega - \omega)}{\kappa/2 + i(\Omega - \omega)}\hat{\mathbf{a}}_{in}(\omega), \tag{5.156}$$

indicating that the transmission function is a Lorentzian of width $\kappa/2$, as expected. D. F. Walls and G. Milburn [6] discuss the application of the input–output formalism to a number of quantum optics situations, including the spectrum of squeezing, the parametric oscillator, and laser fluctuations.

Classical Cavity Driving In many cases, the cavity field, in addition to being coupled to a continuum of modes acting as a reservoir, is also driven by a macroscopically populated field mode that can be treated classically, typically a laser field at some frequency ω. Calling the annihilation and creation operators of this mode $\hat{b}_0$ and $\hat{b}_0^\dagger$, its coupling to the cavity mode is given by the Hamiltonian

$$\hat{V}_0 = \hbar g_0 [\hat{b}_0 \hat{a}^\dagger + \hat{b}_0^\dagger \hat{a}] \,. \tag{5.157}$$

For a classical field, we have $\hat{b}_0 \to \langle \hat{b}_0 \rangle$. In that case, $\hat{V}_0$ takes the form of a displacement operator, and Eq. (5.147) becomes

$$\frac{d\hat{a}(t)}{dt} = \frac{i}{\hbar}[\hat{H}_s, \hat{a}(t)] - \frac{\kappa}{2}\hat{a}(t) + \sqrt{\kappa}\,\alpha_{\text{in}}(t) \,, \tag{5.158}$$

where the classical input field α_{in} has again the units of a flux. This situation will be encountered at several occasions in the following chapters, for example, in Problem 7.1 in the context of cavity QED, in Sect. 9.4 on cavity cooling, and in the discussion of optomechanics of Chap. 11.

Problems

Problem 5.1 In order to highlight the importance of the density of modes on spontaneous emission, derive the analog of the Weisskopf–Wigner spontaneous emission rate in a one-dimensional system and a two-dimensional system.

Problem 5.2 Determine and explain in physical terms the spontaneous emission rates of two two-level atoms whose internal state is $|\psi\rangle(0) = \frac{1}{\sqrt{2}}[|e, g\rangle \pm |g, e\rangle]$, but whose centers of mass are at locations $\mathbf{r}_1$ and $\mathbf{r}_2$, respectively.

Problem 5.3 Derive the equation of motion (5.58) for the mean excitation number of a damped harmonic oscillator

$$\frac{d\langle \hat{a}^\dagger \hat{a} \rangle_s}{dt} = -\gamma \langle \hat{a}^\dagger \hat{a} \rangle_s + \gamma \bar{n} \,.$$

Problem 5.4 Consider the situation discussed in the input–output fields analysis, but with both mirrors of the resonator now semi-transparent, and coupled to separate field reservoirs consisting of the continuous sets of modes $\{\hat{b}(\omega)\}$ and $\{\hat{c}(\omega)\}$, with associated damping rates κ_L and κ_R, see Fig. 5.6. Determine the input–output relations between the fields $\hat{a}$, $\hat{a}_{\text{in, L}}$, $\hat{a}_{\text{out, L}}$, $\hat{a}_{\text{in, R}}$, and $\hat{a}_{\text{out, R}}$ in that case.

Problem 5.5 Follow steps that parallel those leading to the derivation of the master equation for a damped harmonic oscillator to derive the master equation (5.66) for a damped two-level atom.

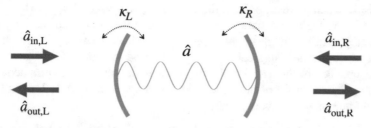

Fig. 5.6 Schematic of the cavity considered in Problem 5.4

Problem 5.6 Determine and explain in physical terms the spontaneous emission rates of two two-level atoms whose internal state is $\hat{\rho} = \frac{1}{2}[|e, e\rangle + |g, g\rangle]$.

Problem 5.7 Carry out the explicit steps to derive equations (5.38), (5.40), and (5.44).

Problem 5.8 Show that the expectation value of the operator

$$\hat{G}_{A^\dagger A}(t) - i \sum_j g_j^* \hat{b}_j^\dagger(t_0) \hat{\Lambda}(t) e^{-i(\Omega - \omega_j)(t - t_0)} + \text{adj}.$$

of Eq. (5.95) is $\langle \hat{G}_{A^\dagger A}(t) \rangle = \gamma \bar{n}$.

Problem 5.9 Derive the Fokker–Planck equation (5.77) from Eq. (5.70), and show that its steady-state solution is

$$P(\alpha) = \frac{1}{\pi \bar{n}} e^{-|\alpha|^2 / \bar{n}}.$$

Problem 5.10 Solve the master equation for a damped two-level atom initially in its excited state $|e\rangle$ numerically by using the quantum trajectories method by (1) decomposing the associated master equation into an effective non-Hermitian Hamiltonian $\hat{H}_{\text{eff}}$ and a quantum jump Liouvillian, (b) writing a code to compute the evolution of a typical stochastic trajectory, and (c) averaging the result over at least 50 trajectories. Show that in that limit, one recovers the average excited state population $p_e(t)$ predicted by an analytical solution of the master equation. What could you expect if you had a lab where you could monitor the decay of individual atoms?

Problem 5.11 The life and death of a photon *The following references are strongly suggested reading in connection with this chapter, in particular with Problems 5.11 and 5.12: "Quantum jumps of light recording the birth and death of a photon in a cavity," by Gleyzes and coworkers [7], as well as Refs. [8] and [9] by the same group. These papers report on a series of remarkable cavity QED experiments that measured and characterized in detail the decay of a single-mode field.*

Consider a damped harmonic oscillator coupled to a reservoir at zero temperature and initially in the mixed state $\hat{\rho} = |3\rangle\langle 3|$. Solve the associated master equation numerically by using the quantum trajectories method, averaging the result over a number of trajectories large enough to obtain good statistics. Determine then the time dependence of the average photon number $\langle n \rangle$ and of the time dependence of the populations $p_n(t)$ of the various levels $|n\rangle$, with $n = 0, 1, 2, 3$, and give a physical interpretation of your result.

Problem 5.12 Carry out a similar simulation as for Problem 5.10, but for the initial state $\hat{\rho} = (1/4)|1\rangle\langle 1| + (1/2)|2\rangle\langle 2| + (1/4)|3\rangle\langle 3|$.

Problem 5.13 Show that for a cavity mode at frequency Ω and a driving field at frequency ω, Eq. (5.147) becomes

$$\frac{d\hat{a}}{dt} = [i(\omega - \Omega) - \kappa/2]\hat{a} + \eta + \sqrt{\kappa}\hat{a}_{in}(t). \qquad (5.159)$$

References

1. R.H. Dicke, Coherence in spontaneous radiation processes. Phys. Rev. **93**, 99 (1954)
2. M. Gross, S. Haroche, Superradiance: an essay on the theory of collective spontaneous emission. Phys. Rep. **93**, 301 (1982)
3. D. Manzano, A short introduction to the Lindblad master equation. AIP Adv. **10**, 025106 (2020)
4. J. Dalibard, J. Dupont-Roc, C. Cohen-Tannoudji, Vacuum fluctuations and radiation reaction: identification of their respective contributions. J. Phys. **43**, 1617 (1982)
5. C.W. Gardiner, M.J. Collett, Input and output in damped quantum systems: Quantum stochastic differential equations and the master equation. Phys. Rev. A **31**(6), 3761 (1985)
6. D.F. Walls, G.J. Milburn, *Quantum Optics* (Springer, Berlin, 2008)
7. S. Gleyzes, S. Kuhr, C. Guerlin, J. Bernu, S. Deléglise, U.B. Hoff, M. Brune, J.-M. Raimond, S. Haroche, Quantum jumps of light recording the birth and death of a photon in a cavity. Nature **447**, 297 (2007)
8. C. Guerlin, J. Bernu, S. Deléglise, C. Sayrin, S. Gleyzes, S. Kuhr, M. Brune, J.-M. Raimond, S. Haroche, Progressive field-state collapse and quantum non-demolition photon counting. Nature **448**, 889 (2007)
9. C. Sayrin, I. Dotsenko, X. Zhou, B. Peaudecerf, T. Rybarczyk, S. Gleyzes, P. Roucon, M. Mirrahimi, H. Amini, M. Brune, J.-M. Raimond, S. Haroche, Real time quantum feedback prepares and stabilizes photon number states. Nature **477**(7362), 73–77 (2011)

Chapter 6
Quantum Measurements

This chapter presents an operational approach to quantum measurements based on the von Neumann projection postulate. After reviewing that postulate we turn to a discussion of measurement back action, leading to an understanding of the standard quantum limit and to the idea of quantum non-demolition measurements. We then extend the idea of projective measurements to positive operator-valued measures, which are then applied to a formulation of weak measurements and to the development of a stochastic Schrödinger equation description of continuous weak measurements. We illustrate this approach with the example of weak continuous measurements of optical fields. The chapter concludes with a brief introduction to the role of environment in establishing a measurement pointer basis.

Quantum measurements are a topic of central importance not just in quantum optics, but in all of quantum mechanics. Because as we have seen the Schrödinger equation predicts that quantum dynamics is reversible, some additional ingredient must be added to the theory to account for the irreversibility and finality of quantum measurements. This is a difficult and somewhat unsettling topic that continues to be the object of much debate, associated at least in part to the interpretation of quantum mechanics. We will not engage in this debate here, but rather limit ourselves to a simple, operational approach based on the von Neumann postulate and its extensions.

The first section briefly reviews that postulate by introducing the concept of projective measurements. We then turn to a discussion of measurement back action, the property that the measurement of some observable of a quantum system usually impacts its subsequent dynamics. This leads to an understanding of the standard quantum limit of measurements and to a discussion of ways to circumvent that limit through the use of the so-called quantum non-demolition measurements [1, 2]. We continue in Sect. 6.3 with an extension of projective measurements to positive operator-valued measures (POVM) and their use in the formulation of weak measurements, more specifically in continuous weak measurements, a type of measurements of considerable practical interest.

© The Author(s), under exclusive license to Springer Nature Switzerland AG 2021
P. Meystre, *Quantum Optics*, Graduate Texts in Physics,
https://doi.org/10.1007/978-3-030-76183-7_6

Because of the irreversible nature of quantum measurements, it is perhaps not surprising that they should be associated with considerations of system–reservoir interactions, which can likewise result in irreversible dynamics as we have seen. With this in mind we conclude this chapter with an introduction to the concept of pointer basis. This shows how specific environments provide a preferred basis of the detector that is immune to environmental decoherence and defines a classical measuring apparatus unambiguously, and closes a loop with the previous chapter.

An excellent and much broader discussion of quantum measurements and their applications, in particular in the detection of classical forces, can be found in the monograph "Quantum Measurement" by V. Braginsky and F. Ya. Khalili [2].

6.1   The von Neumann Postulate

Our starting point is the von Neumann postulate. In its simplest form it deals with the exact measurement of an operator $\hat{A}$ with a (discrete) set of n eigenvectors $\{|n\rangle\}$ and associate eigenvalues $\{a_n\}$. The von Neumann postulate states that if the system is in a state described by the density operator $\hat{\rho}_i$ prior to the measurement, then the result of a measurement of $\hat{A}$ will be one of its eigenvalues $|n\rangle$, with probability

$$p_n = \mathrm{Tr}\left(|n\rangle\langle n|\hat{\rho}_i\right) ,\qquad\qquad (6.1)$$

following which the system will be projected to the pure state $|n\rangle$, with corresponding density operator $\hat{\rho}_f = |n\rangle\langle n|$. For this reason such a measurement is called a *projective measurement*, as its action on the system density operator is characterized by the projection operator

$$\hat{P}_n = |n\rangle\langle n| .\qquad\qquad (6.2)$$

This formulation also applies in the situation by now familiar to us where the system is comprised of several subsystems, and one is interested in observables of one of the subsystems only, call it subsystem A. If the pre-measurement density operator of the full system is $\hat{\rho}_i$, then the n-th possible final state following the measurement of an observable $\hat{A}$ of subsystem A is

$$\hat{\rho}_f = \frac{\hat{P}_n\hat{\rho}_i\hat{P}_n}{\mathrm{Tr}(\hat{P}_n\hat{\rho}_i\hat{P}_n)}\qquad\qquad (6.3)$$

with probability $p(n) = \mathrm{Tr}(\hat{P}_n\hat{\rho}_i\hat{P}_n) = |c_n|^2$. It is readily seen that in case $\hat{A}$ is an observable of the full system rather than a subsystem, then $\hat{\rho}_i$ can be expanded on its complete set of eigenstates as $\hat{\rho}_i = \sum_{j,k} c_{jk}|j\rangle\langle k|$ and Eq. (6.3) reduces to $\hat{\rho}_f = |n\rangle\langle n|$, that is, to the pure state $|n\rangle$ of Eq. (6.1).

Quantum Steering Suppose that a bipartite system is initially prepared in the entangled state $|\psi_i\rangle = \frac{1}{\sqrt{2}} [| \uparrow, \downarrow\rangle + | \downarrow, \uparrow\rangle]_{AB}$, with the corresponding density operator

$$\hat{\rho}_i = \frac{1}{2} [| \uparrow, \downarrow\rangle + | \downarrow, \uparrow\rangle][\langle \uparrow, \downarrow | + \langle \downarrow, \uparrow |]_{AB} . \tag{6.4}$$

Following a measurement projecting one of the subsystems onto one of its eigenstates—say, measuring a qubit A to be in its upper state $| \uparrow\rangle_A$—the post-measurement state of the system will be

$$\hat{\rho}_f = | \uparrow, \downarrow\rangle\langle \uparrow, \downarrow |, \tag{6.5}$$

which is nothing but the pure state $|\psi\rangle_f = | \uparrow, \downarrow\rangle_{AB}$. In contrast, if A is found to be in the lower state $| \downarrow\rangle_A$, we will have $|\psi\rangle_f = | \downarrow, \uparrow\rangle_{AB}$. This result is actually more profound that may appear because it demonstrates that the state of the second spin can be steered by a measurement on a far distant other spin. This *quantum steering* is the result that so bothered Einstein, Podolsky, and Rosen and led to the formulation of the EPR paradox that we discussed in Sect. 4.1.

6.2 Measurement Back Action

An important consequence of performing a measurement on a quantum system is the back action of that measurement on the subsequent system evolution, an effect that is a direct consequence of the Heisenberg uncertainty principle. To gain of intuitive feeling for its origin, consider for a moment the dynamics of a free particle of mass m in one dimension. It is described by the Hamiltonian

$$\hat{H} = \frac{\hat{p}^2}{2m}, \tag{6.6}$$

and its position is

$$\hat{x}(\tau) = \hat{x}(0) + \hat{p}(0)\tau/m . \tag{6.7}$$

Suppose now that at time $t = 0$ we measure $\hat{x}$ with some uncertainty characterized by a variance σ_x^2. From the Heisenberg uncertainty relation $\sigma_x \sigma_p \geq \hbar/2$ and Eq. (6.7) it follows that after a time τ the uncertainty in position will have increased to

$$\sigma_x^2(\tau) \geq \sigma_x^2(0) + \sigma_p^2(0)\tau^2/m^2 \tag{6.8}$$

so that

$$\sigma_x(\tau) \geq \left[\sigma_x^2(0) + \left(\frac{\hbar\tau}{2m\,\sigma_x(0)} \right)^2 \right]^{1/2} \geq \left(\frac{\hbar\tau}{m} \right)^{1/2}. \tag{6.9}$$

This is the so-called *standard quantum limit* (SQL) for free mass position. Physically, the uncertainty in $\hat{p}$ resulting from a first measurement of its conjugate variable $\hat{x}$ imposes an additional uncertainty on subsequent measurements of $\hat{x}$. This back action can be particularly severe if one wishes to perform a sequence of measurements on a system, as we already saw in the simple example of Sect. 3.5.

6.2.1   The Standard Quantum Limit

Since it depends explicitly on the Heisenberg uncertainty relation for the observable under consideration, the standard quantum limit is a function of the specific physical system and type of measurements under consideration. One example of considerable interest is of course the characterization of the ubiquitous harmonic oscillator, which we already encountered in the quantization of the electric field, and will also be central to the discussion of quantum optomechanics in Sect. 11.4.

We limit ourselves for now to a discussion of versions of the standard quantum limit associated with two methods frequently used to characterize quantum harmonic oscillators and that we already encountered in Chap. 2, intensity and quadrature measurements. We will have several opportunities to revisit this problem in other examples in subsequent chapters.

Our starting point is the Hamiltonian,

$$\hat{H} = \frac{\hat{p}^2}{2m} + \frac{1}{2}m\omega^2\hat{x}^2, \tag{6.10}$$

of a quantum harmonic oscillator of mass m, frequency ω, position $\hat{x}$, momentum $\hat{p}$, and their associated creation and annihilation operators

$$\hat{a}^\dagger = \sqrt{\frac{m\omega}{2\hbar}}(\hat{x} - i\hat{p}/m\omega)$$

$$\hat{a} = \sqrt{\frac{m\omega}{2\hbar}}(\hat{x} + i\hat{p}/m\omega) \tag{6.11}$$

where we keep the mass m explicitly since it is important once we stop talking about light fields. We also introduce the quadrature operators

$$\hat{X}_1 = \hat{x}\cos\omega t - (\hat{p}/m\omega)\sin\omega t$$

$$\hat{X}_2 = \hat{x}\sin\omega t + (\hat{p}/m\omega)\cos\omega t, \tag{6.12}$$

or, in terms of $\hat{a}$ and $\hat{a}^\dagger$,

$$\hat{X}_1 = \sqrt{\frac{\hbar}{2m\omega}} \left[\hat{a} e^{i\omega t} + \hat{a}^\dagger e^{-i\omega t} \right],$$

$$\hat{X}_2 = -i \sqrt{\frac{\hbar}{2m\omega}} \left[\hat{a} e^{i\omega t} - \hat{a}^\dagger e^{-i\omega t} \right]. \tag{6.13}$$

Except for a different normalization, they are the same as the operators $\hat{d}_1$ and $\hat{d}_2$ introduced in the discussion of squeezed states in Eq. (2.122). Classically, x and $p/m\omega$ can be thought of as Cartesian coordinates in a phase plane, and X_1 and X_2 as coordinates that rotate clockwise at frequency ω relative to them. Quantum mechanically, these two quadrature operators do not commute,

$$[\hat{X}_1, \hat{X}_2] = [\hat{x}, \hat{p}/m\omega] = i\hbar/m\omega \tag{6.14}$$

so that their variances satisfy

$$\sigma_{X_1} \sigma_{X_2} - \sigma_x \sigma_{(p/\hbar\omega)} \geq \hbar/2m\omega. \tag{6.15}$$

As we discussed in Sect. 2.3.3, and as is also readily apparent from Eqs. (6.13), for a coherent state $|\alpha\rangle$ we have $\sigma_{X_1} = \sigma_{X_2} = \sqrt{\hbar/2m\omega}$, resulting in a circular "error box" in phase space of area $\pi\hbar/2m\omega$. Squeezed coherent states produce the same area, but in the shape of an ellipse with principal axes ΔX_1 and ΔX_2 of different lengths.

Position Measurement Consider then a measurement scheme where the position is measured essentially instantly, that is, in a time short compared to $1/\omega$, with a small uncertainty $\Delta x_0 \ll (\hbar/2m\omega)^{1/2}$. This produces an uncertainty in momentum of $\Delta p_0/m\omega \gg \Delta x_0$, and correspondingly, an uncertainty error box around the oscillator in the form of a strongly elongated ellipse. This ellipse rotates clockwise at frequency ω, see Fig. 6.1, and as a result, the value of a second position measurement will be accompanied by an error Δx_1 that can be anywhere between Δx_0 and $\Delta p_0/m\omega \gg (\hbar/2m\omega)(1/\Delta x_0)$, depending on the precise time at which it is carried out. To guarantee that the maximum possible error is minimized requires therefore that the error box be circular with the minimum allowed radius $\Delta x_0 = \Delta p_0/m\omega = \Delta X_1 = \Delta X_2 = (\hbar/2m\omega)^{1/2}$. An ideal measurement with these uncertainties will drive the oscillator to a minimum uncertainty state that simultaneously minimizes the uncertainties in position and momentum, that is, a coherent state. This uncertainty defines the standard quantum limit of position measurements for a harmonic oscillator.

Amplitude and Phase Measurements Measurements that attempt to determine both the oscillator's amplitude of motion $|X| = (|X_1|^2 + |X_2|^2)^{1/2}$ and its phase $\phi = \tan^{-1}(X_2/X_1)$ are of much interest for a number of applications. An ideal way

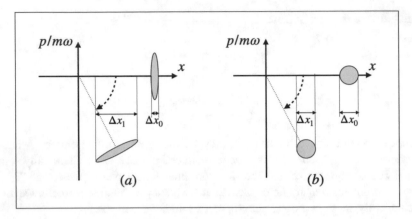

Fig. 6.1 Schematics of the position measurements of a harmonic oscillator, showing the impact of (**a**) an initial measurement with precision $\Delta x_0 \ll (\hbar/2m\omega)^{1/2}$, and (**b**) measurements with precision $\Delta x_1 = \Delta x_2 = (\hbar/2m\omega)^{1/2}$, which minimize the maximum error and drive the system toward a coherent state

to achieve minimum uncertainties in both quantities involves once more the circular error bar characteristic of a coherent state. This results in a Gaussian distribution of measured values of $\langle \hat{X}_1 \rangle$ and $\langle \hat{X}_2 \rangle$ with equal variances $\sigma_{X_1} = \sigma_{X_2} = (\hbar/2m\omega)^{1/2}$. From Eq. (2.108) the standard deviation of a coherent state $|\alpha\rangle$ with mean photon number $\langle n \rangle \equiv \langle \hat{a}^\dagger \hat{a} \rangle = (m\omega/2\hbar)(\langle \hat{X}_1^2 \rangle + \langle \hat{X}_2^2 \rangle) \gg 1$ is

$$\sigma_n = \sqrt{\langle n^2 \rangle - \langle n \rangle^2} \approx \langle n \rangle^{1/2}, \tag{6.16}$$

and

$$\sigma_\phi = \frac{1}{2\sigma_n} = \frac{1}{2\langle n \rangle^{1/2}}. \tag{6.17}$$

These standard deviations represent the *standard quantum limit* of amplitude and phase measurements.

We will return to this topic in Sect. 11.4, which will characterize the noise limits of optical interferometers in some detail. In particular we will show how the optical noise sources acting on the field quadratures $\hat{a}_1$ and $\hat{a}_2$ of Eqs. (2.122), or equivalently on $\hat{X}_1$ and $\hat{X}_2$, are the radiation pressure noise and shot noise, respectively. This understanding plays a central role both in establishing the standard quantum limit of these devices and in finding their optimal point of operation for a specific application, for instance gravitational wave detection. That section will also show how the use of squeezed light permits to circumvent the standard quantum limit in these systems.

Although it is always useful to keep in mind the standard quantum limit as a reference point, it is just as important to also remember that it is not a fundamental measurement precision limit and that it is sometimes possible to evade it. This is the topic of quantum non-demolition measurements, or back action evading measurements, to which we now turn. Substantially more detailed discussions of this topic can be found in the classic book by Braginsky and Khalili [2] and in the review [3] by C. M. Caves and coworkers.

6.2.2 Quantum Non-demolition Measurements

When discussing the back action of a measurement of the position of a free particle, we found that because $\hat{x}(\tau) = \hat{x}(0) + \hat{p}(0)t/m$ the resulting uncertainty in momentum feeds back into the subsequent evolution of $\hat{x}$, limiting the accuracy of successive measurements. The first measurement has "demolished" the possibility of making a second measurement of the same precision. However, this would not be the case if we had chosen to measure $\hat{p}$ instead of $\hat{x}$. Because the momentum is a constant of motion for free particles,

$$\frac{\mathrm{d}\hat{p}}{\mathrm{d}t} = \frac{\mathrm{i}}{\hbar}[\hat{H}, \hat{p}] = 0, \tag{6.18}$$

the uncertainty Δx that results from a first measurement of $\hat{p}$ with precision Δp_0 does not feed back into its subsequent evolution, that is, it does not demolish the possibility to carry out subsequent measurements of $\hat{p}$ with the same precision. Such measurements are called *back action evading*, or quantum non-demolition (QND) measurements. We now derive a necessary and sufficient condition for an observable $\hat{A}$ to be a QND variable.

Let us assume that the observable $\hat{A}$ has a complete set of eigenstates $\{|A, \mu\rangle\}$, where μ labels the states in any degenerate subspace, and that a first measurement of $\hat{A}$ returns some eigenvalue A_0, with a corresponding eigenstate that will in general be a superposition of the form

$$|\psi(t_0)\rangle = \sum_{\mu} c_{\mu}|A_0, \mu\rangle. \tag{6.19}$$

The system evolves then freely until the next measurement at time t_1. During that time interval the Heisenberg picture operator $\hat{A}(t)$ changes, but the state $|\psi(t_0)\rangle$ does not, so that in general it will no longer be an eigenstate of $\hat{A}(t_1)$. However, if the second measurement is to produce a perfectly predictable result, that is, in order to ensure that the free motion of $\hat{A}$ is not altered by the first measurement, then all states $|\psi(A_0, \mu)\rangle$ within a given degenerate subspace must still be eigenstates of $\hat{A}(t_1)$, all with the same eigenvalue—which however needs not in general be equal to A_0. That is, we must have

$$\hat{A}(t_1)|A_0, \mu\rangle = f(A_0)|A_0, \mu\rangle \quad \text{for all } \mu. \tag{6.20}$$

This condition must hold for all eigenvalues A_0, and for all times t_k in a series of k measurements. That is, if the measurements sequence starts with the measurement returning some eigenvalue A_0 of $\hat{A}$, its subsequent free evolution must leave it in an eigenstate of $\hat{A}$ *at each time in the measurement sequence*.

An observable that satisfies Eq. (6.20) at all times is called a *continuous QND observable*, and the simplest way to satisfy that condition is with an observable that is conserved,

$$\frac{\mathrm{d}\hat{A}}{\mathrm{d}t} = -\frac{\mathrm{i}}{\hbar}[\hat{A}, \hat{H}] + \frac{\partial \hat{A}}{\partial t} = 0, \tag{6.21}$$

or

$$[\hat{A}(t), \hat{A}(t')] = 0 \quad \text{for all } t, t'. \tag{6.22}$$

Examples of continuous QND variables that we have encountered so far are the momentum $\hat{p}$ of a free particle, as well as the quadratures $\hat{X}_1$ and $\hat{X}_2$ and excitation number $\hat{N} = \hat{a}^\dagger \hat{a}$ of the harmonic oscillator.

If, on the other hand, the observable satisfies that condition only at some specific times, it is called a *stroboscopic QND observable*, with examples including the position $\hat{x}$ and momentum $\hat{p}$ of the simple harmonic oscillator.

Quantum Mechanics Free Subsystems M. Tsang and C. M. Caves [4] and K. Hammerer and coworkers [5, 6] realized that it is sometimes possible to isolate *quantum mechanics free subsystems* (QMFS) of a quantum system. If this can be achieved, then all observables in these subsystems are by construction QND observables. QMFS may find a number of applications in the detection of feeble forces and fields, including optomechanical sensing, magnetometry, and gravitational wave detection.

A simple setup to implement a QMFS comprises two harmonic oscillators of identical frequencies and opposite masses with Hamiltonian

$$\hat{H} = \frac{\hat{p}^2}{2m} + \frac{1}{2}m\omega^2\hat{q}^2 - \frac{\hat{p}'^2}{2m} - \frac{1}{2}m\omega^2\hat{q}'^2, \tag{6.23}$$

where $[\hat{q}, \hat{p}] = [\hat{q}', \hat{p}'] = \mathrm{i}\hbar$. Considering then the variables

$$\hat{Q} = \hat{q} + \hat{q}' \qquad \hat{P} = \frac{1}{2}(\hat{p} + \hat{p}'),$$

$$\hat{\Phi} = \frac{1}{2}(\hat{q} - \hat{q}') \qquad \hat{\Pi} = \hat{p} - \hat{p}', \tag{6.24}$$

it is easily verified that

$$\dot{\hat{Q}}(t) = \frac{\hat{\Pi}(t)}{m} \qquad \dot{\hat{\Pi}} = -m\omega^2\hat{Q}(t). \tag{6.25}$$

Since $[\hat{Q}, \hat{\Pi}] = 0$, this means that the dynamical pair of observables $\{\hat{Q}, \hat{\Pi}\}$ formed by the collective position $\hat{Q}$ and relative momentum $\hat{\Pi}$ forms a QMFS—and likewise for the pair $\{\hat{\Phi}, \hat{P}\}$.

Implementations of this idea involving for instance two spin ensembles oppositely polarized along an external magnetic field [7] or a hybrid system consisting of a macroscopic mechanical oscillator and a spin oscillator [8] were recently demonstrated. Several other systems have also been proposed, including the use of mechanical oscillators [4] and ultracold atomic systems with an effective negative mass component produced by an optical lattice [9].

Interaction with a Measuring Apparatus While the von Neumann postulate answers the question of what happens to the object during a measurement, it does not address the question of how the measuring device must be designed to achieve a desired measurement. To answer this question one needs to account for the presence of a detector and its interaction with the system to be characterized.

A good measuring apparatus must (obviously) be sensitive to the observable $\hat{X}$ of interest, but ideally it should not be coupled to any other system observable. Its interaction Hamiltonian with the system should therefore be of the form

$$\hat{V} = \hbar g \hat{X} \hat{M}, \tag{6.26}$$

where $\hat{M}$ is an observable of the measuring apparatus and g a coupling constant. It is easily shown that if $\hat{X}$ is a continuous QND observable, then its evolution is not affected by a system–apparatus interaction of the form (6.26).

Broadly speaking one can consider two types of measurements: *direct measurements*, which are quantum measurements where the system interacts directly with a classical device, and *indirect measurements*, which are two-step processes where the object to be characterized interacts directly with another quantum object—a quantum probe—whose state has been carefully prepared, the state of the probe being then reduced through its interaction with a macroscopic detector, for example an amplifier or a photodetector.

Direct measurements are typically characterized by significant randomness in the interaction between the object and the detector, and as a result the detector disturbs the object much more strongly than the minimum required by the Heisenberg uncertainty relations. As a consequence, much better results can be achieved with indirect measurements. Still, since the first step of the measurement process involves the build-up of correlations between the object and the quantum probe, the reduction of the state of the probe results nonetheless in an irreversible back action on the state of the object and an irreversible change in that object as well. This key aspect of the measurement process will be discussed in much more detail in Sect. 6.4.

6.3   Continuous Measurements

Repeated measurements have become increasingly important in quantum optics, where they are routinely used for example to monitor the dynamics of trapped ions or atoms or the optical field in a laser interferometer gravitational wave antenna. We discussed in Sect. 3.5 a crude scheme of repeated measurements of a single-mode field by a stream of two-level atoms and commented on the considerable measurement back action associated with them. QND measurements can mitigate this issue in principle, but they are not necessarily easy to realize. Another way to minimize the effects of measurement back action consists in probing the system so gently as to not significantly perturb its subsequent evolution. This is the approach taken by *non-projective weak measurements*, to which we turn our attention in this section. We will show how in the limit of continuous measurements these can be described in terms of stochastic master equations, or alternatively stochastic Schrödinger equations. To set the stage for this discussion, though, we first revisit briefly the role of back action on continuous projective measurements.

6.3.1   Continuous Projective Measurements

As is clear from their name, continuous measurements extract information from the system continuously. To construct such measurement schemes, time is divided into a sequence of small intervals Δt and a measurement is performed during each of them.

As in the example of Sect. 3.5 we assume that the system, with Hamiltonian $\hat{H}$, is monitored by a stream of j identical and non-interacting particles that successively interact with it during the intervals $t_j \leq t \leq t_{j+1}$, with $t_j = (j-1)\Delta t$, and $\Delta t \to 0$ for continuous monitoring. Information can then be extracted from measuring the state of the successive probes after they exit the system. We denote the system observable to be monitored by the Hermitian operator $\hat{X}$ and assume that the individual system–probe Hamiltonians are of the form (6.26). The associated system–probe evolution operators $\hat{U}_j$ are

$$\hat{U}_j = \exp\left[(-i/\hbar)(\hat{H} + \hbar g \hat{X} \hat{M}_j) \right]$$

$$\approx 1 - \frac{i\Delta t}{\hbar}(\hat{H} + \hbar g \hat{X} \hat{M}_j) - \frac{1}{2}\left(\frac{\Delta t}{\hbar}\right)^2 \left(\hat{H} + \hbar g \hat{X} \hat{M}_j\right)^2 + \ldots \quad (6.27)$$

so that

$$\hat{\varrho}(t_{j+1}) = \hat{U}_j(\Delta t)\hat{\varrho}(t_j)\hat{U}_j^\dagger(\Delta t), \qquad (6.28)$$

where $\hat{\varrho}$ is the combined system–probe density operator. Since the system is initially not correlated with the successive probes, we have that $\hat{\varrho}(t_j) = \hat{\rho}(t_j) \otimes \hat{\rho}_j(t_j)$, where $\hat{\rho}$ and $\hat{\rho}_j$ are the system and probe density operators.

If the state of the probe is not measured at the end of its interaction with the system, that is, if the state of the probe is not collapsed into one of its eigenstates by a direct projective measurement, then the resulting system density operator $\hat{\rho}(t_{j+1}) = \text{Tr}_{\text{probe}}\hat{\varrho}(t_{j+1})$ is, to second order in Δt,

$$
\hat{\rho}(t_{j+1}) = \hat{\rho}(t_j) - \frac{i\Delta t}{\hbar}[\hat{H} + \hbar g \hat{X}\langle M_j \rangle, \hat{\rho}(t_j)] - \frac{1}{2}\left(\frac{\Delta t}{\hbar}\right)^2 \left\{ [\hat{H}, [\hat{H}, \hat{\rho}(t_j)]] \right.
$$
$$
+ \hbar g \langle \hat{M}_j \rangle \left([\hat{H}, [\hat{X}, \hat{\rho}(t_j)]] + [\hat{X}, [\hat{H}, \hat{\rho}(t_j)]] \right)
$$
$$
\left. + (\hbar g)^2 \langle \hat{M}_j^2 \rangle [\hat{X}_j, [\hat{X}_j, \hat{\rho}(t_j)]]^2 \right\}, \tag{6.29}
$$

where $\langle \hat{M}_j \rangle$ and $\langle \hat{M}_j^2 \rangle$ are the mean and mean square of the probe operator $\hat{M}$ before the interaction with the system. The term proportional to $\langle \hat{M}_j \rangle$ is the *direct back action* of the probe on the system, while the term proportional to $\langle \hat{M}_j^2 \rangle$ is its *fluctuational back action*. It introduces random noise in the measurements [2]. Assuming a measurement such that the direct back action can be eliminated, $\langle \hat{M}_j \rangle = 0$, and taking the limit $\Delta t \to 0$ give then

$$
\frac{\partial \hat{\rho}}{\partial t} = -\frac{i}{\hbar}[\hat{H}, \hat{\rho}] - \frac{1}{2\hbar^2}\sigma_M^2[\hat{X}[\hat{X}, \hat{\rho}]], \tag{6.30}
$$

where we have introduced the variance of the probe's back action force on the system

$$
\sigma_M^2 = \hbar^2 g^2 \langle \hat{M}^2 \rangle. \tag{6.31}
$$

It is not surprising that Eq. (6.30) should be of the same form as the master equations that we encountered in the discussion of system–reservoir interactions, since as long as we do not follow the state of the probes, they act effectively as a reservoir that tends to bring the system to a state of equilibrium. Problem 6.4 discusses how this happens for the specific example of a driven two-level system. Of course the situation changes if projective measurements on the successive probe particles are carried out, as we saw in the example of Sect. 3.5. Problem 6.5 explores in particular the quantum Zeno paradox [10] associated with continuous projective measurements where the state of the probe is determined at each step.

6.3.2 Positive Operator-Valued Measures

We indicated that rather than minimizing the effects of projective measurements back action via QND schemes, one can also think of probing the system so gently as to not significantly perturb its subsequent evolution. This is the approach taken by the non-projective, weak measurements to which we now turn. This discussion follows largely the excellent tutorial of Ref. [11], to which the reader is referred for more details.

Measurements other than projective measurements, and weak measurements in particular, are conveniently described by generalizing the idea of projection operators $\hat{P}_n$. Specifically, if we pick any set of $m_{\max}$ operators $\hat{\Omega}_m$ with the restriction

$$\sum_{m=1}^{m_{\max}} \hat{\Omega}_m^\dagger \hat{\Omega}_m = \hat{I} \tag{6.32}$$

where $\hat{I}$ is the identity operator, then it is possible to design a measurement with potential outcomes

$$\hat{\rho}_f = \frac{\hat{\Omega}_m \hat{\rho}_i \hat{\Omega}_m^\dagger}{\mathrm{Tr}[\hat{\Omega}_m \hat{\rho}_i \hat{\Omega}_m^\dagger]} \tag{6.33}$$

occurring with probabilities

$$P(m) = \mathrm{Tr}[\hat{\Omega}_m \hat{\rho}_i \hat{\Omega}_m^\dagger] \tag{6.34}$$

and with the total probability of obtaining a result in the range $[a, b]$ given by

$$P(m \in [a, b]) = \sum_{m=a}^{b} \mathrm{Tr}[\hat{\Omega}_m \hat{\rho} \hat{\Omega}_m^\dagger] = \mathrm{Tr}\left[\sum_{m=a}^{b} \hat{\Omega}_m^\dagger \hat{\Omega}_m \hat{\rho}\right]. \tag{6.35}$$

These generalized measurements are referred to as positive operator-valued measures or POVM. They can be implemented by performing a unitary interaction between the system to be characterized and an auxiliary system and then performing a projective von Neumann measurement on that system. They are of particular interest in the context of weak continuous measurements, to which we now turn, focusing on an outline of the main steps in the derivation of a powerful and elegant description of these measurements in terms of stochastic master equations (or alternatively of stochastic Schrödinger equations).

6.3.3 Weak Continuous Measurements

As in Sect. 6.3.1 we denote the observable to be monitored by the Hermitian operator $\hat{X}$ and assume for simplicity that it has a continuous spectrum of eigenvalues $\{x\}$ with corresponding eigenstates $\{|x\rangle\}$, so that $\langle x|x'\rangle = \delta(x - x')$. Weak measurements of $\hat{X}$ are characterized by a POVM of the form

$$\hat{A}(\alpha) \equiv \left(\frac{4k\Delta t}{\pi}\right)^{1/4} \int_{-\infty}^{+\infty} dx\, e^{-2k\Delta t(x-\alpha)^2} |x\rangle\langle x| , \qquad (6.36)$$

where α is a continuous index that labels the spectrum of measurement results. As we shall see, the parameter k can be understood as a measure of the measurement strength. A continuous measurement results from making a sequence of these measurements and taking the limit $\Delta t \to 0$, or equivalently $\Delta t \to dt$. That is, more measurements are made in any finite time interval, but each is increasingly weak.

A key attribute of $\hat{A}(\alpha)$ is that since it is a Gaussian-weighted sum of projectors onto the eigenstates of $\hat{X}$ peaked at α, it provides only *partial information* about the observable. To see what this information is, consider a generic state $|\psi\rangle = \int dx\, \psi(x)|x\rangle$ of the system, expanded on the eigenstates $\{|x\rangle\}$ of $\hat{X}$. First, by applying Eq. (6.34) to the POVM $\hat{A}(\alpha)$ we have that the probability $P(\alpha)$ to obtain the measurement outcome α is

$$P(\alpha) = \mathrm{Tr}[\hat{A}(\alpha)|\psi\rangle\langle\psi|\hat{A}^\dagger(\alpha)] , \qquad (6.37)$$

from which it follows immediately that

$$\langle\alpha\rangle = \int_{-\infty}^{\infty} d\alpha\, \alpha P(\alpha) = \int_{-\infty}^{\infty} d\alpha\, \alpha \mathrm{Tr}[\hat{A}(\alpha)^\dagger \hat{A}(\alpha)|\psi\rangle\langle\psi|]$$

$$= \sqrt{\frac{4k\Delta t}{\pi}} \int_{-\infty}^{\infty} dx\, |\psi(x)|^2 \int_{-\infty}^{\infty} d\alpha\, e^{-4k\Delta t(x-\alpha)^2}$$

$$= \int_{-\infty}^{\infty} dx\, x|\psi(x)|^2 = \langle\hat{X}\rangle . \qquad (6.38)$$

This shows that the expectation value of α is equal to the expectation value of $\hat{X}$, as should be the case.

Inserting next the definition (6.36) of $\hat{A}(\alpha)$ into Eq. (6.37) gives for the explicit form of $P(\alpha)$

$$P(\alpha) = \sqrt{\frac{4k\Delta t}{\pi}} \int_{-\infty}^{+\infty} dx\, |\psi(x)|^2 e^{-4k\Delta t(\alpha-x)^2} . \qquad (6.39)$$

If Δt is sufficiently small, the Gaussian under the integral is much broader than $\psi(x)$, and it is possible to approximate $|\psi(x)|^2$ by a delta function centered at the expectation value $\langle \hat{X} \rangle = \langle \alpha \rangle$. We then have that

$$P(\alpha) \simeq \sqrt{\frac{4k\Delta t}{\pi}} e^{-4k\Delta t(\alpha - \langle \hat{X} \rangle)^2}, \tag{6.40}$$

which is a Gaussian of variance $\sigma^2 = 1/(8k\Delta t)$ centered at $\langle \hat{X} \rangle$. We can therefore think of α as the *stochastic quantity*

$$\alpha = \langle \hat{X} \rangle + \sigma \frac{\Delta w}{\sqrt{\Delta t}} = \langle \hat{X} \rangle + \frac{\Delta w}{\sqrt{8k\Delta t}}, \tag{6.41}$$

where Δw is a zero-mean Gaussian random variable of variance $\sigma^2_{\Delta w} = \Delta t$ since $4k\Delta t(\alpha - \langle \hat{X} \rangle)^2 = \Delta w^2/2\Delta t$. That is, its root mean square scales as $(\Delta t)^{1/2}$. It is this stochastic nature of α that accounts for the random nature of the successive quantum measurements. Importantly, Eq. (6.41) shows that the larger the parameter k, the smaller the fluctuations in the measurement outcomes. This justifies associating it with the measurement strength.

In the infinitesimal limit $\Delta t \to dt$ and $\Delta w \to dw$ the stochastic variable $w(t)$ is referred to as a Wiener process, a random walk with arbitrary small, independent steps taken arbitrarily often. Importantly, one needs to keep in mind that the Wiener differential dw satisfies the Itô rule $dw^2 = dt$. This might appear surprising since while dw is a stochastic quantity dt is not, and therefore neither is dw^2. This subtle point is discussed in a pedagogical way in Section 5 of Ref. [11].[1]

These results permit to numerically determine at each time step the evolution of the wave function $|\psi(t)\rangle$, subject to measurements characterized by the POVM $\hat{A}(\alpha)$. The infinitesimal, stochastic change in the quantum state following a single measurement is given (before normalization) by

$$|\psi(t + \Delta t)\rangle \propto \hat{A}(\alpha)|\psi(t)\rangle \propto e^{-2k\Delta t(\alpha - \hat{X})^2}|\psi(t)\rangle. \tag{6.42}$$

Inserting the expression (6.41) for α into this equation, expanding the exponential to first order in $\Delta t \to dt$, but keeping terms up to second order in the Wiener process dw, a necessary step since $dw^2 = dt$, give

$$|\psi(t + dt)\rangle = \left[1 - 2k\hat{X}^2 dt + \hat{X}\left(4k\langle \hat{X} \rangle dt + \sqrt{2k}\,dw + k\hat{X}\,dw^2\right)\right]|\psi(t)\rangle. \tag{6.43}$$

---

[1] The basic rule of Itô calculus is that $dw^2 = dt$, and $dt^2 = dt\,dw = 0$. To use this calculus, count the increment dw as if it were equivalent to $\sqrt{dt}$. As an example, if $dx = \alpha dt + \beta dw$ and $y = \exp(x)$, then $dy = \exp(x + dx) - \exp(x) = \exp(x)[\exp(dx) - 1]$ is obtained by expanding $\exp(dx)$ to second order and keeping terms up to first order in dt but second order in dw, with the result $dy = y[\alpha dt + \beta dw + (\beta^2/2)dt]$, see Ref. [11].

Finally normalizing $|\psi(t + dt)\rangle$ results in the stochastic Schrödinger equation

$$d|\psi\rangle = [-k(\hat{X} - \langle\hat{X}\rangle)^2 dt + \sqrt{2k}(\hat{X} - \langle\hat{X}\rangle)dw]|\psi(t)\rangle, \qquad (6.44)$$

where $d|\psi\rangle = |\psi(t+dt)\rangle - |\psi(t)\rangle$ and the expectation values are taken using $|\psi(t)\rangle$. This equation describes the evolution of the system conditioned on a specific stream of random measurement results

$$dy = \langle\hat{X}\rangle dt + \frac{dw}{\sqrt{8k}} \qquad (6.45)$$

in that time interval. The successive measurements give the expected value $\langle\hat{X}\rangle$ plus a random component due to the width of $P(\alpha)$. The measurements record is called a quantum trajectory, just like in Sect. 5.4.1, albeit for a very different type of measurement. In the present situation it describes the result of weak continuous measurements, while the quantum jumps characteristic of Monte Carlo trajectories considered in that earlier section could be interpreted as resulting from projective measurements effectively performed by the environment.

The stochastic Schrödinger equation (6.44) can also be written in terms of a density operator $\hat{\rho}$ as the *stochastic master equation*

$$d\hat{\rho} = \left(d|\psi\rangle\langle\psi| + |\psi\rangle(d\langle\psi|) + \left(d|\psi\rangle(d\langle\psi|)\right)\right.$$
$$= -k[\hat{X}[\hat{X}, \hat{\rho}]]dt + \sqrt{2k}\left(\hat{X}\hat{\rho} + \hat{\rho}\hat{X} - 2\langle\hat{X}\rangle\hat{\rho}\right)dw, \qquad (6.46)$$

where we have kept all terms proportional to dw^2.

Importantly, Eqs. (6.44) and (6.46) only account for the effect of the continuous weak measurements. A full description of the system must include also its unitary evolution from the Hamiltonian $\hat{H}$ not associated with the measurement process,

$$d|\psi\rangle = -\frac{i}{\hbar}\hat{H}|\psi\rangle dt \quad \text{or} \quad d\hat{\rho} = -\frac{i}{\hbar}[\hat{H}, \hat{\rho}]dt .$$

6.3.4   Continuous Field Measurements

As an illustration of continuous weak measurements we revisit the example of Sect. 3.5, where we considered the repeated measurements of a single-mode field confined in an optical cavity by a sequence of two-state atoms acting as probes. We now modify this scheme so that the interaction of the field with the successive atoms, taken to form a continuous stream, is described by a weak measurement POVM. As a result the field dynamics is then governed by a stochastic Schrödinger equation of the general form (6.44).

We consider both the case where the atom–field interaction is resonant and described by the Jaynes–Cummings Hamiltonian (3.1)

$$\hat{H}_a = \tfrac{1}{2}\hbar\omega_0\hat{\sigma}_z + \hbar\omega\hat{a}^\dagger\hat{a} + \hbar g\left(\hat{\sigma}_+\hat{a} + \hat{a}^\dagger\hat{\sigma}_-\right) \qquad (6.47)$$

and the off-resonant regime described by its dispersive limit (3.14),

$$\hat{H}_d \approx \tfrac{1}{2}\hbar\omega_0\hat{\sigma}_z + \hbar\omega\hat{a}^\dagger\hat{a} + \hbar g_d\hat{a}^\dagger\hat{a}\,\hat{\sigma}_z . \qquad (6.48)$$

Here $g_d = g^2/\Delta$ and the vacuum induced light shift $(\hbar g^2/\Delta)$ of state $|e\rangle$ has been absorbed in the atomic Hamiltonian, $\omega_e \rightarrow \omega_e + \hbar g^2/\Delta$, as discussed in Chap. 3. Measuring the state of the successive atoms as they exit the resonator provides information on the field, following well-established methods of cavity QED that will be discussed in the next chapter.

Absorptive Measurements We first consider the resonant situation $\omega_0 = \omega$ and assume that the successive atoms are prepared in their ground state $|g\rangle$. Their interaction with the optical field induces Rabi changes in the atomic state that can then be monitored by indirect projective measurements on the atoms after they exit the cavity. Each series of measurements results in a stochastic measurement sequence, or quantum trajectory "j" described by the stochastic wave function $|\psi_j(t)\rangle$ of Eq. (6.44), adapted to the system at hand. Since the Hamiltonian $\hat{H}_a$ corresponds to measurements of the amplitude of the single-mode field the relevant system observable is $\hat{X} = (\hat{a} + \hat{a}^\dagger)$ and one finds readily, with $\hat{\sigma}_-|g\rangle = 0$, [11, 12]

$$d|\psi(t)\rangle = \left\{\left[-\frac{i}{\hbar}\hat{H}_a - \frac{1}{2}\lambda_a\left(\hat{a}^\dagger\hat{a} - \langle\hat{a} + \hat{a}^\dagger\rangle\hat{a} + \frac{\langle\hat{a} + \hat{a}^\dagger\rangle^2}{4}\right)\right]dt \right.$$
$$\left. + \sqrt{\lambda_a}\left(\hat{a} - \frac{\langle\hat{a} + \hat{a}^\dagger\rangle}{2}\right)dw\right\}|\psi(t)\rangle , \qquad (6.49)$$

where the measurement strength $\lambda_a = g\tau$ can be controlled by varying the atomic transit time τ. The term proportional to λ_a on the right-hand side of Eq. (6.49) accounts for the measurement-induced dissipation of the intracavity field, a consequence of the absorption of photons by the successive probe atoms. The stochastic term proportional to $\sqrt{\lambda_a}$ and to the Wiener process dw describes the stochastic changes of the intracavity field about its expected value $\langle\hat{a} + \hat{a}^\dagger\rangle$ resulting from the measurement outcomes.

Dispersive Measurements In this second measurement scheme the atoms are far off-resonant from the optical field, and we assume that they are prepared in the coherent superposition $|+\rangle = (|e\rangle + |g\rangle)/\sqrt{2}$ of the ground and excited states before entering the resonator. Information on the intracavity field is then inferred from a change in the phase of the atomic state. The effect of the measurements on the optical field is now described by the stochastic Schrödinger equation

$$d|\psi_j(t)\rangle = \left\{\left[-\frac{i}{\hbar}\hat{H}_d - \frac{1}{2}\lambda_d\left(\hat{n} - \langle\hat{n}\rangle\right)^2\right]dt + \sqrt{\lambda_d}(\hat{n} - \langle\hat{n}\rangle)dw\right\}|\psi_j(t)\rangle,$$

$$(6.50)$$

where $\lambda_d = g_d^2\tau$ [13].

Like Eq. (6.49) this equation comprises two contributions, but the underlying physics that they describe accounts for the important differences in the back action of the two measurement schemes. Specifically, because the non-resonant atom–field coupling $\hat{V}_d = \hbar g_d\hat{a}^\dagger\hat{a}\hat{\sigma}_z$ of Eq. (6.48) is a QND interaction for the photon number $\hat{n}_a = \hat{a}^\dagger\hat{a}$, the dissipative channel of Eq. (6.49) is replaced by a number conserving term, resulting therefore in additional damping of the phase of the optical field.

Figure 6.2 shows several stochastic trajectories corresponding to these QND measurements. Each path corresponds to a different initial seed value for the random noise generator and traces an individual continuous QND measurement of $\hat{n}$. The estimated photon number fluctuates immediately after the measurement starts, but these fluctuations diminish as the measurement sequence proceeds. The distribution of final estimated values of $\langle\hat{n}\rangle$ reflects the stochastic nature of light, and the computer simulations confirm that the distribution of estimates of the photon number n tends to coincide with the initial photon number distribution.

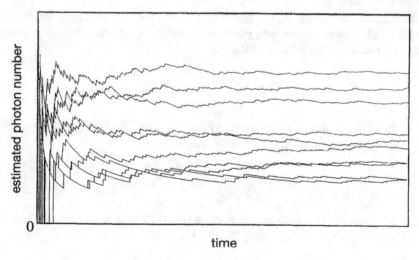

Fig. 6.2 Typical stochastic trajectories for the dispersive measurements of the photon number in a field initially in a coherent state. Arbitrary units. (Adapted from Ref. [13])

6.4   The Pointer Basis

We now return to a point briefly brought up in Sect. 6.2.2 and related to the importance of the environment on the measurement process. When discussing quantum measurements we indicated that the system to be measured is coupled to a detector, or probe, itself a quantum object, and that this probe is in turn coupled to the environment, a quantum object as well. So far we have however not really addressed the reason for this additional coupling.

The key point here is that the Schrödinger equation predicts a fully reversible time evolution, and yet a specific measurement results in a single outcome that is a classical quantity, "cast in stone." To quote John Wheeler [14] "No elementary quantum phenomenon is a phenomenon until it is a registered ('observed, indelibly recorded') phenomenon, brought to a close by an irreversible act of amplification." This brings up the fundamental question of what introduces classicality in an otherwise quantum world. If we assume that quantum physics is the fundamental theory of the Universe, then the Universe is itself a quantum object, and since it is a closed system (by definition of the Universe) its evolution must keep it in a pure state. Classicality must therefore be an emerging property, associated somehow with the ignorance associated with considering only subsystems of the full Universe. As we have seen in Chap. 5, the coupling of such subsystems to their environment can for all practical purposes introduce irreversibility in their otherwise reversible behavior, so it makes considerable operational sense to include them as the final element of the measurement process.

To further illustrate the difficulty associated with the lack of irreversibility when coupling a quantum system to a probe consider the simplest quantum system, a single two-state system initially in the state

$$|\psi\rangle_S = \frac{1}{\sqrt{2}}[|1\rangle + |0\rangle]_S , \qquad (6.51)$$

and a probe that is likewise a two-state system. Assume for concreteness that their interaction leaves the combined system in the state

$$|\psi\rangle = \frac{1}{\sqrt{2}}[|1\rangle_S |1\rangle_P + |0\rangle_S |0\rangle_P] \qquad (6.52)$$

that maps the state $|1\rangle_S$ of the system to the state $|1\rangle_P$ of the probe, and its state $|0\rangle_S$ to the state $|0\rangle_P$. The corresponding density operator is simply

$$\hat{\rho}_{SP} = \frac{1}{2}\big[|1, 1\rangle\langle 1, 1| + |1, 1\rangle\langle 0, 0| + |0, 0\rangle\langle 1, 1| + |0, 0\rangle\langle 0, 0| \big]_{SP} . \qquad (6.53)$$

We might (carelessly) decide to associate the mapping of the system to the probe with the diagonal elements of $\hat{\rho}_{SP}$ and the probabilities that they provide. However, this would be not just careless, but also incorrect. This is because we could just as well describe the state (6.52) in the $\{|+\rangle, |-\rangle\}$ basis of the system and probe, with $|\pm\rangle = \frac{1}{2}[\,|1\rangle \pm |0\rangle\,]$. It is easy to verify that the state of the combined system would then read

$$|\psi\rangle = \frac{1}{\sqrt{2}}[\,|+\rangle_S\,|+\rangle_P + |-\rangle_S\,|-\rangle_P\,]. \tag{6.54}$$

However, the diagonal elements of the corresponding density operator

$$\hat{\rho}_{SP} = \frac{1}{2}[\,|+, +\rangle\langle+, +| + |+, +\rangle\langle-, -| + |-, -\rangle\langle+, +| + |-, -\rangle\langle-, -|\,]_{SP} \tag{6.55}$$

are completely different from those of the density operator (6.53), although they both describe precisely the same state! Why then should we not associate the mapping of the system to the probe with the diagonal elements in the $\{|+\rangle, |-\rangle\}$ basis instead? There is nothing at this point that allows us to favor one representation over the other, and it is indeed incorrect in general to associate the diagonal elements of $\hat{\rho}$ with *classical* probabilities.

Since there are an infinite number of bases on which to expand the state of the system, there are also an infinite number of equivalent density operators that contain the same information, with different elements on their diagonal. There is therefore no unambiguous way to associate pure states with classical probabilities: In order for the measurement to produce a *classical* result, pure states must be somehow transformed into mixtures. This was already realized by J. von Neumann when he postulated the nonunitary "reduction of the wave function" in order to project pure states into an appropriate mixture that depends on the measurement being carried out.

As stated by W. H. Zurek in his analysis of the measurement problem [15, 16], "the role of measurements is to convert quantum states and quantum correlations into classical, definite outcomes." Since a unitary transformation is involved in the interaction between the system and the probe, and hence the purity of pure states is preserved, a way must be found to eliminate the maximum quantum information characteristic of that state, or perhaps more accurately, to disperse it to a place where it is not accessible to the detector. The considerable merit of the *pointer basis* approach is the additional role played by the environment: in addition to providing the loss of information and irreversible system dynamics that we are already familiar with, it plays in addition a fundamental role in determining a *preferred basis* for the detector. This pointer basis is immune to environmental decoherence and defines a classical measuring apparatus unambiguously.

To see how this works more concretely, let us include, in addition of the coupling of the system to the probe, a coupling of the probe to the environment, so that they become correlated either through an interaction $\hat{V}_{PE}^{(1)}$, resulting in a state of the form

$$|\psi\rangle = \frac{1}{\sqrt{2}}\left[|1, 1\rangle_{SP} \otimes |1\rangle_E + |0, 0\rangle_{SP} \otimes |0\rangle_E\right], \qquad (6.56)$$

or alternatively through an interaction $V_{PE}^{(2)}$ resulting in the state

$$|\varphi\rangle = \frac{1}{\sqrt{2}}\left[|+, +\rangle_{SP} \otimes |+\rangle_E + |-, -\rangle_{SP} \otimes |-\rangle_E\right]. \qquad (6.57)$$

Tracing over the environment results in the first case in the system–probe density operator

$$\hat{\rho}_{SP}^{(1)} = \frac{1}{2}[\,|1, 1\rangle\langle 1, 1| + |0, 0\rangle\langle 0, 0|\,]_{SP}, \qquad (6.58)$$

and in the second case to

$$\hat{\rho}_{SP}^{(2)} = \frac{1}{2}[\,|+, +\rangle\langle+, +| + |-, -\rangle\langle-, -|\,]_{SP}, \qquad (6.59)$$

a result reminiscent of the discussion of entanglement monogamy, see for instance Eq. (4.14). The coupling $V_{PE}^{(1)}$ of the probe to the environment results in the selection of the $\{|1\rangle, |0\rangle\}$ pointer basis after tracing over the environment, and hence in the unambiguous measurement of the classical probabilities to find the system in states $|1\rangle$ or $|0\rangle$. For the probe–environment coupling $V_{PE}^{(2)}$, in contrast, the resulting pointer basis is the set of states $\{|+\rangle, |-\rangle\}$, with the probe measuring then the classical probabilities of finding the system in either state $|+\rangle$ or $|-\rangle$. That is, while the system–probe coupling develops the quantum correlations needed to learn about the system, it is the coupling of the probe to the environment that develops the additional correlations that, when traced over the environment, select the observable measured by the probe.

The results and insight gained from this simple example can easily be generalized to any system–detector combination and to a variety of environments. Environments have typically a very large number of degrees of freedom, and their interaction with the probe must be described carefully, see in particular Refs. [15, 17]. The main conclusion of this analysis is that the detector–environment coupling should preserve the relevant diagonal elements of the system–detector density operator. This will occur only if it leaves the diagonal terms of the system–detector density operator invariant, that is, if it commutes with the projection operators that appear in that diagonal.

Calling $\hat{V}_{PE}$ the probe–environment coupling Hamiltonian and $\{|B_k\rangle\}$ a complete orthonormal set of desired pointer basis states, this implies that

$$\sum_k p_k \left[\hat{V}_{PE}, |B_k\rangle\langle B_k|\right] = 0, \qquad (6.60)$$

for all p_k real with $\sum_k p_k = 1$, that is, not surprisingly the operator $\sum_k |B_k\rangle\langle B_k|$ must be a quantum non-demolition operator with respect to its coupling to the environment.

Problems

Problem 6.1 Consider a system described by the Hamiltonian $\hat{H}(\hat{A}, \hat{B})$, where the operators $\hat{A}$ and $\hat{B}$ are conjugate variables. Can they be simultaneous QND observables of the system, and why?

Problem 6.2 Show that the quadratures

$$\hat{X}_1 = \sqrt{\frac{\hbar}{2m\omega}}\left[\hat{a}e^{i\omega t} + \hat{a}^\dagger e^{-i\omega t}\right] \; ; \; \hat{X}_2 = -i\sqrt{\frac{\hbar}{2m\omega}}\left[\hat{a}e^{i\omega t} - \hat{a}^\dagger e^{-i\omega t}\right]$$

are QND variables of the harmonic oscillator.

Problem 6.3 Carry out the steps that lead from Eq. (6.27) to Eq. (6.30),

$$\frac{\partial\hat{\rho}}{\partial t} = -\frac{i}{\hbar}[\hat{H}, \hat{\rho}] - \frac{1}{2\hbar^2}\sigma_M^2[\hat{X}[\hat{X}, \hat{\rho}]].$$

Problem 6.4 Consider a system consisting of a two-level atom driven by a classical field, and described in the rotating wave approximation by the Hamiltonian (1.60),

$$\hat{H} = \frac{1}{2}\hbar\omega_0\hat{\sigma}_z - d\left[\hat{\sigma}_+ E^+(\mathbf{R}, t) + \hat{\sigma}_- E^-(\mathbf{R}, t)\right].$$

For an atom initially in its excited state $|e\rangle$ determine the dynamics of this system under the combined effects of the classical field $E^+(\mathbf{R}, t)$ and of continuous measurements on the dynamics of this system, assuming that the state of the probe is not measured at the end of its interaction with the system. Show that the measurements result in the evolution of the system toward a stationary state, and determine both that state and the rate at which it is reached.

Problem 6.5 The quantum Zeno paradox— *This problem addresses the quantum Zeno paradox, the fact that the evolution of quantum systems can be inhibited by measuring it frequently. It was first discussed theoretically by E. C. G. Sudarshan and B. Misra [10] and first demonstrated experimentally by W. Itano and colleagues [18].*

We consider an observable $\hat{A}$ of some system described by the Hamiltonian $\hat{H}$ that is monitored by a series of instantaneous measurements, separated by an interval Δt, that put the system in one of the eigenstates $|\psi_m\rangle$ of $\hat{A}$. We assume that the measurements are "exact," in the sense that they barely induce transitions from one to some other eigenstate of $\hat{A}$ and do not affect the free evolution in the intervals between measurements.

(a) Assuming that the system is left in the eigenstate $|\psi_n\rangle$ after a first measurement, show that just after the next measurement its probability to still be in that state is, to lowest order in Δt,

$$p_n \approx 1 - \left(\frac{\Delta t}{\hbar}\right)^2 \sigma_{E_n}^2 ,$$

where $\sigma_{E_n}^2 = \left[\langle\psi_n|\hat{H}^2|\psi_n\rangle - (\langle\psi_n|\hat{H}|\psi_n\rangle)^2\right]$ is the variance in energy of the initial state $|\psi_n\rangle$.

(b) Show then that after sequence of k measurements over an interval $\tau = k\Delta t$ and the limit of continuous monitoring, $\Delta t \to 0$ and $k \to \infty$, the probability to still be in the initial state $|\psi_n\rangle$ at time τ becomes

$$p_n(\tau) \to \exp\left[-\left(\frac{\tau\Delta t}{\hbar^2}\right)^2 \sigma_{E_n}^2\right]$$

so that in the limit $\Delta t \to 0$ we have $p_n(\tau) = 1$, that is, the system remains in its initial state, no matter how long the total measurement time τ. This effect is known as the *quantum Zeno paradox* [10].

Hint: Remember the identity $e^x = \lim_{n\to\infty}(1 + x/n)^n$.

Problem 6.6 Carry out the steps that lead from Eq. (6.42) to the stochastic differential equation (6.44).

Problem 6.7 Show that for the absorptive measurement of a single-mode intra-cavity field amplitude $\hat{X} = \hat{a} + \hat{a}^\dagger$ the stochastic equation (6.44) results in the form (6.50),

$$d|\psi_j(t)\rangle = \left\{\left[-\frac{i}{\hbar}\hat{H}_d - \frac{1}{2}\lambda_d\left(\hat{n} - \langle\hat{n}\rangle\right)^2\right]dt + \sqrt{\lambda_d}(\hat{n} - \langle\hat{n}\rangle)dw\right\}|\psi_j(t)\rangle.$$

Problem 6.8 Write a program to simulate the stochastic equation of Problem 6.6 numerically for (a) a field initially in a coherent state $|\alpha\rangle$ and (b) a number state, both with $\langle n\rangle = 9$. Plot a sample of several characteristic resulting stochastic trajectories as a function of time.

References

1. V. Braginsky, Y.I. Voronstsov, K.S. Thorne, Quantum nondemolition measurements. Science **209**, 547 (1980)
2. V. Braginski, F.Ya. Khalili, *Quantum Measurement* (Cambridge University, Cambridge, 1992), ed. by K. Thorne
3. C.M. Caves, K.S. Thorne, R.W.P. Drever, V.D. Sandberg, M. Zimmerman, On the measurement of a weak classical force coupled to a quantum-mechanical oscillator. Rev. Mod. Phys. **52**, 341 (1980)
4. M. Tsang, C.M. Caves, Evading quantum mechanics: Engineering a classical subsystem within a quantum environment. Phys. Rev. X **2**, 031016 (2012)
5. K. Hammerer, M. Aspelmeyer, E.S. Polzik, P. Zoller, Establishing Einstein-Poldosky-Rosen channels between nanomechanics and atomic ensembles. Phys. Rev. Lett. **102**, 020501 (2009)
6. E. Polzik, K. Hammerer, Trajectories without quantum uncertainties. Ann. Phys. **527**, A15 (2015)
7. W. Wasilewski, K. Jensen, H. Krauter, J.J. Renema, M.V. Balabas, E.S. Polzik, Quantum noise limited and entanglement assisted magnetometry. Phys. Rev. Lett. **104**, 133601 (2010)
8. C.B. Møller, R.A. Thomas, G. Vasilakis, E. Zeuthen, Y. Tsaturyan, M. Balabas, K. Jensen, A. Schliesser, K. Hammerer, E.S. Polzik, Quantum back action evading measurement of motion in a negative mass reference frame. Nature **547**, 191 (2017)
9. K. Zhang, P. Meystre, W. Zhang, Back-action-free quantum optomechanics with negative-mass Bose-Einstein condensates. Phys. Rev. A **88**, 043632 (2013)
10. B. Misra, E.C.G. Sudarshan, The Zeno's paradox in quantum theory. J. Math. Phys. **18**, 756 (1977)
11. K. Jacobs, D.A. Steck, A straightforward introduction to continuous quantum measurement. Contemp. Phys. **47**, 279 (2006)
12. N. Imoto, M. Ueda, T. Ogawa, Microscopic theory of the continuous measurement of photon number. Phys. Rev. A **41**, 4127(R) (1990)
13. M. Ueda, N. Imoto, N. Nagaoka, T. Ogawa, Continuous quantum non-demolition measurement of photon number. Phys. Rev. A **46**, 2859 (1992)
14. J.A. Wheeler, Law without Law, in *Quantum Theory and Measurement*, ed. by J.A. Wheeler, W.H. Zurek (Princeton University, Princeton, 1983), p. 182
15. W.H. Zurek, Pointer basis of quantum apparatus: Into what mixture does the wave packet collapse? Phys. Rev. D **24**, 1516 (1981)
16. W.H. Zurek, Decoherence and the transition from quantum to classical. Los Alamos Sci. **27**, 2 (2002)
17. W.H. Zurek, Environment induced superselection rules. Phys. Rev. D **26**, 1862 (1982)
18. W. Itano, D. Heinzen, J. Bollinger, D. Wineland, Quantum Zeno effect. Phys. Rev. A **41**, 2295 (1990)

Chapter 7
Tailoring the Environment—Cavity QED

It is possible to create electromagnetic environments where spontaneous emission can be enhanced or inhibited, or even becomes reversible. This is the domain of cavity QED. Following a review of its basic aspects we discuss the micromaser, where a tailored environment allows to produce strongly nonclassical radiation. We also consider the off-resonant situation where a single atom operates as a dispersive medium. We then transition to the quantization of LC electric circuits and superconducting qubits, which permit to extend cavity QED to the emerging area of many-body circuit QED and the possibility to reach extremely high atom–field couplings. The chapter concludes with a brief introduction to the Casimir effect.

As apparent from the previous chapters, quantum optics deals with the dynamics of coupled quantum systems that can be as small as one or two atoms or a single mode of the electromagnetic field, or very large like an electromagnetic continuum of modes, an atomic vapor, or perhaps a solid in which a handful of qubits are embedded. Their coupling may be under superb control, as for instance in quantum cryptography experiments, when trying to detect extraordinarily faint signals, or when making measurements of exquisite precision. In other cases it is much less so, in particular when dealing with the unavoidable coupling of a system to its environment, a coupling that can also lead as we have seen to the emergence of classical dynamics.

As early as 1946, E. M. Purcell et al. [1] observed an increased spontaneous emission rate of several orders of magnitude when an atom was surrounded by a cavity tuned to the transition frequency of the atom. This observation was later further elaborated upon by D. Kleppner [2], who discussed theoretically the possibilities of enhanced and inhibited spontaneous emission. With hindsight the fact that controlling the rate of spontaneous emission should be possible is not surprising: the Weisskopf–Wigner theory of Sect. 5.1 shows that the spontaneous decay rate of an atom depends explicitly on the mode density of the electromagnetic field. It follows that it must be possible to modify it by manipulating the density of modes, that is, by controlling the electromagnetic environment. Still, it took a

© The Author(s), under exclusive license to Springer Nature Switzerland AG 2021
P. Meystre, *Quantum Optics*, Graduate Texts in Physics,
https://doi.org/10.1007/978-3-030-76183-7_7

few more years before experimental capabilities permitted to achieve this goal in
a controlled way, first in the microwave regime, and later in optical resonators,
opening up the field of cavity quantum electrodynamics, or cavity QED. Progress
in this direction has now reached the point where it is even possible to create
electromagnetic environments where spontaneous emission becomes reversible, the
environment ceasing to behave as a Markovian reservoir where energy is forever
dissipated.

Following a review of the basic aspects of cavity QED this chapter discusses
in some detail the micromaser, where a tailored environment allows to produce
radiation that is strongly nonclassical, in contrast to the situation in conventional
lasers. We also consider the off-resonant situation where a single atom operates as a
dispersive medium that can for instance be exploited to generate optical Schrödinger
cats, or perhaps more accurately Schrödinger kitten. We then transition to the use of
"artificial atoms" in the form of superconducting qubits that permit to extend cavity
QED to the emerging area of circuit QED. These systems raise the possibility to
reach atom–field coupling constants in the ultrastrong coupling regime introduced
in Sect. 3.6, in addition to offering much promise for quantum information science
and technology applications.

The chapter concludes with an introduction to the Casimir effect. In 1948 H.
Casimir realized that the vacuum forces are dependent on the geometry of the
system in which they are contained, resulting in particular in an attractive force
between two conducting plates facing each other due to the simple presence of
the electromagnetic vacuum. An important physical quantity when discussing the
Casimir force is the radiation pressure exerted from light on massive objects. As
such this force, in addition to being arguably the simplest manifestation of a tailored
vacuum, also provides a natural bridge to the discussion of the mechanical effects
of light that comprise much of the next three chapters.

For a much more complete presentation of cavity QED the reader is referred to
Chap. 5 of "Exploring the Quantum—Atoms, Cavities and Photons," by S. Haroche
and J.-M. Raimond [3], a wonderful text that covers this topic in considerably more
detail and depth than can be done in this short chapter.

7.1   Enhanced and Inhibited Spontaneous Emission

7.1.1   Master Equation for the Atom–Cavity System

A simple theoretical model that permits to identify the three main regimes of
spontaneous emission in tailored environments consists of a single two-level atom of
transition frequency ω_0 trapped in an open-sided Fabry–Pérot cavity and effectively
interacting with a single cavity mode of frequency $\omega \approx \omega_0$. This situation can
be realized if the cavity length L is short enough that the axial mode frequency
separation $c/2L$ is large compared to the dipole coupling frequency between the

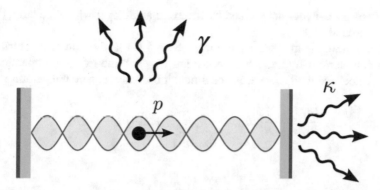

Fig. 7.1 Fabry-Pérot cavity with a standing wave mode of frequency ω interacting with an atom of center-of-mass momentum **p**. The rates γ and κ account for atomic losses into the free space background and cavity losses due to imperfect mirrors and diffraction, respectively

atom and that mode, their detuning $|\Delta| = |\omega_0 - \omega|$, as well as the cavity decay rate—or, in the microwave regime, if $c/2L$ is large compared to ω and ω_0.

Under these conditions, and neglecting the effects of transverse modes for now,[1] the subsystem consisting of the atom and field mode can then be described by the Jaynes–Cummings Hamiltonian (3.1) with spatially dependent dipole coupling

$$g(z) = -(d\mathcal{E}_\omega/2\hbar)\cos Kz, \tag{7.1}$$

where z is the position of the atom along the resonator axis.

As we have already discussed, it is impossible in practice to perfectly isolate it from the external world. In contrast to the situations that we have encountered so far, it should be thought of as interacting with (at least) two reservoirs: The first one is associated with the coupling of the atom to the free space electromagnetic background through the sides of the resonator, and the second one to the coupling of the cavity mode to the outside world via mirror losses and diffraction, as illustrated in Fig. 7.1. The first process is important in open cavities and cannot be ignored then. It results in an incoherent decay of the excited atomic state *a la* Weisskopf–Wigner, but at a rate γ that can be significantly different from the free space rate Γ of Eq. (5.18). The second reservoir accounts for the irreversible escape of cavity photons.

Due to the additive nature of the two decay mechanisms the single reservoir master equation of Chap. 5 can readily be expanded to a form whose non-Hermitian component now comprises two contributions. The first one, given by Eq. (5.57), accounts for the dissipation of the cavity field mode at rate κ and the second one, given by Eq. (5.66), for a spontaneous emission of the atom at rate γ into the subset

---

[1] We will return to this important point later in this section.

of the free space modes determined by the solid angle over which the atom "sees" that background.

Decomposing the master equation as in Sect. 5.4.1 as the sum of a contribution from a non-Hermitian effective Hamiltonian $\hat{H}_{\text{eff}}$ and the two jump operators $\hat{\mathcal{L}}_{\kappa}$ and $\hat{\mathcal{L}}_{\gamma}$ associated with these decay channels it becomes, at zero temperature,

$$\frac{d\hat{\rho}_s}{dt} = -\frac{i}{\hbar}[\hat{H}_{\text{eff}}\hat{\rho}_s - \hat{\rho}_s\hat{H}_{\text{eff}}^{\dagger}] + \hat{\mathcal{L}}_{\kappa}[\hat{\rho}] + \hat{\mathcal{L}}_{\gamma}[\hat{\rho}]. \tag{7.2}$$

Here

$$\hat{H}_{\text{eff}} = \hat{H}_s + \hat{H}_{\text{loss}}, \tag{7.3}$$

where $\hat{H}_s$ is the Jaynes–Cummings Hamiltonian (3.1) and

$$\hat{H}_{\text{loss}} = -\frac{i\hbar}{2}[\gamma\hat{\sigma}_+\hat{\sigma}_- + \kappa\hat{a}^{\dagger}\hat{a}]. \tag{7.4}$$

The Liouvillian

$$\hat{\mathcal{L}}_{\kappa}[\hat{\rho}] = \kappa\hat{a}\hat{\rho}_s\hat{a}^{\dagger} \tag{7.5}$$

accounts for quantum jumps associated with the dissipation of the cavity field mode at rate κ and

$$\hat{\mathcal{L}}_{\gamma}[\hat{\rho}] = \gamma\hat{\sigma}_-\hat{\rho}_s\hat{\sigma}_+ \tag{7.6}$$

accounts for those associated with the spontaneous decay of the excited atomic state $|e\rangle$.

We are interested in the spontaneous emission of the excited atom in the cavity environment, or more precisely perhaps, in the dynamics of the small atom–cavity mode subsystem initially in the state

$$|\psi(0)\rangle = |e, 0\rangle \tag{7.7}$$

and coupled to its two reservoirs. As we have seen in the discussion of the Jaynes-Cummings model, in the absence of dissipation, the total number of excitations in the atom–field mode subsystem, one in the present case, remains constant. However, the coupling to the reservoirs involves the loss of excitation from the small system. Consequently there are now three relevant states involved in its dynamics: the "one-quantum" states $|e, 0\rangle$ and $|g, 1\rangle$, and the "zero-quantum" state $|g, 0\rangle$.

Following the Monte Carlo wave function approach of Sect. 5.4 we proceed by introducing the *unnormalized* one-quantum state

$$|\psi(t)\rangle = C_e\, e^{-i\Delta/2}|e, 0\rangle + C_g\, e^{i\Delta/2}|g, 1\rangle, \tag{7.8}$$

where $\Delta = \omega_0 - \omega$. It is governed by the effective non-Hermitian Schrödinger equation

$$i\hbar \frac{d|\psi(t)\rangle}{dt} = \hat{H}_{\text{eff}}|\psi(t)\rangle, \tag{7.9}$$

which describes the evolution of the system within the one-excitation manifold. In addition, the transitions to the zero-quantum state $|g, 0\rangle$ are driven by the quantum jumps

$$|\psi\rangle \rightarrow \sqrt{\kappa}\,\hat{a}|\psi\rangle \quad \text{or} \quad |\psi\rangle \rightarrow \sqrt{\gamma}\,\hat{\sigma}_-|\psi\rangle \tag{7.10}$$

for events resulting in cavity field mode and atomic energy dissipation, respectively.

The Schrödinger-like equation (7.9) yields the probability amplitude equations of motion

$$\frac{dC_e(t)}{dt} = -(\gamma/2)C_e(t) - igC_g(t), \tag{7.11}$$

$$\frac{dC_g(t)}{dt} = (i\Delta - \kappa/2)C_g(t) - igC_e(t). \tag{7.12}$$

At this point it is useful to distinguish between two qualitatively different regimes. In the first one the irreversible decay rates κ and γ dominate over the dipole interaction between the atom and the cavity mode, whose strength is given by g. This is traditionally called the *weak coupling regime*, or bad cavity limit. In contrast, the *strong coupling regime*, or good cavity limit, is characterized by the fact that the coherent interaction between the atom and the cavity mode dominates over the irreversible decay mechanisms. In the closed superconducting cavities sometimes used in microwave experiments we have $\gamma \simeq 0$, so that the strong coupling regime corresponds to $g \gg \kappa$ and the weak coupling regime to $g \ll \kappa$. In contrast, most optical cavities encompass only a small fraction of the free space solid angle 4π, so that $\gamma \lesssim \Gamma$. In this case, the strong coupling regime corresponds to $g \gg \{\Gamma, \kappa\}$ and the weak coupling regime to $g \ll \{\Gamma, \kappa\}$.

7.1.2 Weak Coupling Regime

Formally integrating Eq. (7.12) gives readily

$$C_g(t) = -ig \int_0^t dt'\, C_e(t')e^{(i\Delta - \kappa/2)(t-t')}. \tag{7.13}$$

Assuming consistently with the weak coupling condition $g \ll \{\Gamma, \kappa\}$ that $C_e(t')$ varies slowly compared to $1/(|\Delta| + |\kappa|/2)$ it can be taken outside the integral and evaluated at $t' = t$, a step that we recognize from the discussion of the Markov approximation. That is, all memory effects in its evolution are effectively washed out by the dissipation of the intracavity field and the evolution of the one-excitation manifold becomes Markovian. From Eqs. (7.11) and (7.12), this requires that g and γ are small compared to $|\Delta| + |\kappa|/2$. The remaining integral gives, for $t \gg \kappa^{-1}$,

$$C_g(t) = \frac{ig}{i\Delta - \kappa/2} C_e(t) \,, \tag{7.14}$$

and, after substitution of this expression into Eq. (7.11),

$$\frac{dC_e(t)}{dt} = -\left[(\gamma/2) + \frac{g^2(\kappa/2 + i\Delta)}{\Delta^2 + \kappa^2/4}\right] C_e(t) \,. \tag{7.15}$$

Hence, the upper electronic state population $p_e(t)$ undergoes an exponential decay at the rate

$$\gamma_{\text{eff}} = \gamma + \gamma_c \,, \tag{7.16}$$

where the term

$$\gamma_c = \left(\frac{2g^2}{\kappa}\right) \frac{1}{1 + (2\Delta/\kappa)^2} \tag{7.17}$$

accounts for a contribution to the upper atomic state damping from the dissipation of the intracavity field. In the free space limit $\kappa \to \infty$ and $\gamma \to \Gamma$, γ_{eff} reduces as it should to the Weisskopf–Wigner result Γ.

Closed Cavity Consider now the case of a closed cavity, $\gamma = 0$, so that $\gamma_{\text{eff}} = \gamma_c$. Assume also that the atom is at an antinode of the field mode, so that $\cos(Kz) = 1$. Expressing the dipole matrix element g in terms of the free space decay rate Γ using Eq. (5.18) with $\mathcal{E}_\omega = \sqrt{\hbar\omega/\epsilon_0 V}$ and introducing the *quality factor* of the resonator $Q \equiv \omega/\kappa$ we find at resonance $\Delta = 0$

$$\gamma_c \to \gamma_{\max} = \frac{3Q}{4\pi^2}\left(\frac{\lambda^3}{V}\right)\Gamma \,, \tag{7.18}$$

where $\lambda = 2\pi c/\omega$. The enhancement factor

$$\frac{\gamma_{\max}}{\Gamma} = \frac{3Q}{4\pi^2}\left(\frac{\lambda^3}{V}\right) \tag{7.19}$$

is called the *Purcell factor*. For sufficiently high quality factors Q and transition wavelengths comparable to the cavity size, this expression predicts a considerable *enhancement* of the spontaneous emission rate as compared to its free space value.

This is essentially the effect observed by E. M. Purcell and first quantitatively verified by R. Goy et al. [4].

Equation (7.17) also predicts an *inhibition* of spontaneous emission for atoms far detuned from the cavity resonance frequency ω. For instance, for $|\Delta| = \omega$ we have for $Q \gg 1$

$$\gamma_c \simeq \gamma_{max} \left(\frac{1}{4Q^2} \right) = \frac{3}{16\pi^2 Q} \left(\frac{\lambda^3}{V} \right) \Gamma \ . \tag{7.20}$$

For large quality factors it is therefore possible to almost completely switch off spontaneous emission, an effect first demonstrated by R. Hulet and colleagues [5].

We note that Eqs. (7.18) and (7.20) seem to imply that a transition wavelength comparable to the cavity size is necessary to obtain a significant enhancement or inhibition of spontaneous emission. This turns out to be incorrect, however, and results from an oversimplified description of the cavity modes that neglects transverse effects. In particular, in the case of a confocal resonator of length L and for Gaussian modes of waist $w_0 = \sqrt{L\lambda/\pi}$ the possible wavelengths are given by $L = (q + 1/4)\lambda/2$, where q is an integer, and the mode volume is $v = \pi w_0^2 L/4 = (q + 1/4)^2 \lambda^3/16$. In this case, the wavelength dependence in Eqs. (7.18) and (7.20) largely disappears, demonstrating that wavelength-size cavities are not required in general to observe enhanced or inhibited spontaneous emission.

Propagation Effects It may be useful at this point to make a general comment on the dependence of the spontaneous emission rate on the cavity density of modes. The mode structure depends on the boundary conditions imposed by the cavity, and one may wonder how the atom can initially "know" that it is inside a cavity rather than in free space. Is there some instantaneous action at a distance involved, and if not, what is the mechanism through which the atom learns about its environment? The single-mode theory presented in this section does not permit to answer this question, since it cannot account for the propagation of wave packets along the cavity axis. Using a proper multimode theory, J. Parker and C. Stroud [6] and R. J. Cook and P. W. Milonni [7] showed that there is a simple answer to that question. In a real cavity, the initially excited atom starts to decay while radiating a wave packet in the form of a multimode field that propagates away from it. Eventually, this field encounters the cavity walls, which reflect it. The reflected field acts back on the atom, carrying information about the cavity walls as well as about the state of the atom itself at earlier times. Depending upon the phase of this field relative to that of the atomic polarization, it will either accelerate or prevent the further atomic decay. But for times shorter than the transit time between the atom and the cavity walls and back, it always decays at its free space rate.

7.1.3　Strong Coupling Regime

We now turn to the strong coupling regime, characterized by the fact that the coupling g between the atom and the cavity mode is now large enough for a photon emitted into the cavity to have a significant probability of being reabsorbed before it escapes the resonator. To analyze this regime we first consider the general solution of Eqs. (7.11) and (7.12) for arbitrary values of γ, κ, and g,

$$C_e(t) = C_{e1}\, e^{\alpha_1 t} + C_{e2}\, e^{\alpha_2 t} \,, \qquad (7.21)$$

where

$$\alpha_{1,2} = \frac{1}{2}\left(\frac{\gamma}{2} + \frac{\kappa}{2} - i\Delta\right) \pm \frac{1}{2}\left[\left(\frac{\gamma}{2} + \frac{\kappa}{2} - i\Delta\right)^2 - 4g^2\right]^{1/2} \qquad (7.22)$$

and the constants C_{e1} and C_{e2} are determined from the initial conditions $C_e(0) = 1$ and $C_g(0) = 0$.

In the strong coupling regime $g \gg \gamma$, κ these exponents reduce to

$$\alpha_{1,2} = -\frac{1}{2}\left(\frac{\gamma}{2} + \frac{\kappa}{2} - i\Delta\right) \pm ig \,, \qquad (7.23)$$

and the amplitude of their imaginary part is much larger than that of the real part. As a result the evolution of the upper state population consists now of slowly decaying oscillations at the vacuum Rabi frequency $2g$, see Fig. 7.2. These *vacuum Rabi oscillations* were first observed in the microwave regime by M. Brune et al. [8]. In this regime, the spectrum of spontaneous emission consists of a doublet of Lorentzian lines of equal widths $(\gamma + \kappa)/4$ and split by the vacuum Rabi frequency $2g$, rather than the familiar Lorentzian associated with free space exponential decay.

Fig. 7.2 Atomic excited state probability as a function of time, in dimensionless units. The dashed exponentially decaying curve is for the weak coupling regime, and the solid damped oscillations correspond to the strong coupling regime

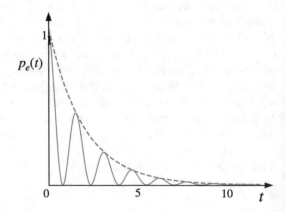

A simple physical interpretation of that spectrum can be obtained from the Jaynes–Cummings dressed energy spectrum (3.5) of the atom–cavity mode system,

$$E_{1n} = \hbar(n + \tfrac{1}{2})\omega + \hbar\Omega_n$$

$$E_{2n} = \hbar(n + \tfrac{1}{2})\omega - \hbar\Omega_n \,, \qquad (7.24)$$

where

$$\Omega_n = \frac{1}{2}\sqrt{\Delta^2 + 4g^2(n + 1)} \,. \qquad (7.25)$$

In the case of spontaneous emission the atom–cavity system is initially in the one-quantum manifold, and there are only two allowed transitions, $|1, 0\rangle \rightarrow |g, 0\rangle$ and $|2, 0\rangle \rightarrow |g, 0\rangle$. The frequencies of these transitions are $-\Delta/2 \pm 2g$, consistent with the result of Eq. (7.23).

7.2 The Micromaser

So far we have considered the spontaneous decay of a single atom at rest inside the optical or microwave resonator. An interesting extension consists in injecting a beam of atoms transversally through the cavity, at a rate low enough that only one atom at a time is present inside the resonator—at least if one wishes to avoid collective effects such as sub- or superradiance, see Fig. 7.3. This situation is reminiscent of the discussions of Sect. 3.5, where a stream of atoms was used to gain information on an intracavity field, and of Sect. 6.3.4, where this measurement scheme was improved upon by using weak continuous measurements of the field, again using a stream of atoms as probes.

Our present goal is however different: rather than using the atoms as field sensors we now exploit them to build a specific intracavity field in the presence of weak dissipation. We have seen in the previous section that in the strong coupling regime atoms injected in their upper state $|e\rangle$ undergo damped vacuum Rabi oscillations. One can therefore expect that if the time the atoms remain inside the resonator and the rate at which they are injected are just right, they will tend to deposit their energy inside the resonator and build the intracavity field. As such this system is reminiscent of traditional lasers or masers, but with the field built up "one photon at a time." Since in this system only one atom at a time is inside the resonator, instead of the vast numbers characteristic of usual lasers and masers, this system is called a "one-atom maser," or *micromaser*. This is a deceptively simple idea, but not surprisingly its experimental realization, first achieved by H. Walther and collaborators [9], is much less so.

Theoretical Model If the cavity transit time of the individual atoms is much shorter than both the spontaneous emission time $1/\gamma$ and the cavity mode decay time $1/\kappa$,

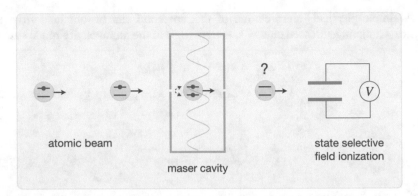

atomic beam

maser cavity

state selective
field ionization

Fig. 7.3 Schematic of a micromaser, with a beam of two-level atoms traversing a high-Q single-mode microwave cavity at a rate such that at most one atom is present in the cavity at a time. If desired, information on the state of the field can be inferred from atomic state measurements after they exit the resonator, in many cases in the form of state-selective field ionization. This approach is often favored in microwave cavity QED due to the absence of good single-photon detectors in that wavelength regime

one can separate the evolution of the atom–field system into alternating intervals, where an interval during which an atom is inside the resonator and the dynamics is governed by the Jaynes–Cummings Hamiltonian is followed by a period where the resonator is empty and the field dynamics follows the master equation (5.57) for a damped harmonic oscillator.

Consider then the evolution of the system, starting from the time t_i when the ith atom enters the cavity. At that instant the atom and the field are uncorrelated, so that the state of the combined atom–field system is

$$\hat{\rho}_{af}(t_i) = \hat{\rho}_a(t_i) \otimes \hat{\rho}_f(t_i) \,, \tag{7.26}$$

with $\hat{\rho}_a(t_i)$ and $\hat{\rho}_f(t_i)$ the atomic and field mode density operators, respectively. After a transit time τ the atom exits the resonator, leaving the field in the state

$$\rho(t_i + \tau) = \mathrm{Tr}_{\mathrm{atom}}\{\hat{U}(\tau)\hat{\rho}_{af}(t_i)\hat{U}^\dagger(\tau)\} \equiv \hat{F}(\tau)[\hat{\rho}(t_i)] \,, \tag{7.27}$$

where $\hat{U}(t) = \exp(-i\hat{H}t/\hbar)$ and $\hat{H}$ is the Jaynes–Cummings Hamiltonian (3.1). This equation also defines the operator $\hat{F}(\tau)$, which we use in the following to simplify the notation.

Importantly, in contrast with the repeated field measurements of Sects. 3.5 and 6.3.4, or the Schrödinger cats generation to be discussed in Sect. 7.3, we focus here on the situation where the internal state of the successive atoms is not measured after they exit the resonator. Between $t_i + \tau$ and the time t_{i+1} at which the next atom is injected, the field then simply dissipates energy at rate $\kappa = \omega/Q$, with its density operator governed by the master equation (5.57)

$$\frac{d\hat{\rho}}{dt} \equiv \hat{\mathcal{L}}[\hat{\rho}] = -\frac{\omega}{2Q}(n_{\text{th}} + 1)[\hat{a}^\dagger \hat{a}\hat{\rho}(t) - \hat{a}\hat{\rho}(t)\hat{a}^\dagger]$$

$$-\frac{\omega}{2Q}n_{\text{th}}[\hat{\rho}(t)\hat{a}\hat{a}^\dagger - \hat{a}^\dagger\hat{\rho}(t)\hat{a}] + \text{h.c.} \qquad (7.28)$$

If waiting long enough it would reach a thermal steady state with a temperature-dependent mean number n_{th} of thermal photons[2] so clearly the next atom needs to be injected at a time t_{i+1} before this can happen. At that time the field density operator is simply given by

$$\hat{\rho}(t_{i+1}) = \exp(\hat{\mathcal{L}}t_p)\hat{F}(\tau)\hat{\rho}(t_i). \qquad (7.29)$$

More concretely, if the successive atoms enter the cavity in their excited state $|e\rangle$ and the field is initially diagonal in energy, the reduced field density operator is easily verified to remain diagonal at all times and it is sufficient to concentrate on its diagonal elements $p_n = \langle n|\hat{\rho}_f|n\rangle$. The atom–field density operator as the ith atom enters the cavity is then

$$\rho_{af}(t_i) = |e\rangle\langle e| \otimes \sum_n p_n(t_i)|n\rangle\langle n|, \qquad (7.30)$$

and in the resonant case $\omega = \omega_0$ the reduced field density operator becomes simply, with Eqs. (3.20) and (3.21),

$$\hat{\rho}(t_i + \tau) = \sum_n p_n(t_i)\cos^2(g\sqrt{n}\tau)|n\rangle\langle n| + \sin^2(g\sqrt{n+1}\tau)|n+1\rangle\langle n+1|.$$

$$(7.31)$$

The diagonality of the field is preserved during its decay, so that the master equation (7.28) can be restricted to its diagonal elements

$$\frac{dp_n}{dt} = -\kappa(n_{\text{th}} + 1)[np_n - (n+1)p_{n+1}] - \kappa n_{\text{th}}[(n+1)p_n - np_{n-1}]. \qquad (7.32)$$

---

[2]Instead of labeling the mean number of thermal photons by $\bar{n}$ as in Chap. 5, we use here the less compact but more descriptive notation n_{th}, while "$\langle n \rangle$" denotes the average photon numbers over the successive atoms driving the micromaser.

Under these conditions, successive iterations of the Jaynes–Cummings and field dissipation sequence eventually yield a diagonal steady-state field density matrix ρ_{st}, which is the solution of this equation with $\hat{\rho}(t_{i+1}) = \hat{\rho}(t_i)$.[3]

As a final step we assume that the atoms enter the cavity according to a Poisson process with mean spacing $1/R$ between events, where R is the atomic flux, and average over the random times t_p between events. Since $\hat{\rho}(t_i)$ depends only on earlier time intervals, it is statistically independent of the current $\exp(\hat{\mathcal{L}}t_p)$, and we can factor the average of $\hat{\rho}(t_{i+1})$ as

$$\langle \hat{\rho}(t_{i+1}) \rangle = \langle \exp(\hat{\mathcal{L}}t_p) \rangle \hat{F}(\tau) \langle \hat{\rho}(t_i) \rangle$$

$$= R \int_0^\infty dt_p \exp[-(R - \hat{\mathcal{L}})t_p] \hat{F}(\tau) \langle \hat{\rho}(t_i) \rangle$$

$$= \frac{R}{R - \hat{\mathcal{L}}} \hat{F}(\tau) \langle \hat{\rho}(t_i) \rangle . \tag{7.33}$$

In the last step we have averaged the damping operator $\exp(\hat{\mathcal{L}}t_p)$ over an exponential distribution of intervals between atoms with average injection rate R. In steady state $\langle \hat{\rho}(t_{i+1}) \rangle = \langle \hat{\rho}(t_i) \rangle \equiv \bar{\rho}$, so that

$$R[1 - \hat{F}(\tau)]\bar{\rho} = \hat{\mathcal{L}}\bar{\rho} . \tag{7.34}$$

This leads, with Eqs. (7.31) and (7.32) and after some straightforward algebra, to

$$\bar{p}_n = \frac{n_{th}\kappa + R \sin^2(g\sqrt{n}\tau)}{(n_{th} + 1)\kappa} \bar{p}_{n-1} , \tag{7.35}$$

that is,

$$\bar{p}_n = \bar{p}_0 \prod_{k=1}^n \frac{n_{th}\kappa + R \sin^2(g\sqrt{k}\tau)}{(n_{th} + 1)\kappa}$$

$$= \bar{p}_0 \left(\frac{n_{th}}{1 + n_{th}}\right)^n \prod_{k=1}^n \left[1 + \left(\frac{N}{n_{th}}\right)\sin^2(\sqrt{k/N}\Theta)\right] \tag{7.36}$$

with $\bar{p}_0$ determined by the normalization condition $\sum_n \bar{p}_n = 1$. In the second line we have introduced the dimensionless parameters

$$N = R/\kappa , \tag{7.37}$$

---

[3]Note that this is not a "true" steady state, but rather a stroboscopic steady state. Physically, it corresponds to a situation where the same field state repeats at the precise instants when successive atoms exit the cavity.

which is the number of atoms injected by cavity decay time $1/\kappa$, as well as the effective "pump parameter"

$$\Theta = \sqrt{N}g\tau . \tag{7.38}$$

Features of the Photon Statistics Since the intracavity field always remains diagonal, the photon statistics (7.36) contain all information about the single-time statistical properties of the steady-state field reached by the micromaser.

The left side of Fig. 7.4 shows the average photon number

$$\langle n \rangle = \sum_n n\bar{p}_n , \tag{7.39}$$

normalized to N as a function of the pump parameter Θ. Figure 7.4a, c corresponds to $N = 20$ and 200, respectively, with a number of thermal photons $n_{th} = 0.1$. A feature common to all cases is that $\langle n \rangle$ is nearly zero for small Θ, but a finite value of $\langle n \rangle$ emerges at the threshold value $\Theta = 1$. For Θ increasing past this point, $\langle n \rangle$ first grows rapidly but then decreases to reach a minimum at about $\Theta \simeq 2\pi$, where

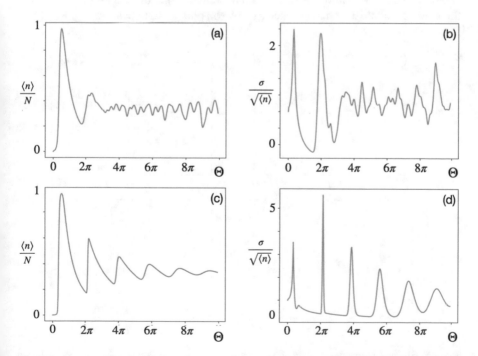

Fig. 7.4 Average photon number $\langle n \rangle$, normalized to N, as a function of the pump parameter Θ for (**a**) $N = 20$ and (**c**) $N = 200$, and corresponding normalized standard deviation $\sigma/\sqrt{\langle n \rangle} = [(\langle n^2 \rangle - \langle n \rangle^2)/\langle n \rangle]^{1/2}$ for (**b**) $N = 20$ and (**d**) $N = 200$. The number of thermal photons is $n_{th} = 0.1$

the field abruptly jumps to a higher intensity. This general behavior recurs roughly at integer multiples of 2π, although it becomes less pronounced for increasing Θ. Finally, a stationary regime with $\langle n \rangle$ nearly independent of Θ is reached. Outside the time scale of the figure there is an additional structure somewhat reminiscent of the Jaynes–Cummings revivals.

The number and sharpness of the features in the mean photon number depend on N. At the onset of the field around $\Theta = 1$, $\langle n \rangle$ is essentially independent of N for $N \gg 1$, but the subsequent transitions become sharper for increasing N although the micromaser remains largely dominated by thermal noise for small N, as illustrated in Fig. 7.4a, c. In the limit $N \rightarrow \infty$, this behavior hints at an interpretation of the first transition, the analog of the threshold in conventional lasers, in terms of a continuous phase transition, while subsequent transitions near $\Theta \approx 2n\pi$ are similar to first-order phase transitions [10]. As expected these transitions are characterized by the onset of bimodal photon statistics, as illustrated for the case of the first such transition in Fig. 7.5. They are also responsible for the sharp peaks in the normalized standard deviation $\sigma / \sqrt{\langle n \rangle} = [(\langle n^2 \rangle - \langle n \rangle^2) / \langle n \rangle]^{1/2}$ of $\langle n \rangle$, which is shown in Fig. 7.4b, d for the two cases $N = 20$ and $N = 200$.

Both the photon statistics and its second moment show that the micromaser field is characterized by features alien to ordinary single-mode masers and lasers, which far above threshold are characterized by Poisson photon statistics, see e.g. Ref. [11].

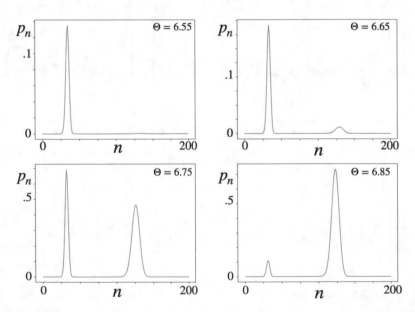

Fig. 7.5 Changes in the steady-state photon statistics $\bar{p}_n$ for $N = 200$ and $\bar{n} = 0.1$, as the pump parameter Θ is varied across the values near $\Theta = 2\pi$ where the micromaser undergoes its first phase-like transition, and $\bar{p}_n$ acquires a bimodal nature. The values of Θ corresponding to the cases just below, at, and just above the transition are indicated on the respective plots

The field is typically strongly "nonclassical," where by classical we mean a field with a positive definite $P(\alpha)$ distribution. And it has no particular tendency, even far above threshold, of being Poissonian, in which case we would have as we have seen $\sigma = \sqrt{\langle n \rangle}$.

These differences with conventional lasers originate from the fact that the micromaser possesses *less* stochasticity and noise than macroscopic masers and lasers, for which the atom–field interaction is terminated by exponential atomic decays rather than a transit time. As a result, the coherence of the quantum mechanical light-matter interaction is averaged over in conventional lasers, and the purely quantum mechanical features appearing in micromasers are largely lost.

Temperature Dependence The temperature dependence of the micromaser steady state illustrates particularly clearly the difference between the effects of quantum and thermal noise. To see this, note that the n-dependence of the quantum Rabi oscillations implies the existence of number states $|n_q\rangle$ that cause successive atoms to experience $2q\pi$ pulses, where q is an integer, during their transit time τ through the cavity. At resonance $\omega = \omega_0$ this happens for $g(n_q + 1)^{1/2}\tau = q\pi$. That is, for these states the atom entering the resonator in its excited state will exit it in that same state, thereby prohibiting the growth of the cavity field past them.

Competing with the blocking effect of these *trapping states* [12] is dissipation, which leads to an incoherent transfer of population both up and down the ladder of states of the cavity mode. Hence, thermal fluctuations allow the micromaser to jump past the trapping states and rapidly wash out their effect. In the limit $T \to 0$, however, $n_{\text{th}} \to 0$ and Eq. (7.32) reduces to

$$\dot{p}_n = -\frac{\omega}{Q}[np_n - (n+1)p_{n+1}],\qquad(7.40)$$

so that in the absence of thermal effects dissipation only causes downward transitions. In contrast to thermal fluctuations, vacuum fluctuations do not permit the growth of the maser past the trapping states. In this limit, we can expect remnants of these states to appear in the steady-state properties of the maser.

Figure 7.6 shows the steady-state mean photon number $\langle n \rangle$, normalized to $N = 200$ and for $n_{\text{th}} = 10^{-6}$ thermal photons, as a function of the micromaser pump parameter Θ. The "narrow resonances" are easily interpreted in terms of the trapping condition that becomes, in terms of the parameters N and Θ,

$$\frac{N}{\Theta^2} = \frac{n_q + 1}{q^2\pi^2}.\qquad(7.41)$$

For fixed N, the successive resonances correspond to values of Θ where decreasing Fock states $|n_1\rangle$ become trapping states for $q = 1$, see Ref. [12] for more details. The existence of these trapping states was experimentally demonstrated by M. Weidinger et al. [13].

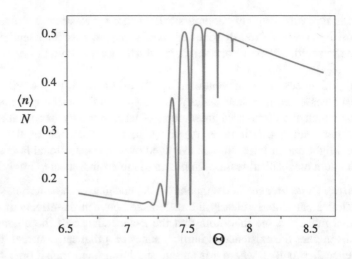

Fig. 7.6 Mean photon number $\langle n \rangle$, normalized to N, as a function of the pump parameter Θ for $N = 200$ and $n_{\mathrm{th}} = 10^{-6}$. For these parameters and $q = 1$ the Fock state $|n_{q=1} = 38\rangle$ becomes a dark state for $\Theta \approx 7.2$, the state $|n_1 = 37\rangle$ at $\Theta \approx 7.3$, $|n_1 = 36\rangle$ at $\Theta \approx 7.4$, $|n_1 = 35\rangle$ at $\Theta \approx 7.5$, and $|n_1 = 34\rangle$ at $\Theta \approx 7.6$

7.3 Dispersive Regime

So far we have concentrated on the resonant regime $\Delta = 0$ of the Jaynes–Cummings model. We now turn to the dispersive regime, where the field frequency is far off-resonant from the atomic transition frequency so that

$$|\Delta| = |\omega_0 - \omega| \gg g\sqrt{n+1} \tag{7.42}$$

for all relevant number states $|n\rangle$. We have seen in Sect. 3.1 that in this limit the eigenenergies (3.5) reduce to

$$E_{1n} = \hbar(n + 1/2)\omega + \frac{1}{2}\hbar\Delta + \frac{\hbar g^2(n+1)}{\Delta}$$

$$E_{2n} = \hbar(n + 1/2)\omega - \frac{1}{2}\hbar\Delta - \frac{\hbar g^2(n+1)}{\Delta}, \tag{7.43}$$

and the Jaynes–Cummings dressed states approach the bare states of the atom–field system,

$$|1, n\rangle \rightarrow |e, n\rangle$$

$$|2, n\rangle \rightarrow |g, n + 1\rangle \tag{7.44}$$

for $\Delta > 0$ and

$$|1, n\rangle \rightarrow |g, n+1\rangle$$

$$|2, n\rangle \rightarrow |e, n\rangle \tag{7.45}$$

for $\Delta < 0$. In that limit the Jaynes–Cummings Hamiltonian simplifies to the effective Hamiltonian

$$\hat{H}_{\mathrm{JC, eff}} = \frac{1}{2}\hbar\omega_0\hat{\sigma}_z + \hbar\omega\hat{a}^\dagger\hat{a} + \frac{\hbar g^2}{\Delta}\left[(\hat{a}^\dagger\hat{a} + 1)|e\rangle\langle e| - \hat{a}^\dagger\hat{a}|g\rangle\langle g|\right]$$

$$\approx \frac{1}{2}\hbar\omega_0\hat{\sigma}_z + \hbar\omega\hat{a}^\dagger\hat{a} + \frac{\hbar g^2}{\Delta}\hat{a}^\dagger\hat{a}\,\hat{\sigma}_z, \tag{7.46}$$

where the last term is the intensity dependent light shift of Eq. (3.15). In the second line of Eq. (7.46) we have incorporated the vacuum induced shift $s_0 = g^2/\Delta$ in the upper state frequency, $\omega_e \rightarrow \omega_e' = \omega_e + s_0$, and used the fact that for large detunings $s_0 \ll \omega_e$, so that $\omega_0' \equiv \omega_e' - \omega_g \approx \omega_0$, an approximation that we already encountered in Eqs. (3.14) and (6.48). It follows from the discussion of Sect. 6.2.2 that in this limit the number operator $\hat{N} = \hat{a}^\dagger\hat{a}$ is a QND variable since it commutes with $\hat{H}_{\mathrm{JC, eff}}$, and so is $\hat{\sigma}_z$.

Inverse Stern–Gerlach Effect Consider then an atom injected transversally into a resonator along some trajectory $\mathbf{r}(t)$. It will experience a spatially dependent vacuum Rabi frequency $g(\mathbf{r})$, a simple extension of the Rabi frequency expression (3.2) that accounts for the transverse profile of the cavity mode.[4] If in addition the atom is slow enough that it will not undergo any nonadiabatic transition between $|e\rangle$ and $|g\rangle$ during its transit, that is, if [14]

$$\frac{1}{\Delta^2}\frac{dg(\mathbf{r}(t))}{dt} < 1, \tag{7.47}$$

then all that happens is that it acquires a number state $|n\rangle$ and position-dependent light shift $-(\hbar g(\mathbf{r})^2/\Delta)n$ for a ground state atom or $(\hbar g(\mathbf{r})^2/\Delta)(n+1)$ for an excited atom.

Since both $\hat{a}^\dagger\hat{a}$ and $\hat{\sigma}_z$ are constants of motion of the Hamiltonian (7.46), one can think of these light shifts as potentials acting on the center-of-mass motion of the atom only. The n-dependence of the associated forces, given by the negative of their derivatives, implies that the atomic center-of-mass wave function will split into

---

[4]This spatial dependence was ignored for simplicity in the discussion of the micromaser.

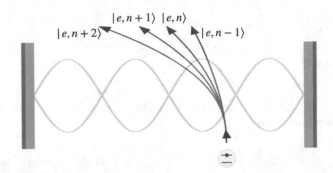

Fig. 7.7 Schematic of the inverse Stern–Gerlach effect, with the atom, taken to be in its excited state $|e\rangle$, exiting the cavity in a superposition of $|n\rangle$ dependent trajectories with probabilities reflecting the photon statistics p_n of the intracavity field

partial waves whose trajectories reflect the state of the field and its photon statistics.[5] In particular the vacuum field $|n\rangle = |0\rangle$ only produces a force on excited atoms, but not on ground state atoms, as follows from the discussion of the vacuum ac Stark shift of Eq. (3.15).

Let us assume for concreteness that a very slow atom enters the cavity in its excited state $|e\rangle$ with wave packet $|\psi_e(\mathbf{r}, t)\rangle$, and that the field is in a superposition $|\Phi\rangle = \sum_n c_n |n\rangle$, so that the initial state of the atom–field system is

$$|\Psi(\mathbf{r}, t_i)\rangle = \sum_n c_n |\psi_e(\mathbf{r}, t_i)\rangle \otimes |n\rangle \equiv \sum_n c_n |\Psi_{e,n}(\mathbf{r}, t_i)\rangle . \qquad (7.48)$$

As the atom enters the resonator and starts interacting with the field, and assuming a positive atom–field detuning $\Delta > 0$, we have that $|\Psi_e(\mathbf{r}, t_i)\rangle \rightarrow |\Psi_1(\mathbf{r}, t_i)\rangle = \sum_n |\Psi_{1,n}(\mathbf{r}, t_i)\rangle$, with the individual components all subject to different n-dependent optical potentials and their evolution governed by the corresponding Schrödinger equations

$$i\hbar \frac{\partial |\Psi_{1,n}(\mathbf{r}, t)\rangle}{\partial t_f} = \left[-\frac{\hbar^2}{2m} \nabla^2 + E_{1n}(\mathbf{r}) \right] |\Psi_{1,n}(\mathbf{r}, t)\rangle . \qquad (7.49)$$

As a result they will have acquired an n-dependent spatial component when the atom exits the resonator at time t_f and the dressed states of the system are mapped back to the bare states, $|\Psi_{1,n}(\mathbf{r}, t)\rangle \rightarrow |\Psi_{e,n}(\mathbf{r}, t_f)\rangle$, as schematically shown in Fig. 7.7.

The spatially resolved detection of the atom at a given point and time will therefore collapse the field wave function to a single number state $|n\rangle$, in analogy

---

[5]To properly account for these forces and the mechanical effects of light on atoms requires that the Hamiltonian (7.46) be properly modified to include their kinetic energy. This step will be carried out in some detail in the following chapters.

to the way a particle detection on a traditional Stern–Gerlach apparatus reduces its spin to a single value. For this reason this effect has been dubbed the *inverse Stern Gerlach effect* by S. Haroche and J. M. Raimond [15].

We will return to the mechanical effects of light in considerably more detail in the discussions of atom optics, laser cooling, and cavity optomechanics of Chaps. 8–11. In particular, Sect. 8.4.3 will show how the optical inverse Stern–Gerlach effect just discussed is formally closely related to the *matter-wave* Stern Gerlach effect.

Dispersive Schrödinger Cats In addition to resulting in a number-state-dependent force on the atomic center of mass, the effective Hamiltonian (7.46) indicates that as an atom traverses the resonator at velocity v along some trajectory it imprints on the field an additional state-dependent phase

$$\phi_1(n) = \frac{1}{v} \int dx\, E_{1n}(x)/\hbar - \Delta \tag{7.50}$$

for the state $|1, n\rangle$ and similarly for the state $|2, n\rangle$, where the spatial dependence of the eigenenergies $E_{1,n}(x)$ results as before from the spatial dependence of the vacuum Rabi frequency $g(x)$ along the atomic trajectory.

Consider then an excited atom interacting with an intracavity field in a coherent state $|\alpha\rangle$ with mean photon number $|\alpha|^2 = \bar{n}$.[6] As we recall, its photon statistics is a Poisson distribution with width $\sqrt{\bar{n}}$, so that the initial state of the atom–field system is

$$|\Psi_e(t_i)\rangle = e^{-|\alpha|^2/2} \sum_n \frac{\alpha^n}{\sqrt{n!}} |n\rangle |e\rangle. \tag{7.51}$$

For large $\bar{n}$ we can expand the accumulated phase (7.50) about its value for $n = \bar{n}$ as

$$\phi(n) = \phi(\bar{n}) + (n - \bar{n})\phi'(\bar{n}) + O(1/\bar{n}), \tag{7.52}$$

where, with the form (3.5) of $E_{1,n}$,

$$\phi'(\bar{n}) = \frac{1}{4v} \int \frac{4g^2(x)}{\sqrt{\Delta^2 + 4g^2(\bar{n} + 1)}}. \tag{7.53}$$

The phase (7.52) has two components of very different natures [3]: The first one,

$$\psi_e(\bar{n}) \equiv \phi(\bar{n}) - \bar{n}\phi'(\bar{n}), \tag{7.54}$$

---

[6]To keep the notation from becoming unnecessarily cumbersome we call the mean photon number $\bar{n}$ rather than $\langle n \rangle$ in this section.

is an overall phase that depends on $\bar{n}$ only and is accumulated by the state of the system as a whole. In contrast, the phase

$$\phi_n \equiv n\phi'(\bar{n}) \tag{7.55}$$

is a number state-dependent phase. It follows that ϕ_n effectively adds a classical phase $\Phi(\bar{n}) = \phi'(\bar{n})$ to α, so that $\alpha \to \alpha \exp[i\phi'(\bar{n})]$ and the coherent field becomes

$$|\alpha\rangle \to e^{-|\alpha|^2/2} \sum_n \frac{\alpha^n e^{in\phi'(\bar{n})}}{\sqrt{n!}} |n\rangle \equiv |\alpha e^{i\Phi(\bar{n})}\rangle . \tag{7.56}$$

For positive detuning Δ, as the atom exits the cavity at time t_f, the dressed state $|1, n\rangle$ returns adiabatically back to the bare state $|e, n\rangle$ so that the input state (7.51) has evolved to

$$|\Psi_e(t_f)\rangle = e^{-i\psi_e(\bar{n})} |e, \alpha e^{-i\Phi(\bar{n})}\rangle . \tag{7.57}$$

Similarly, if the atom had been prepared in its ground state $|g\rangle$, the final state of the atom–field system would be

$$|\Psi_g(t_f)\rangle = e^{-i\psi_g(\bar{n})} |g, \alpha e^{+i\Phi(\bar{n})}\rangle \tag{7.58}$$

with

$$\psi_g(\bar{n}) \equiv \phi(\bar{n} - 1) + \bar{n}\phi'(\bar{n} - 1) , \tag{7.59}$$

the $(\bar{n} - 1)$ factor accounting for the fact that $|g, n\rangle$ is a superposition of the dressed states $|1, n - 1\rangle$ and $|2, n - 1\rangle$, see Eq. (3.3).

Under many circumstances the global phase of a quantum state is irrelevant, but this is of course not so if two states with different global phases are made to interfere. Such a situation can be realized here if the atoms enter the resonator in the coherent superposition $(|e\rangle + |g\rangle)/\sqrt{2}$, in which case the state of the atom–field system evolves to the entangled state

$$|\Psi(t_f)\rangle = \frac{1}{\sqrt{2}} \left[e^{-i\psi_e(\bar{n})} |e, \alpha e^{-i\Phi(\bar{n})}\rangle + e^{-i\psi_g(\bar{n})} |g, \alpha e^{+i\Phi(\bar{n})}\rangle \right] . \tag{7.60}$$

After exiting the resonator, the atomic states can be subjected to the unitary transformation

$$|e\rangle \to \frac{1}{\sqrt{2}} \left[|e\rangle + e^{i\varphi} |g\rangle \right] \quad ; \quad |g\rangle \to \frac{1}{\sqrt{2}} \left[|g\rangle - e^{-i\varphi} |e\rangle \right] , \tag{7.61}$$

for example through a sequence of two spatially separated $\pi/2$ pulses with an adjustable relative phase, a so-called Ramsey interferometer. Following this step the state of the atom–field system becomes

$$|\Psi\rangle = \frac{1}{2}|e\rangle \otimes \left[e^{-i\psi_e(\bar{n})}|\alpha e^{-i\Phi(\bar{n})}\rangle + e^{-i[\psi_g(\bar{n})+\varphi]}|\alpha e^{+i\Phi(\bar{n})}\rangle \right]$$

$$+ \frac{1}{2}|g\rangle \otimes \left[e^{-i[\psi_e(\bar{n})-\varphi]}|\alpha e^{-i\Phi(\bar{n})}\rangle - e^{-i\psi_g(\bar{n})}|\alpha e^{+i\Phi(\bar{n})}\rangle \right]. \quad (7.62)$$

A final detection of the atomic state then projects the field into one of the two cat states

$$|\Psi\rangle_{\text{cat, e}} = \frac{e^{-i\psi_e(\bar{n})}}{\sqrt{2}} \left[|\alpha e^{-i\Phi(\bar{n})}\rangle + |\alpha e^{+i\Phi(\bar{n})}\rangle \right] \quad (7.63)$$

or

$$|\Psi\rangle_{\text{cat, g}} = \frac{e^{-i\psi_g(\bar{n})}}{\sqrt{2}} \left[|\alpha e^{-i\Phi(\bar{n})}\rangle - |\alpha e^{+i\Phi(\bar{n})}\rangle \right], \quad (7.64)$$

where we have set the adjustable phase to $\varphi = \psi_e(\bar{n}) - \psi_g(\bar{n})$. These states are coherent superpositions of two coherent states centered at the angles $\pm\Phi(\bar{n})$ in the $\{\text{Re}(\alpha), \text{Im}(\alpha)\}$ phase plane. Because coherent states can be relatively large quantum objects, the states (7.63) and (7.64) can therefore be considered as Schrödinger cats—or perhaps more accurately Schrödinger kittens, see Fig. 7.8. These optical Schrödinger cats were first experimentally demonstrated by S. Deléglise et al. [16].

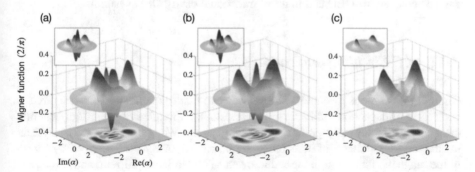

Fig. 7.8 Schrödinger cat and decoherence: (**a**) Reconstruction of the Wigner function of a cat state with $\langle n \rangle = 3.5$ photons, a short time (1.3 ms) after its preparation by a single atom crossing the cavity. (**b**) and (**c**) The same cat state after 4.3 and 16 ms: the vanishing of the fringe interfering features is a manifestation of decoherence. The inserts are the corresponding theoretical Wigner functions. (From Ref. [16])

7.4   Circuit QED

Circuit QED (cQED) is an extension of cavity QED that uses artificial atoms instead of atoms to achieve many of the same objectives, but with differences that open up promising new opportunities. The artificial atoms most frequently used in these systems comprise several varieties of superconducting qubits. They are coupled to microwave fields, most often in one-dimensional transmission line resonators. The qubit–field interactions can be orders of magnitude larger than in atomic systems, see Table 7.1, a result of the extremely small mode volumes that can be achieved in one-dimensional transmission lines. In addition these systems can be fabricated on microchips and as such offer an elegant route toward the solid-state realization of multiple qubit systems of interest for quantum information science, quantum communications, and quantum metrology. However, potential experimental challenges arise from the fact that they must operate in cryogenic environments and that artificial atoms can suffer from atom to atom variations, in contrast to the situation for atoms—all sodium atoms are the same!

This section gives a brief overview of these systems, starting with the field quantization of LC circuits and extending then the discussion to multimode one-dimensional transmission line resonators. As it turns out, LC circuits quantization is also useful in the description of artificial atoms, since they are likewise generated by electric circuits, with the nonlinearity required to simulate atoms provided by superconducting Josephson junctions that act as nonlinear inductances. As an example we will discuss in some detail one specific type of superconducting artificial atom, the Cooper pair box, and will show how its coupling to the field results under appropriate conditions in a realization of the Jaynes–Cummings model. We conclude the section with a brief comparison of typical parameters that can be achieved in cQED and in more traditional cavity QED systems.

7.4.1   LC Circuit Quantization

Circuit QED exploits resonators and microwave fields generated in electric circuits consisting of capacitors, inductors, and resistors, although since resistors generate dissipation it is normally preferable to avoid them. We consider therefore first an undamped LC oscillator of inductance L and capacitance C, where $L = \Phi/I$ with Φ the magnetic flux through the inductor and I the inductor current, and $C = q/V$ with V the voltage across the capacitor and q the capacitor charge.

A possible Lagrangian $\mathcal{L}$ for the LC oscillator can be obtained from the difference between the energy $LI^2/2$ stored in the inductor and the potential energy $q^2/2C$ inside the capacitor. Making use of charge conservation $I = \dot{q}$, it takes the form

$$\mathcal{L} = \frac{1}{2}L\dot{q}^2 - \frac{1}{2C}q^2 \,. \tag{7.65}$$

The associated Euler–Lagrange equation

$$\frac{\partial \mathcal{L}}{\partial q} - \frac{d}{dt}\frac{\partial \mathcal{L}}{\partial \dot{q}} = 0 \tag{7.66}$$

yields the familiar harmonic oscillator equation of motion

$$\ddot{q} + \omega^2 q = 0, \tag{7.67}$$

with $\omega = \sqrt{LC}$, which is the known equation of motion for the charge in an LC oscillator. This justifies a posteriori the choice of the Lagrangian (7.65). The conjugate momentum of the charge q,

$$\frac{\partial \mathcal{L}}{\partial \dot{q}} = L\dot{q} = LI = \Phi, \tag{7.68}$$

is the magnetic flux Φ though the inductor. The Hamiltonian of the circuit is therefore

$$H = \Phi \dot{q} - \mathcal{L} = \frac{1}{2L}\Phi^2 + \frac{1}{2C}q^2, \tag{7.69}$$

with associated Hamilton equations of motion

$$\dot{q} = \frac{\partial H}{\partial \Phi} = \frac{\Phi}{L} \quad ; \quad \dot{\Phi} = -\frac{\partial H}{\partial q} = -\frac{q}{C} = V, \tag{7.70}$$

where V is the voltage at the node connecting the inductor with the capacitor.

In analogy with the harmonic oscillator of Sect. 2.1 we quantize this system by promoting q and Φ to quantum operators subject to the canonical commutation relation

$$[\hat{q}, \hat{\Phi}] = i\hbar, \tag{7.71}$$

and introduce the creation and annihilation operators

$$\hat{a} = \frac{1}{\sqrt{2\hbar\omega C}}\hat{q} + i\frac{1}{\sqrt{2\hbar\omega L}}\hat{\Phi}, \tag{7.72}$$

$$\hat{a}^\dagger = \frac{1}{\sqrt{2\hbar\omega C}}\hat{q} - i\frac{1}{\sqrt{2\hbar\omega L}}\hat{\Phi}, \tag{7.73}$$

with $[\hat{a}, \hat{a}^\dagger] = 1$, so that

$$\hat{H} = \hbar\omega\left(\hat{a}^\dagger\hat{a} + \tfrac{1}{2}\right). \tag{7.74}$$

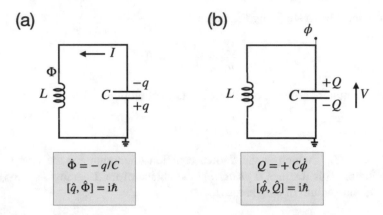

Fig. 7.9 Lossless LC electrical oscillator. (**a**) The coordinate is taken to be q and the conjugate momentum is Φ; (**b**) the coordinate is taken to be ϕ and the conjugate momentum is Q. Note the important sign change in the definitions q and Q of the charge between the two approaches, needed to maintain the canonical commutation relation between momentum and coordinate. (Adapted from Ref. [17])

As discussed by S. Girvin in his comprehensive Les Houches Lecture Notes on cQED [17], this description amounts to considering $\hat{q}$ as the coordinate, so that the inductance plays effectively the same role as the mass in the mechanical oscillator, and C takes the role of the inverse spring constant. Alternatively, instead of considering q as the "coordinate" and Φ as the momentum, it is also possible to reverse their roles and choose the integral of the voltage across the capacitor, the node flux ϕ, as the "coordinate" and Q as the associated "momentum." This point of view, which amounts to exchanging the roles of the "kinetic" and "potential" energies as compared with the Lagrangian (7.65), is useful in the description of superconducting qubits, which behave as nonlinear inductors, see Fig. 7.11. In that case it is better, or at least intuitively more appealing, to think of the energy stored in the inductor as a potential energy.

To proceed in this way we note that the node flux at the location indicated in Fig. 7.9b is

$$\phi(t) = \int_{-\infty}^{t} d\tau\, V(\tau), \qquad (7.75)$$

or $V = \dot{\phi}$, so that the potential energy stored in the capacitor is $U = C\dot{\phi}^2/2$, except that with the new choice of coordinates it now looks like a "kinetic energy." At the same time, the kinetic energy $\phi^2/2L$ stored in the inductor now looks like a "potential energy," so that the Lagrangian of the LC circuit takes the form

$$\mathcal{L} = \frac{C}{2}\dot{\phi}^2 - \frac{1}{2L}\phi^2, \qquad (7.76)$$

compare this expression with Eq. (7.65). In this approach C plays therefore the role of the mass, and L of the inverse spring constant.

The conjugate variable of Φ is now

$$Q = \frac{\partial \mathcal{L}}{\partial \dot{\phi}} = +C\dot{\phi}, \tag{7.77}$$

so that the Hamiltonian associated with the Lagrangian (7.76) is

$$H = Q\dot{\phi} - \mathcal{L} = \frac{1}{2C}Q^2 + \frac{1}{2L}\phi^2, \tag{7.78}$$

with Hamilton equations of motion

$$\dot{\phi} = \frac{\partial H}{\partial Q} = \frac{Q}{C} = V \quad ; \quad \dot{Q} = -\frac{\partial H}{\partial \phi} = -\frac{\phi}{L}. \tag{7.79}$$

Comparing these equations with Eq. (7.70) shows an important sign change in the definition of the charge, $Q = -q$. This is required to maintain the canonical commutation relation between momentum and coordinate, which is now $[\hat{\phi}, \hat{Q}] = i\hbar$.

With this choice of coordinates, and when expressed in terms of annihilation and creation operators, the quantized Hamiltonian of the LC circuit still takes the form (7.74), but now with

$$\hat{a} = \frac{1}{\sqrt{2\hbar\omega L}}\hat{\phi} + i\frac{1}{\sqrt{2\hbar\omega C}}\hat{Q}, \tag{7.80}$$

$$\hat{a}^\dagger = \frac{1}{\sqrt{2\hbar\omega L}}\hat{\phi} - i\frac{1}{\sqrt{2\hbar\omega C}}\hat{Q}. \tag{7.81}$$

With $\hat{V} = d\hat{\phi}/dt = \hat{Q}/C$ this gives

$$\hat{V} = -i\sqrt{\frac{\hbar\omega}{2C}}(\hat{a} - \hat{a}^\dagger). \tag{7.82}$$

One-dimensional Transmission Line Resonator We now extend these results to the quantization of a one-dimensional transmission line resonator of length L, following again Ref. [17]. In that case the flux (7.75) becomes

$$\Phi(x,t) = \int_{-\infty}^{t} d\tau\, V(x,\tau), \tag{7.83}$$

and the local voltage on the transmission line is

$$V(x,t) = \partial_t \Phi(x,t), \tag{7.84}$$

so that the flux through the inductance ℓdx along a segment of the line of length dx is $-[\partial_x \Phi(x, t)]dx$, the voltage drop is $\partial_x[\partial_t \Phi(x, t)]dx$, and the local value of the current is

$$I(x, t) = -\frac{1}{\ell}\partial_x \Phi(x, t).$$ (7.85)

The Lagrangian (7.76) is therefore replaced by

$$\mathcal{L} = \int_0^L dx \left[\frac{c}{2}(\partial_t \Phi)^2 - \frac{1}{2\ell}(\partial_x \Phi)^2 \right],$$ (7.86)

where ℓ and c are the inductance and capacitance per unit length.[7] The conjugate momentum to $\Phi(x)$ is now the charge density

$$\frac{\delta \mathcal{L}}{\delta(\partial_t \Phi)} = q(x, t) = c \, \partial_t \Phi(x, t) = cV(x, t),$$ (7.87)

and the Hamiltonian becomes

$$\hat{H} = \int_0^L dx \left[\frac{1}{2c}q^2 + \frac{1}{2\ell}(\partial_x \Phi)^2 \right].$$ (7.88)

Very much like in the multimode field quantization of Sect. 2.2 we now expand $\Phi(x, t)$ in a set of orthonormal modes as

$$\Phi(x, t) = \sum_{n=0}^{\infty} \xi_n(t)u_n(x).$$ (7.89)

We consider open boundary conditions such that the current (but not the voltage) vanishes at the ends of the resonator. We then have

$$u_n(x) = \sqrt{2}\cos(k_n x)$$ (7.90)

and $k_n = n\pi/L$ so that

$$\frac{1}{L}\int_0^L dx \, u_n(x)u_m(x) = \delta_{nm}$$

$$\frac{1}{L}\int_0^L dx \, |\partial_x u_n(x)||\partial_x u_m(x)| = k_n^2 \delta_{nm}.$$

---

[7]Importantly, keep in mind that L is now the length of the transmission line, rather than an induction as was the case in the discussion of the LC circuit, and c is an inductance per unit length, not the speed of light.

The Lagrangian (7.86) becomes the sum over modes

$$\mathcal{L} = \frac{L}{2}c\sum_n \left[|\partial_t \xi_n|^2 - \omega_n^2 \xi_n^2\right],\tag{7.91}$$

where $\omega_n = v_p k_n$ and $v_p = 1/\sqrt{\ell c}$. All modes are then individually quantized, resulting after introduction of associated creation and annihilation operators in the multimode Hamiltonian

$$\hat{H} = \frac{1}{2}\sum_n \left[\frac{1}{Lc}\hat{q}_n^2 + Lc\omega_n^2 \xi_n^2\right] = \sum_n \hbar\omega_n \left(\hat{a}_n^\dagger \hat{a}_n + \tfrac{1}{2}\right)\tag{7.92}$$

with

$$\hat{\xi}_n = \sqrt{\frac{\hbar}{2\omega_n Lc}}\left(\hat{a}_n + \hat{a}_n^\dagger\right),\tag{7.93}$$

$$\hat{q}_n = -i\sqrt{\frac{\hbar\omega_n Lc}{2}}\left(\hat{a}_n - \hat{a}_n^\dagger\right).\tag{7.94}$$

The voltage operator $\hat{V}(x) = \hat{q}(x)/c$, which will be needed in determining the coupling Hamiltonian between the transmission line and the superconducting qubit, is therefore

$$\hat{V}(x) = \frac{1}{Lc}\sum_n u_n(x)\hat{q}_n = -i\sum_n \sqrt{\frac{\hbar\omega_n}{2Lc}}\left(\hat{a}_n - \hat{a}_n^\dagger\right)u_n(x).\tag{7.95}$$

7.4.2   Superconducting Qubits

The idea of cQED is to couple superconducting artificial atoms to microwave fields supported by LC transmission lines. One significant advantage of these systems is that in addition to single qubit–field systems, which can as we shall see be described by the Jaynes–Cummings Hamiltonian under appropriate conditions, it is also relatively straightforward to set up multi-qubit configurations, for example to investigate collective and many-body effects and/or to develop modules such as quantum gates for quantum information applications. In addition, as already mentioned, the small mode volume of the field in one-dimensional transmission lines can result in vacuum Rabi frequencies orders of magnitude larger than can be achieved in more traditional cavity QED configurations.

Superconductivity is the phenomenon whereby below a critical temperature the electrical resistance of a material vanishes and magnetic flux fields are expelled from it. The physical effect underlying superconductivity is that the effective

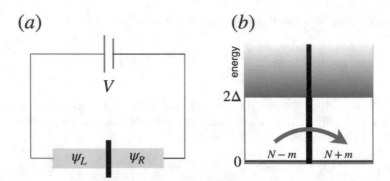

Fig. 7.10 (a) Schematic of a Josephson junction, with the wave functions ψ_L and ψ_R of the superconducting states at the left and right of the junction. (b) Sketch of the energy diagram of a Josephson junction, showing in red the ground state of its left and right superconducting elements, with all electrons bound in Cooper pairs (for an even number of electrons). This state is separated by the continuum of not fully paired states, shown in grey, by an energy gap 2Δ. The blue arrow illustrates the tunneling of Cooper pairs between the two sides of the junction

attractive interaction resulting from virtual phonon exchange leads to the pairing of electrons of opposite spins into *Cooper pairs*, which are composite spin $S = 0$ bosons.[8] If the number of electrons in the electrode is even, then the quantum ground state of the system is characterized by all of the electrons paired up. The energy required to break a pair is the energy gap 2Δ. The remarkable properties of superconducting materials have now been exploited to develop a variety of artificial atoms, with properties described in detail in several excellent reviews [17–19]. Here we concentrate on just one system, the so-called Cooper pair box.

Just like this book is not a book on atomic physics, it is also not the place to provide a detailed description of superconductivity and Josephson junctions. These can be found in numerous texts, for example the classic book by Tinkham [20]. Instead, we limit ourselves to those few elements that are essential for a phenomenological understanding of these systems and the way they can operate as two-state systems, starting with a brief review of Josephson junctions.

Josephson Junction Josephson junctions are devices that consist of two superconductors coupled by a thin insulating barrier or a short section of non-superconducting metal (Fig. 7.10). At low enough temperatures the ground state of the junction corresponds to the two superconductors being occupied by macroscopic numbers N_L and N_R of Cooper pairs with $N_R + N_L = N$, each with charge $-2e$. At the simplest level, it is possible to ignore the continuum of non-fully paired states, which are separated from the ground state by an energy gap 2Δ, and to describe the

---

[8]Unconventional Cooper pairing resulting in spin $S = 1$ composite bosons is also possible.

junction in terms of the effective single-particle wave function

$$|\psi\rangle = \psi_L|L\rangle + \psi_R|R\rangle, \tag{7.96}$$

with the number of pairs on each side of the junction given by

$$N_i = N|\psi_i|^2 \ , \quad i = \{L, R\}. \tag{7.97}$$

The tunneling of pairs between the two sides of the junction can be described by a hopping Hamiltonian

$$\hat{H} = -\begin{pmatrix} 2eV_L & w \\ w & 2eV_R \end{pmatrix}, \tag{7.98}$$

where

$$|L\rangle = \begin{pmatrix} 1 \\ 0 \end{pmatrix} \quad ; \quad |R\rangle = \begin{pmatrix} 0 \\ 1 \end{pmatrix}, \tag{7.99}$$

with corresponding equations of motion

$$i\hbar \frac{\partial \psi_L}{\partial t} = -2eV_L\psi_L - w\psi_R$$
$$i\hbar \frac{\partial \psi_R}{\partial t} = -2eV_R\psi_R - w\psi_L. \tag{7.100}$$

Here V_L and V_R are the external electric potentials on the left and right sides of the junction and w is a constant that is characteristic of the junction and accounts for quantum tunneling across it. If the electric potential difference across the junction is V, then $2e(V_R - V_L) = 2e\,V$.

Introducing the so-called Ginzburg–Landau order parameter

$$\psi_i = \sqrt{n_i}e^{i\varphi_i} \tag{7.101}$$

where $i = (R, L)$ and n_i are the densities of Cooper pairs on both sides of the junction, it is easily found from Eq. (7.100) that

$$\frac{\partial n_L}{\partial t} = -\frac{\partial n_R}{\partial t} = -\frac{2w}{\hbar}\sqrt{n_L n_R}\sin(\varphi_L - \varphi_R), \tag{7.102}$$

$$\frac{\partial}{\partial t}(\varphi_R - \varphi_L) = \left(\frac{2e}{\hbar}\right)(V_R - V_L). \tag{7.103}$$

With Eq. (7.97) and introducing the Cooper pair current $I = -2e(\partial N_L/\partial t)$ and $\varphi = \varphi_L - \varphi_R$ finally gives the first and second *Josephson relations*. Noting that the variations in N_R and N_L remain small they take the form

$$I(\varphi) = I_c \sin \varphi \,, \qquad (7.104)$$

$$\frac{\partial \varphi}{\partial t} = \frac{2eV(t)}{\hbar} \,, \qquad (7.105)$$

where the critical current is

$$I_c = \frac{2e}{\hbar} E_J \,. \qquad (7.106)$$

The parameter E_J, given in this model by

$$E_J = 2w\sqrt{N_L N_R} \,, \qquad (7.107)$$

is called the Josephson coupling energy. It is a measure of the ability of Cooper pairs to tunnel through the junction. The Josephson relations show that a DC current can be drawn through the junction even without a voltage drop V, as long as the current is smaller than I_c. This is the so-called DC Josephson effect. But if a voltage V or a DC current larger than I_C is applied to the junction, Eq. (7.105) shows that the Josephson current will oscillate at the frequency $(2e/\hbar)V$, the AC Josephson effect.

The Cooper Pair Box A Cooper pair box consists of a superconducting island that is connected via Josephson junctions to a grounded reservoir, so that Cooper pairs can tunnel into and out of the island, see Fig. 7.11. It is modeled by a capacitance

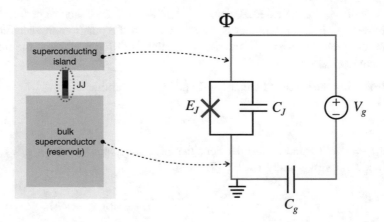

Fig. 7.11 Schematic of Cooper pair box, illustrating the coupling of a superconducting island to a bulk superconductor via a Josephson junction modeled by a capacitance C_J in parallel with an effective nonlinear inductor characterized by the Josephson coupling energy E_J. The charging energy required to add another Cooper pair to the system is $E_C = e^2/2(C_J + C_g)$. The applied gate potential V_g capacitively coupled to the island provides a gate charge $n_g = C_g V_g/2e$ that acts as a control parameter

C_J in parallel with an effective nonlinear inductor characterized by the Josephson coupling energy E_J, and in addition by the charging energy

$$E_C = \frac{e^2}{2(C_J + C_g)} \equiv \frac{e^2}{2C_\Sigma} \tag{7.108}$$

required to add another Cooper pair to the island. Finally, the applied gate potential V_g capacitively coupled to the island provides a gate charge $n_g = C_g V_g / 2e$ that acts as a control parameter.

We model this device by considering two superconducting electrodes with nominal numbers of Cooper pairs N_L and N_R and separated by a tunnel junction that permits Cooper pairs to be transferred from one to the other, see Fig. 7.11. The tunneling process can be described by the phenomenological Hamiltonian

$$\hat{H}_T = -\frac{1}{2}E_J \sum_n [\,|n\rangle\langle n+1| + |n+1\rangle\langle n|\,] , \tag{7.109}$$

where $|n\rangle = |N_L - m, N_R + m\rangle$ is the state of the system with m pairs having been transferred from the nominal values $\{N_L, N_R\}$ through the junction from the "left" to the "right" electrode, and m can either increase or decrease by unity corresponding to the tunneling of a pair to the right or the left. It is easily verified by direct substitution that its (unnormalized) eigenstates are

$$|\varphi\rangle = \sum_{m=-\infty}^{\infty} e^{im\varphi}|m\rangle , \tag{7.110}$$

with eigenvalues

$$\hat{H}_T|\varphi\rangle = -E_J \cos\varphi|\varphi\rangle . \tag{7.111}$$

Introducing the number of pairs operator

$$\hat{n} = \sum_n n|n\rangle\langle n| \tag{7.112}$$

and the associated current operator $\hat{I} = 2e\, d\hat{n}/dt$, given by

$$\hat{I} = 2e\frac{i}{\hbar}[\hat{H}_T, \hat{n}] = -\frac{eE_J}{\hbar} \sum_{m=-\infty}^{\infty} [\,|m\rangle\langle m+1| - |m+1\rangle\langle m|\,] , \tag{7.113}$$

it is easily shown that the eigenstates $|\phi\rangle$ of $\hat{H}_T$ are also eigenstates of $\hat{I}$ with

$$\hat{I}|\varphi\rangle = I_c \sin\varphi|\varphi\rangle , \tag{7.114}$$

which recovers the first Josephson relation (7.104).

A full description of the Cooper pair dynamics must also account for the Coulomb energy

$$U = 4E_C(n - n_g)^2 \tag{7.115}$$

required to transfer n pairs of charge $2e$ across the junction. Here the gate charge $n_g = C_g V_g / 2e$ is a continuously variable control parameter. It accounts for the effect of an externally applied gate voltage V_g to a nearby electrode that is capacitively coupled to the island. The total effective Hamiltonian describing the Cooper pair box is therefore

$$\hat{H} = \sum_n \left[4E_C(n - n_g)^2 |n\rangle\langle n| - \frac{1}{2} E_J |n\rangle\langle n + 1| + |n + 1\rangle\langle n| \right], \tag{7.116}$$

which shows explicitly the anharmonicity resulting from the presence of the Josephson junction.

Two-level Approximation In the regime $4E_c \gg E_J$ and for most values of the external voltage, the eigenenergies of the Hamiltonian (7.116) are dominated by the charging part $\sum_n 4E_C(n - n_g)^2 |n\rangle\langle n|$, with the tunnel coupling acting as a small perturbation. If the controllable gate charge $n_g \in [0, 1]$ it is easily verified that the box has two low energy levels $|0\rangle$ and $|1\rangle$ with energies $E_0 \approx E_1 \approx E_C/4$ (with equal signs for $E_J = 0$), all other levels having much higher energies. More generally, for $n_g = n + 1/2$ the two levels of $\hat{H}$ of energies E_n and E_{n+1} become approximately degenerate with $E_n \approx E_{n+1} \approx E_C$—and exactly so for $E_J \to 0$—while the next neighboring levels have much higher energies, $E_{n-1} \approx E_{n+2} \approx 9E_C/4$. That is, there are two close energy levels that are well separated from all other levels, making this a good two-state system for an appropriate choice of driving field frequencies.

Concentrating for concreteness on the case $n = 0$ and discarding all terms that involve states other than the two states $|0\rangle$ and $|1\rangle$ reduce the Cooper pair box to an effective two-level system described by the qubit Hamiltonian

$$\hat{H} = -\frac{E_{el}}{2}\hat{\sigma}_z - \frac{E_J}{2}\hat{\sigma}_x, \tag{7.117}$$

where $E_{el} = 4E_C(1 - 2n_g)$. The eigenstates $|e\rangle$ and $|g\rangle$ of this Hamiltonian are easily found to be

$$|e\rangle = \cos\theta|1\rangle - \sin\theta|0\rangle,$$
$$|g\rangle = \sin\theta|1\rangle + \cos\theta|0\rangle, \tag{7.118}$$

with eigenenergies

$$E_e = -E_g = \frac{1}{2}\sqrt{E_{\text{el}}^2 + E_J^2}$$ (7.119)

and $\tan(2\theta) = E_J/E_{\text{el}} = E_J/4E_C(1 - 2n_g)$.

Other Superconducting Qubits Following the realization of the Cooper pair box a number of other superconducting qubits have been invented, characterized by various ratios E_J/E_C of Josephson to charging energy, number of Josephson junctions involved, and topology of the circuits in which they are embedded. These include in particular the quantronium [21], the transmon [22], and the fluxonium [23]. It is beyond the scope of this brief section to review these developments, which are discussed in some detail in the Lecture Notes [17] and the recent reviews [18] and [19].

7.4.3 Field–Qubit Coupling

We have seen in Sect. 7.4.1 that a one-dimensional transmission line produces the voltage (7.95)

$$\hat{V}(x) = \frac{1}{Lc}\sum_n u_n(x)\hat{q}_n = -i\sum_n \sqrt{\frac{\hbar\omega_n}{2Lc}}\left(\hat{a}_n - \hat{a}_n^\dagger\right)u_n(x),$$ (7.120)

so that the Cooper box will be driven both by the DC voltage V_g and the AC field $\hat{V}(x)$. For a single-mode field and a Cooper box at a location x inside the resonator, see Fig. 7.12, the Hamiltonian of the full system is then [24]

$$\hat{H} = \frac{1}{2}\hbar\Omega\hat{\sigma}_z + \hbar\omega\hat{a}^\dagger\hat{a} - ie\frac{C_g}{C_\Sigma}\sqrt{\frac{\hbar\omega}{Lc}}(\hat{a} - \hat{a}^\dagger)\left[1 - 2n_g - \cos(2\theta)\hat{\sigma}_z + \sin(2\theta)\hat{\sigma}_x\right],$$ (7.121)

where the Pauli matrices $\hat{\sigma}_x$ and $\hat{\sigma}_z$ are now in the $\{|e\rangle, |g\rangle\}$ basis. For a gate charge $n_g = 1/2$ and $2\theta = \pi/2$ this reduces to the Jaynes–Cummings Hamiltonian

$$\hat{H} = \frac{1}{2}\hbar\Omega\hat{\sigma}_z + \hbar\omega\hat{a}^\dagger\hat{a} - i\hbar g(\hat{a} - \hat{a}^\dagger)\hat{\sigma}_x$$ (7.122)

with vacuum Rabi frequency

$$g = \frac{e\beta}{\hbar}\sqrt{\frac{\hbar\omega}{cL}}$$ (7.123)

and $\beta = C_g/C_\Sigma$.

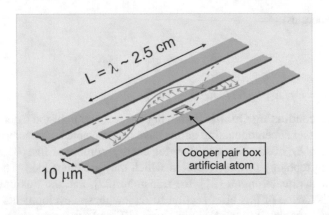

Fig. 7.12 Schematic of a 1-D transmission line resonator with Cooper pair box placed at a maximum of the voltage standing wave. The transmission line resonator consists of a full-wave section of superconducting coplanar waveguide. Multiple qubits can be placed at different antinodes of the standing wave across distances of several millimeters. Such 1-D configurations are characterized by extremely small mode volumes, of the order of $10^{-6}\lambda$, resulting in vacuum Rabi frequencies that can be orders of magnitude larger than achievable in atomic systems, see Table 7.1. (Adapted from Ref. [24])

Table 7.1 Typical rates and parameters for 3-D cavity QED optical and microwave atomic systems, compared to 1-D systems using superconducting circuits and artificial atoms

Parameter	Symbol	Optical CQED	Microwave CQED	Circuit QED
Transition frequency	ω_0	350 THz	50 GHz	10 GHz
Vacuum Rabi frequency	$g/2\pi$	220 MHz	50 KHz	100 MHz
Cavity lifetime	$1/\kappa$	10 ns	> 10 ms	150 ns
Atomic transition lifetime	$1/\Gamma$	60 ns	30ms	> 2 μs
Atomic transit time	τ	>50 μs	100 μs	∞
Single atom cooperativity	$C = g^2/2\kappa\Gamma$	50	$1.5 \cdot 10^8$	$6 \cdot 10^4$
Critical photon number	$n_0 = \Gamma^2/8g^2$	10^{-4}	$5 \cdot 10^{-8}$	$2.5 \cdot 10^{-6}$

Microwave cavity QED systems operate typically on transitions between neighboring atomic Rydberg states of high principal quantum number n and orbital quantum number $\ell = n - 1$ (the so-called circular Rydberg states), for which the electric dipole moment scales approximately as n^2 and the radiative lifetime as n^3. Cavity lifetimes of tens of milliseconds have been achieved in the superconducting cavities used in these systems. The superconducting cQED parameters are for a full-wave $L = \lambda$ resonator, a relatively low $Q = 10^4$, and $C_g/C_\Sigma = 0.1$. See Problem 7.1 for the physical interpretation of the single atom cooperativity C, critical photon number n_0, and "critical atom number" $N_0 \equiv 1/C$. (Adapted from Ref. [24])

As already indicated, superconducting qubit-based circuit QED can present considerable advantages over real atom-based cavity QED, as summarized in Table 7.1. In particular, these systems can have considerably larger vacuum Rabi frequencies, longer lifetimes, and of course, infinite transit times through the resonator. For this reason they are of much interest both in fundamental studies, as well as in potential applications in quantum information science.

7.5   The Casimir Force

The previous sections discussed how the electromagnetic vacuum can be tailored to qualitatively and quantitatively control the radiative properties of atoms. As it turns out, tailoring the electromagnetic vacuum results in observable effects even in the absence of atoms or other radiators. The most famous of these manifestations is the attractive Casimir force between two perfectly conducting plates in a vacuum. This force, which was predicted as early as 1948 [25, 26], was long considered an academic curiosity, but it now plays an increasing role in nanophotonics and other nanoscience applications.

The origin of the Casimir force is the zero-point energy of the electromagnetic field, which we first encountered in the Hamiltonian $\hat{H} = \hbar\omega(\hat{a}^\dagger\hat{a}+\frac{1}{2})$ of the single-mode electromagnetic field. We determined then, in the quantization of multimode fields, that every mode contains such a zero-point energy contribution. Since free space contains an infinite number of modes, we conclude that this energy must be infinite. Since it is also a constant, we have largely ignored it so far, but as a wise physicist once explained to me, "it is not because a quantity is infinite that you can simply set it as equal to zero!" ... The Casimir force is a perfect example in point, although it is minute and can be ignored in most everyday situations. It was first observed by M. J. Sparnaay [27], but it is not until 1997 that S. K. Lamoreaux [28] carried out its first precision measurement between a spherical lens and an optical quartz plate connected to a torsion pendulum. He was able to demonstrate that the Casimir force pulled the two objects together and caused the pendulum to twist, in agreement with theory. His results permitted in addition to test the inverse square law of gravitational attraction at distances much shorter than had been possible until then, a point to which we will return in Sect. 12.1.

That this force should be attractive can be understood by a simple argument: if instead of being separated by d the plates were separated by a larger distance $d + \Delta d$, then the zero-point energy would be larger, since at a larger separation the system enclosed by the plates can support more modes of the electromagnetic field. The minimal energy configuration must therefore be for zero separation, and hence, the force must be attractive.

Consider then an empty box with perfectly conducting walls and of transverse dimensions $L_x = L_y = L$ and longitudinal length $d \ll L$, see Fig. 7.13. The total zero-point energy of the system is the sum of the contributions of all field modes supported by the cavity

$$E(d) = \sum_\alpha \frac{1}{2}\hbar\omega_\alpha = \frac{\hbar c}{2} \sum_\alpha |\mathbf{k}_\perp| \tag{7.124}$$

$$= \frac{\hbar c}{2} \left(\frac{L}{2\pi}\right)^2 \iint d^2k_\perp \left(|\mathbf{k}_\perp| + 2\sum_{n=1}^{\infty} \sqrt{|\mathbf{k}_\perp|^2 + n^2\pi^2/d^2}\right),$$

Fig. 7.13 Schematic
illustration of the origin of the
Casimir force, with the
zero-point energy from the
outside modes exerting a
larger radiation pressure force
on the plates than the
zero-point energy from the
intracavity modes

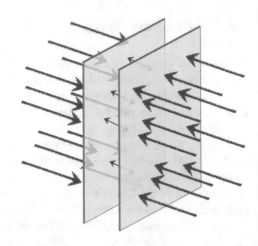

where we have used the fact that each wave vector, except for $k_z = 0$, supports
two field polarizations. In the case of free space quantization we have instead,
considering the same volume,

$$
E_0 = \frac{\hbar c}{2} \left(\frac{L}{2\pi} \right)^2 \iint d^2 k_\perp \int_0^\infty dn\, 2\sqrt{|\mathbf{k}_\perp|^2 + n^2\pi^2/d^2}
$$

$$
= \frac{\hbar c}{2} \frac{d}{2\pi} \left(\frac{L}{2\pi} \right)^2 \iint d^2 k_\perp \int_{-\infty}^\infty dk_z\, 2\sqrt{|\mathbf{k}_\perp|^2 + k_z^2}. \tag{7.125}
$$

The difference in energies per unit area is therefore, when expressed in polar
coordinates,

$$
\Delta E = \frac{E(d) - E_0}{L^2} \tag{7.126}
$$

$$
= \frac{\hbar c}{2\pi} \int_0^\infty k\, dk \left(\frac{k}{2} + \sum_{n=1}^\infty \sqrt{k^2 + n^2\pi^2/d^2} - \int_0^\infty dn\sqrt{k^2 + n^2\pi^2/d^2} \right).
$$

The integral in this expression diverges for $k \to \infty$; however, it is important to
keep in mind that our analysis is not really valid in this regime, in part because the
model of perfectly conducting plates is not realistic above the plasma frequency,
where they become effectively transparent. For this reason we introduce a cutoff
momentum $k_{\max}$ and require that

$$
f(k) = \begin{cases} 1 & \text{for } k < k_{\max}, \\ 0 & \text{for } k \geq k_{\max}. \end{cases} \tag{7.127}
$$

The specific value of $k_{\max}$ is not important as long as it is large enough since as
we shall see it disappears from the final answer. With the change of variable $u =$

$d^2 k^2 / \pi^2$, ΔE takes then the form

$$\Delta E = \hbar c \frac{\pi^2}{4d^3} \left[\frac{1}{2} F(0) + \sum_{n=1}^{\infty} F(n) - \int_0^{\infty} dn \, F(n) \right], \qquad (7.128)$$

where we have introduced the function

$$F(n) \equiv \int_0^{\infty} du \sqrt{u + n^2} f \left(\frac{\pi}{d} \sqrt{u + n^2} \right). \qquad (7.129)$$

This expression can be evaluated using the Euler–Maclaurin resummation formula

$$\frac{1}{2} F(0) + \sum_{n=1}^{\infty} F(n) - \int_0^{\infty} dn \, F(n) = -\frac{1}{2!} B_2 F'(0) - \frac{1}{4!} B_4 F'''(0) + \ldots, \qquad (7.130)$$

where the coefficients B_n are Bernoulli numbers with $B_2 = 1/6$ and $B_4 = -1/30$. Rewriting $F(n)$ as

$$F(n) = \int_{n^2}^{\infty} dv \sqrt{v} f(\pi \sqrt{v}/d), \qquad (7.131)$$

we find readily

$$F'(n) = -2n^2 f(n\pi/d) \quad ; \quad F'''(n) = -4, \qquad (7.132)$$

with all further derivatives vanishing. Inserting these expressions into Eq. (7.128) gives finally

$$\Delta E = -\frac{\pi^2}{720} \frac{\hbar c}{d^3}, \qquad (7.133)$$

a result independent of the cutoff momentum k_{max}, as advertised. The Casimir force per unit area is

$$F = -\frac{\partial}{\partial d} \Delta E = -\frac{\pi^2}{240} \frac{\hbar c}{d^4}. \qquad (7.134)$$

As expected from our discussion of its physical origin, this is a purely quantum mechanical effect, as evidenced by its disappearance for $\hbar = 0$. For a plate separation of $d = 1 \mu$m this attractive force is just $F = -1.3 \cdot 10^{-3} \text{Nm}^{-2}$.

Despite its weak magnitude, it can be argued that the Casimir force is perhaps the most fundamental consequence of tailoring the electromagnetic vacuum, since it manifests itself with no atom or other radiating elements between the conducting plates. The Casimir force also offers a fascinating link between the topic of this chapter and the mechanical effects of light to which we will turn next. This is because it can be thought of as resulting from the radiation pressure exerted by the vacuum field on the cavity end plates. While at the resonance frequency of the resonator the radiation pressure inside the cavity is stronger than outside, and the mirrors can therefore be pushed apart, out of resonance, the radiation pressure inside the cavity is smaller than outside and the mirrors are drawn toward each other. On balance, the attractive components have a slightly stronger impact than the repulsive ones. For two perfect plane parallel mirrors the radiation pressure of the Casimir force is therefore attractive and the mirrors are pulled together.

Problems

Problem 7.1 Driven-damped Jaynes–Cummings model—*This important exercise discusses the situation where the optical cavity is driven by an external field. It introduces two important parameters frequently encountered in cavity QED and circuit QED, the single atom cooperativity parameter C and the critical photon number n_0.*

We consider a two-level atom placed in a cavity and subject to the master equation

$$\frac{d\hat{\rho}_s}{dt} = -\frac{i}{\hbar}[\hat{H}, \hat{\rho}_s] + \hat{\mathcal{L}}_\kappa[\hat{\rho}] + \hat{\mathcal{L}}_\gamma[\hat{\rho}],$$

where

$$\hat{H} = \frac{1}{2}\hbar\Delta\hat{\sigma}_z + \hbar\delta\hat{a}^\dagger\hat{a} + \hbar g(\hat{a}\hat{\sigma}_+ + \hat{a}\hat{\sigma}_-) + \hbar\sqrt{\kappa}\mathcal{E}_L(\hat{a} + \hat{a}^\dagger)$$

is the Jaynes–Cummings Hamiltonian, with an extra term accounting for the fact that the intracavity field is driven by a classical field $\mathcal{E}_L$ of frequency ω_L. Here $\Delta = \omega_0 - \omega_L$ and $\delta = \omega_c - \omega_L$, with ω_0 the atomic transition frequency and ω_c the cavity mode frequency. In addition,

$$\hat{\mathcal{L}}_\kappa[\hat{\rho}] = -\frac{\kappa}{2}[\hat{a}^\dagger\hat{a}\hat{\rho}(t) - \hat{a}\hat{\rho}(t)\hat{a}^\dagger] + \text{adj.}$$

is the Liouvillian describing the dissipation of the cavity mode at rate κ, see Eq. (5.57), and

$$\hat{\mathcal{L}}_\Gamma[\hat{\rho}] = -\frac{\Gamma}{2}[\hat{\sigma}_+\hat{\sigma}_-\hat{\rho}(t) - \hat{\sigma}_-\hat{\rho}(t)\hat{\sigma}_+] + \text{adj.}$$

is the Liouvillian that accounts for atomic spontaneous emission, see Eq. (5.66). Both reservoirs are assumed to be at zero temperature.

(a) Show that the equations of motion for $\langle \hat{a} \rangle$, $\langle \sigma_z \rangle$, and $\langle \sigma_- \rangle$ are then

$$\frac{d}{dt} \langle \hat{a} \rangle = - (\kappa/2 + i\delta) \langle \hat{a} \rangle - i\sqrt{\kappa} \mathcal{E}_L - ig \langle \hat{\sigma}_- \rangle,$$

$$\frac{d}{dt} \langle \hat{\sigma}_- \rangle = - (\Gamma/2 + i\Delta) \langle \hat{\sigma}_- \rangle + ig \langle \hat{a} \hat{\sigma}_- \rangle,$$

$$\frac{d}{dt} \langle \hat{\sigma}_z \rangle = -\Gamma(\langle \sigma_z \rangle + 1) - 2ig[\langle \hat{a} \hat{\sigma}_+ \rangle - \langle \hat{a}^\dagger \hat{\sigma}_- \rangle].$$

(b) What additional information/equations would be required to solve this problem exactly?

(c) Introducing a semiclassical approximation where $\hat{a} \to \alpha$, so that $\langle \hat{a} \hat{\sigma}_- \rangle \to \alpha \langle \hat{\sigma}_- \rangle$ and $\langle \hat{a} \hat{\sigma}_+ \rangle \to \alpha^* \langle \hat{\sigma}_+ \rangle$, show that in steady state and for $\Delta = \delta = 0$ we have then

$$\alpha = - \frac{2i\mathcal{E}_L}{\sqrt{\kappa}} \left[1 + \frac{2C}{1 + n/n_0} \right]^{-1},$$

$$\langle \hat{\sigma}_z \rangle = - \frac{1}{1 + n/n_0},$$

$$\langle \hat{\sigma}_- \rangle = \frac{2ig}{\Gamma} \alpha \langle \hat{\sigma}_z \rangle,$$

where $n = |\alpha|^2$ and

$$C = 2g^2/\kappa\Gamma \quad ; \quad n_0 = \Gamma^2/8g^2. \tag{7.135}$$

C is called the *single atom cooperativity* and n_0 is the *critical photon parameter*. Some authors also introduce the *critical atom number* $N_0 = 1/C$.

(d) Discuss the interpretation and physical significance of these parameters, in particular in terms of the strong coupling regime and the saturation of the atomic transition.

Problem 7.2 Consider a micromaser system pumped by a mixture of excited and ground state atoms injected inside the cavity at rate R_e for atoms in state $|e\rangle$, and R_g for atoms in the ground state $|g\rangle$. Determine and plot the resulting steady-state photon statistics as a function of the pump parameter Θ and a function of the ratio R_e/R_g for $n_{th} = 0.1$.

Problem 7.3 Consider a micromaser operated in the dispersive limit, where the atom–field interaction is described by the effective Jaynes–Cummings Hamiltonian

$$H = \frac{1}{2}\hbar\omega_0 \hat{\sigma}_z + \hbar\omega \hat{a}^\dagger \hat{a} + \frac{\hbar g^2}{\Delta} \hat{a}^\dagger \hat{a} [|e\rangle\langle e| - |g\rangle\langle g|].$$

Following an approach that parallels the analysis leading to Eq. (7.36), determine the stationary photon statistics $\bar{p}_n$ as a function of $\Theta\sqrt{n}g\tau$ in that regime.

Problem 7.4 Evaluate and plot the $P(\alpha)$ distribution of the cat states

$$|\psi\rangle_\pm = \frac{1}{\sqrt{2}}(|\alpha\rangle \pm |-\alpha\rangle),$$

where $|\alpha\rangle$ is a coherent state with $\alpha = 2$.

Problem 7.5 Evaluate and plot the Wigner function of these same two cat states as in Problem 7.4.

Problem 7.6 Show that the charge and flux operators $\hat{Q}$ and $\hat{\phi}$ of an LC circuit can be expressed as

$$\hat{Q} = -iQ_{\text{ZPF}}(\hat{a} - \hat{a}^\dagger) \quad ; \quad \hat{\phi} = \Phi_{\text{ZPF}}(\hat{a} + \hat{a}^\dagger),$$

with $Q_{\text{ZPF}} = (\hbar/2Z)^{1/2}$ and $\Phi_{\text{ZPF}} = (\hbar Z/2)^{1/2}$, where Z is the characteristic impedance $Z = \sqrt{L/C}$ of the circuit. Express also these quantities in terms of the superconducting resistance quantum $R_Q = h/(2e)^2$ and the superconducting flux quantum $\Phi_0 = h/2e$.

Problem 7.7

(a) Show that the eigenstates and eigenenergies of the Hamiltonian

$$\hat{H}_T = -\frac{1}{2}E_J \sum_n [\,|n\rangle\langle n+1| + |n+1\rangle\langle n|\,]$$

are

$$|\phi\rangle = \sum_{m=-\infty}^{\infty} e^{im\varphi}|m\rangle$$

with eigenvalues $-E_J \cos\varphi$, that is, $\hat{H}_T|\varphi\rangle = -E_J\cos\varphi|\varphi\rangle$.

(b) Show also that

$$\hat{I} = 2e\frac{i}{\hbar}[\hat{H}_T, \hat{n}] = -\frac{eE_J}{\hbar}\sum_{m=-\infty}^{\infty} [|m\rangle\langle m+1| - |m+1\rangle\langle m|]\,,$$

where $\hat{n} = \sum_n n|n\rangle\langle n|$.

References

1. E.M. Purcell, H.C. Torrey, R.V. Pound, Resonance absorption by nuclear magnetic moments in a solid. Phys. Rev. **69**, 37 (1946)
2. D. Kleppner, Inhibited spontaneous emission. Phys. Rev. Lett. **47**, 233 (1981)
3. S. Haroche, J.-M. Raimond, *Exploring the Quantum—Atoms, Cavities and Photons* (Oxford, New York, 2006)
4. P. Goy, J.M. Raimond, M. Gross, S. Haroche, Observation of cavity-enhanced single-atom spontaneous emission. Phys. Rev. Lett. **50**, 1903 (1983)
5. R.G. Hulet, E.S. Hilfer, D. Kleppner, Inhibited spontaneous emission by a Rydberg atom. Phys. Rev. Lett. **55**, 2137 (1985)
6. J. Parker, C.R. Stroud, Transient theory of cavity-modified spontaneous emission. Phys. Rev. A **35**, 4226 (1987)
7. R.J. Cook, P.W. Milonni, Quantum theory of an atom near partially reflecting walls. Phys. Rev. A **35**, 5081 (1987)
8. M. Brune, F. Schmidt-Kaler, A. Maali, J. Dreyer, E. Hagley, J.M. Raimond, S. Haroche, Quantum Rabi oscillation: A direct test of field quantization in a cavity. Phys. Rev. Lett. **76**, 1800 (1996)
9. D. Meschede, H. Walther, G. Müller, One-atom maser. Phys. Rev. Lett. **54**, 551 (1985)
10. A.M. Guzman, P. Meystre, E.M. Wright, Semiclassical theory of a micromaser. Phys. Rev. A **40**, 2471 (1989)
11. P. Meystre, M. Sargent III, *Elements of Quantum Optics*, 4th edn. (Springer, Berlin, 2007)
12. P. Meystre, G. Rempe, H. Walther, Very-low temperature of a micromaser. Optics Lett. **13**, 1078 (1988)
13. M. Weidinger, B.T.H. Varcoe, R. Heerlein, H. Walther, Trapping states in the micromaser. Phys. Rev. Lett. **82**, 3795 (1999)
14. A. Messiah, *Quantum Mechanics* (North Holland, Amsterdam, 1961)
15. S. Haroche, J.-M. Raimond, *Manipulation of Nonclassical Field States in a Cavity by Atom Interferometry: In Cavity Quantum Electrodynamics*, ed. by P. Berman (Academic Press, Boston, 1993), p. 123
16. S. Deléglise, I. Dotsenko, C. Sayrin, J. Bernu, M. Brune, J.-M. Raimond, S. Haroche, Reconstruction of non-classical cavity field states with snapshots of their decoherence. Nature **455**, 510 (2008)
17. S. Girvin, Circuit QED: superconducting qubits coupled to microwave photons, in *Proceedings of the 2011 Les Houches Summer School on Quantum Machines*, ed. by M.H. Devoret, B. Huard, R.J. Schoelkopf, B.L.F. Cugliandolo (Oxford University, Oxford, 2014)
18. P. Krantz, M. Kjaergaard, F. Yan, T.P. Orlando, S. Gustavsson, W.D. Oliver, A quantum engineer's guide to superconducting qubits. Appl. Phys. Review **6**, 021318 (2019)
19. A. Blais, A.L. Grimsmo, S.M. Girvin, A. Wallraff, Circuit quantum electrodynamics. Rev. Mod. Phys. **93**, 025005 (2021)
20. M. Tinkham, *Introduction to Superconductivity* (Dover, New York, 1996)
21. D. Vion, A. Aassime, A. Cottet, P. Joyez, H. Pothier, C. Urbina, D. Esteve, M.H. Devoret, Manipulating the quantum state of an electrical circuit. Science **296**, 886 (2002)
22. J. Koch, T.M. Yu, J. Gambetta, A.A. Houck, D.I. Schuster, J. Majer, A. Blais, M.H. Devoret, S.M. Girvin, R.J. Schoelkopf, Charge-insensitive qubit design derived from the cooper-pair box. Phys. Rev. A **76**, 042319 (2007)
23. V.E. Manucharyan, J. Kock, L.I. Glazman, M.H. Devoret, Fluxonium: single cooper-pair circuit free of charge offsets. Science **326**, 113 (2009)
24. A. Blais, R.-S. Huang, A. Wallraff, S.M. Girvin, R.J. Schoelkopf, Cavity quantum electrodynamics for superconducting electrical circuits: An architecture for quantum computation. Phys. Rev. A **69**, 062320 (2004)
25. H.B.G. Casimir, On the attraction between two perfectly conducting plates. Proc. Kon. Ned. Akad. Wet. **51**, 793 (1948)

26. H.B.G. Casimir, D. Polder, The influence of retardation on the London-van der Waals forces. Phys. Rev. **73**, 360 (1948)
27. M.J. Sparnaay, Measurements of attractive forces between flat plates. Physica **24**, 6 (1958)
28. S.K. Lamoreaux, Demonstration of the Casimir force in the 0.6–6μm range. Phys. Rev. Lett. **78**, 5 (1997)

Chapter 8
Mechanical Effects of Light

This chapter takes a first quantitative look at the way light can modify atomic trajectories. Except for spontaneous emission, which is treated phenomenologically, we describe the optical fields classically, an approximation that is sufficient to introduce two key components of the light force, the gradient (or dipole) force and the dissipative radiation pressure force. We then introduce several aspects of atomic diffraction by light fields, including the Raman–Nath, Bragg, and Stern–Gerlach regimes. The chapter concludes with an introduction to atom interferometry.

The idea that light carries momentum and hence can influence the trajectory of massive particles goes back to Johannes Kepler, who observed that the tail of comets always points away from the sun and concluded that "The direct rays of the Sun strike upon it [the comet], penetrate its substance, draw away with them a portion of this matter, and issue thence to form the track of light we call the tail." (J. Kepler, as quoted in "A Comet Called Halley," by I. Ridpath [1].) This idea was elaborated upon by Newton, but of course it is Maxwell's theory of electromagnetism that put it on a solid theoretical footing.

In the previous chapter we briefly encountered mechanical effects of light in the discussion of the inverse Stern–Gerlach effect and the realization of optical Schrödinger cats, and then again in the discussion of the Casimir force. Our goal is now to put this analysis on a more solid footing, concentrating first on two-level atoms. The next chapters will then move on to laser cooling and the remarkable new directions of research that it has opened, most importantly perhaps with the realization of quantum degenerate atomic systems and of quantum simulators of many-body solid-state systems. We will finally turn to quantum optomechanics, where the cooling of mesoscopic and macroscopic objects offers enormous promise in quantum metrology and quantum information. As such we are now redirecting our focus toward the analysis of optical ways to bring the motion of massive systems deep into the quantum regime and to optically control and manipulate that motion.

© The Author(s), under exclusive license to Springer Nature Switzerland AG 2021   229
P. Meystre, *Quantum Optics*, Graduate Texts in Physics,
https://doi.org/10.1007/978-3-030-76183-7_8

8.1   Semiclassical Atom–Field Interaction Revisited

Our purpose in this chapter is to take a first quantitative look at the way light can modify the trajectory of two-level atoms. At this stage we describe the optical field classically, an approximation that is sufficient to introduce two key components of the light force, the gradient (or dipole) force and the radiation pressure force. It is characterized by a slowly varying amplitude $\mathcal{E}(z)$, frequency ω, and polarization $\vec{\epsilon}(z)$,

$$\mathbf{E}(\hat{z}, t) = \vec{\epsilon}(\hat{z})\mathcal{E}(\hat{z}) \cos[\omega t + \Phi(\hat{z})]. \tag{8.1}$$

Note that when expressed in this form the phase $\Phi(z)$ is *not* slowly varying, as it includes the usual kz spatial dependence of the quasi-monochromatic field $\mathbf{E}(\hat{z}, t)$. Spontaneous emission is introduced phenomenologically, and we assume that it leads to the decay of the atom from its excited state $|e\rangle$ to the ground state $|g\rangle$ at the Weisskopf–Wigner rate (5.18),

$$\Gamma = \frac{1}{4\pi\epsilon_0} \frac{4\omega_0^2 |d|^2}{3\hbar c^2}. \tag{8.2}$$

This is essentially the level of approximation of Chap. 1, the new element being that we now account explicitly for the changes in the center-of-mass motion of the atom resulting from its interaction with the field.

Working for now in one spatial dimension only we proceed by quantizing the atomic center-of-mass position and momentum z and p, which become the canonically conjugate operators $\hat{z}$ and $\hat{p}$. It is therefore no longer sufficient to describe the atoms with a density operator $\hat{\rho}$ that characterizes their internal atomic state only. Rather, the matrix elements ρ_{ij}, with $i, j = \{e, g\}$, are now operator-valued quantities $\hat{\rho}_{ij}(z)$ or $\hat{\rho}_{ij}(p)$, depending on the representation—coordinate or momentum—selected for the description of the center-of-mass dynamics.

After adding the kinetic energy of a two-level atom of mass m to the Hamiltonian (1.61) the atom–field system is then described in the electric dipole and rotating wave approximations by

$$\hat{H} = \frac{\hat{p}^2}{2m} + \hbar\omega_0 |e\rangle\langle e| - \frac{\hbar\Omega_r(\hat{z})}{2} \left[e^{-i[\omega t + \Phi(\hat{z})]} |e\rangle\langle g| + \text{h.c.} \right], \tag{8.3}$$

where we have taken the energy $\hbar\omega_g$ of the ground electronic level $|g\rangle$ as the zero of energy. The electric field is evaluated at the position $\hat{z}$ of the atom, and the spatially dependent resonant Rabi frequency (1.62) has been slightly generalized to account for a potentially spatially varying field polarization. It reads now

$$\Omega_r(\hat{z}) = d[\vec{\epsilon}_z \cdot \vec{\epsilon}(\hat{z})]\mathcal{E}(\hat{z})/\hbar, \tag{8.4}$$

where $\vec{\epsilon}_z$ is a unit vector along the quantization axis. In addition, the atom is subject to spontaneous emission.

8.2   Gradient and Radiation Pressure Forces

The change in momentum $\hat{p}$ of the atom under the influence of the Hamiltonian (8.3) is given by the Heisenberg equation of motion

$$\frac{d\hat{p}}{dt} = \frac{i}{\hbar}[\hat{H}, \hat{p}],\tag{8.5}$$

which readily yields

$$\begin{aligned}\frac{d\hat{p}}{dt} &= \frac{i}{\hbar}\left[-\frac{\hbar\Omega_r(\hat{z})}{2}\left(e^{-i\Phi(\hat{z})}e^{-i\omega t}|e\rangle\langle g| + \text{h.c.}\right), \hat{p}\right]\\ &= \frac{\hbar}{2}|e\rangle\langle g|\nabla\left[\Omega_r(z)e^{-i\Phi(z)}e^{-i\omega t}\right] + \text{h.c.},\end{aligned}\tag{8.6}$$

where we have used in the second equality the coordinate representation commutation relation

$$[f(\hat{x}), \hat{p}] = i\hbar\nabla f(\hat{x}).\tag{8.7}$$

Although it might appear that the kinetic energy part of the atomic Hamiltonian plays no role here since it commutes with $\hat{p}$, this is not the case since in addition to Eq. (8.6) we need to consider the equation of motion for the center-of-mass position,

$$\frac{d\hat{x}}{dt} = \frac{\hat{p}}{m}.\tag{8.8}$$

The full description of the influence of light on atomic trajectories requires the simultaneous solution of the coupled operator equations (8.6) and (8.8), but for now we just consider the expectation value of Eq. (8.6), which can be thought of as a form of Newton's law for the mean atomic momentum. In this spirit, the right-hand side of that equation can be interpreted as a light force acting on the atomic center of mass. Its explicit form is

$$F(z) = \left\langle\frac{d\hat{p}}{dt}\right\rangle = \frac{\hbar}{2}\left\langle|e\rangle\langle g|\nabla\left[\Omega_r(z)e^{-i\Phi(z)}e^{-i\omega t}\right] + \text{c.c.}\right\rangle,\tag{8.9}$$

where the expectation value is taken on both the internal degrees of freedom and the center-of-mass state of the atom.

The recoil momentum imparted on the atom by the absorption or stimulated emission of a photon is $\hbar k$, and the associated frequency is the *recoil frequency*

$$\omega_{\text{rec}} = \frac{\hbar k^2}{2m}.$$ (8.10)

Its inverse defines a characteristic time ω_{rec}^{-1} for the center-of-mass dynamics. It should be compared to the characteristic time for the internal dynamics, which is of the order of the spontaneous lifetime Γ^{-1} of the transition. In many cases these two time scales are vastly different, with ω_{rec} typically of the order of 10–500 s^{-1} and Γ of the order of $10^6 - 10^9$ s^{-1}. If that is the case, the internal state of the atoms can be assumed to be in a quasi-steady-state relative to that of the center of mass, and the internal and external contributions to the force (8.9) may be factorized as

$$F(z) = \frac{\hbar}{2} \langle |e\rangle \langle g| \rangle_{\text{internal}} \left\langle \boldsymbol{\nabla}[\Omega_r(z)e^{-i\Phi(z)}]e^{-i\omega t} \right\rangle_{\text{external}} + \text{c.c.}$$ (8.11)

It is important however to keep in mind that this factorization scheme is not always justified. In particular it ignores any possible quantum entanglement between the internal and center-of-mass motion of the atom. This is an important pitfall in many of the most interesting applications of atom optics, which oftentimes exploit such entanglements as we already saw in the discussion of optical Schrödinger cats of Sect. 7.3.

A further difficulty arises when trying to evaluate the center-of-mass expectation value of the operator $\boldsymbol{\nabla}\left[\Omega(z)\exp(-i\Phi(z))\right]$ because it is generally a complicated function of z. For ultracold particles, in particular, there is no obvious way to evaluate this expression short of determining the center-of-mass wave function $\psi(z, t)$. For well-localized particles, however, one can approximate this wave function by a δ-function located at some location $z_0(t)$. In this case, and keeping in mind these limitations, Eq. (8.11) reduces to

$$F(z) \simeq \frac{\hbar}{2} \langle |e\rangle \langle g| \rangle_{\text{internal}} \boldsymbol{\nabla}[\Omega_r(z)e^{-i\Phi(z)}e^{-i\omega t}]_{z=z_0} + \text{c.c.},$$ (8.12)

where $z_0(t)$ is the classical center-of-mass location of the atom. Because this approximation is reminiscent to the ray optics limit of conventional optics we adopt the same usage here and call it the *ray atom optics limit* of the light force. We will first encounter the wave atom optics regime in the discussion of atomic diffraction of Sect. 8.4.

In the limit where $\langle |e\rangle \langle g| \rangle_{\text{internal}}$ can be evaluated in steady state we have finally

$$F(z) = \frac{\hbar \Omega_r(z)}{2} \left[U_{\text{st}}\boldsymbol{\alpha}(z) + V_{\text{st}}\boldsymbol{\beta}(z)\right],$$ (8.13)

where we have introduced the parameters

$$\alpha(\mathbf{r}) \equiv \frac{\nabla \Omega_r(\mathbf{r})}{\Omega_r(\mathbf{r})} \tag{8.14}$$

$$\beta(\mathbf{r}) \equiv \nabla \Phi(\mathbf{r}), \tag{8.15}$$

or more precisely their one-dimensional version, and expressed the density matrix elements ρ_{eg} and ρ_{ge} in terms of the steady-state Bloch vector components of Eqs. (1.89) and (1.90),

$$U_{\text{st}} = -\frac{2\Delta}{\Omega_r} \left(\frac{s}{1+s} \right) \quad ; \quad V_{\text{st}} = \frac{\Gamma}{\Omega_r} \left(\frac{s}{1+s} \right),$$

with

$$s = \frac{\Omega_r^2/2}{\Gamma^2/4 + \Delta^2}$$

the saturation parameter. We also dropped the subscript in $\mathbf{r}_0$ for notational clarity.

We recall from Chap. 1 that the U-component of the Bloch vector is responsible for dispersive effects, while the V-component is responsible for absorption and emission. This is apparent from the dispersive form of $U_{\text{st}}(\Delta)$, which should be contrasted to the Lorentzian absorption/emission profile associated with $V_{\text{st}}(\Delta)$. This naturally leads to the decomposition of the force F into two components as

$$F(z) = F_{\text{rp}}(z) + F_{\text{gr}}(z), \tag{8.16}$$

where $F_{\text{rp}}(z)$ is the *radiation pressure* force

$$F_{\text{rp}}(z) = \frac{1}{2} \hbar \Omega_r V_{\text{st}} \beta = \frac{\hbar \Gamma}{2} \left(\frac{s}{1+s} \right) \nabla \Phi(z), \tag{8.17}$$

which we already briefly encountered in the discussion of the Casimir force of Sect. 7.5, and $F_{\text{gr}}(z)$ is a reactive force known as the *dipole* or *gradient* force[1]

$$F_{\text{gr}}(z) = \frac{1}{2} \hbar \Omega_r U_{\text{st}} \alpha = -\hbar \Delta \left(\frac{s}{1+s} \right) \frac{\nabla \Omega_r(z)}{\Omega_r(z)}. \tag{8.18}$$

Radiation Pressure Force   The expression (8.17) shows that the radiation pressure force is nonvanishing provided the laser field exhibits a phase gradient. This force is central to the Doppler cooling technique that will be discussed in the next chapter.

---

[1]Remember when comparing different publications that many authors use the alternative definition of detuning $\delta = \omega - \omega_0 = -\Delta$.

In contrast, the dipole (or gradient) force requires a field amplitude gradient. Note also the change in sign of the dipole force as the laser is tuned across the atomic resonance, a direct consequence of its dispersive nature. This property can be used to achieve state-selective atomic mirrors and optical dipole traps, which play an important role in the study of ultracold and quantum degenerate atomic and molecular systems, as we will see in Chap. 10.

For a monochromatic running wave

$$\mathbf{E}(z, t) = \vec{\epsilon} \, \mathcal{E} \cos(\omega t - kz) , \qquad (8.19)$$

the radiation pressure force becomes

$$F_{\text{rp}}(z) = \hbar k \frac{\Gamma}{2} \left(\frac{\Omega_r^2 / 2}{\Omega_r^2 / 2 + \Delta^2 + (\Gamma/2)^2} \right) . \qquad (8.20)$$

It has the familiar power-broadened Lorentzian line shape associated with absorption in two-level systems, see Fig. 8.1. Note that as the Rabi frequency Ω is increased, F_{rp} saturates to the value $F_{\text{rp}} \to \hbar \Gamma k/2$.

Since the reemission of a photon into the laser mode from which it was absorbed does not change the momentum of the field, it follows that the atomic momentum must remain constant as well. It follows that the radiation pressure force must result from the absorption of a photon from the laser beam, with associated momentum

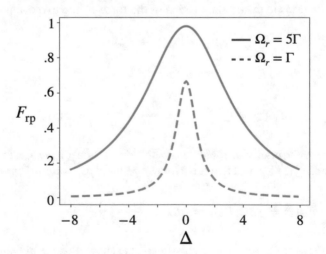

Fig. 8.1 Radiation pressure force, in units of $\hbar k \Gamma/2$, as a function of the laser-atom detuning $\Delta = \omega_0 - \omega$ for a monochromatic running wave. All frequencies are in units of the damping rate Γ. Note that if the atomic center-of-mass motion is treated classically, the velocity dependence of the force is readily obtained by including the Doppler shift, $\Delta \to \Delta \pm kv$, where the "+" sign corresponds to an atom moving in the opposite direction of propagation of the field and the "−" sign to an atom moving in the direction of propagation of the field

Fig. 8.2 Gradient (or dipole) force $F_{gr}(z, \Delta)$ resulting from a standing wave along the z-axis as a function of the detuning $\Delta = \omega_0 - \omega$, in units of Γ

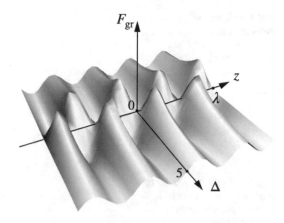

transfer $\hbar k$ to the atom, followed by its spontaneous reemission into the continuum of vacuum modes, a process that *on the average* does not result in any change in atomic momentum, but see Sect. 8.3 for a more careful discussion of this point.

Gradient Force In contrast to the radiation pressure force the gradient force (8.18) vanishes in the case of a plane running wave since such a wave does not exhibit a gradient in its field amplitude. But the situation is of course different for a standing wave or for a general superposition of plane waves. In that case we have, in one dimension,

$$F_{gr}(z) = \frac{\hbar \Delta}{4} \left(\frac{\nabla \Omega_r^2(z)}{\Omega_r^2(z)/2 + \Delta^2 + (\Gamma/2)^2} \right). \tag{8.21}$$

This force is plotted in Fig. 8.2 as a function of Δ for a standing wave along the z-axis.

Because $F_{gr}(z)$ is proportional to the U-component of the Bloch vector, it does not involve the absorption of energy from the field. Rather, it is due solely to the redistribution of momentum between the various plane waves composing that field by the atom. More specifically, an optical beam with a spatial inhomogeneity $\Omega(\mathbf{r})$ is comprised of a superposition of many plane waves propagating within the divergence angle of the beam, and the elementary process underlying the gradient force is the absorption by the atom of a photon from one of these plane waves and its subsequent stimulated emission into another, a point discussed in detail in Ref. [2]. For lasers tuned to the red of the atomic transition frequency, $\Delta > 0$, the atom is "strong field seeking," in that the dipole force directs the atoms toward regions of stronger fields. For blue detunings $\Delta < 0$, this force is repulsive and leads the atoms to regions of weak laser intensity. Note that in contrast to F_{rp}, F_{gr} does not saturate for increasing Rabi frequencies. Also, in addition to having different physical origins, F_{rp} and F_{gr} are also in general in different directions. This is illustrated in Fig. 8.3 for the case of a Gaussian beam profile.

Fig. 8.3 Schematic of the
radiation force $F = F_{rp} + F_{gr}$
on a two-level atom in a
focused Gaussian laser beam

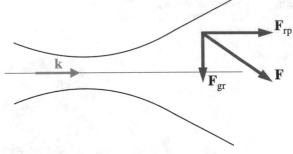

Fig. 8.4 Optical potential (in
arbitrary units) associated
with the dipole force of
Fig. 8.2. Detuning
$\Delta = \omega_0 = \omega$ in units of Γ

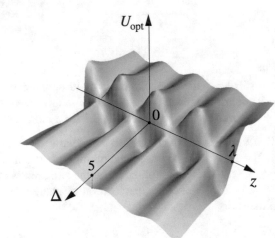

Since in contrast to the radiation pressure force the gradient force is associated
with a conservative process it can be described in terms of a potential U_{opt} (not to be
confused with the U-component of the Bloch vector), so that with $F_{gr} = -\nabla U_{opt}$ it
is possible to interpret the gradient force F_{gr} as deriving from the "optical potential"

$$U_{opt}(z) = -\frac{\hbar\Delta}{2} \ln\left(1 + \frac{\Omega_r^2(z)/2}{\Delta^2 + (\Gamma/2)^2}\right), \tag{8.22}$$

which is plotted in Fig. 8.4 for the example of Fig. 8.2. At large detunings, $|\Delta| \gg$
Γ, Ω_r is reduced simply to

$$U_{opt}(z) \approx -\frac{\hbar\Omega_r^2(z)}{4\Delta}. \tag{8.23}$$

The gradient and radiation pressure forces are central to a number of applications
in AMO physics and quantum optics, ranging from optical tweezers to optical
lattices, and from laser cooling and atom interferometry to the study of quantum
degenerate atomic systems. In many cases, though, it is necessary as we shall see

to go past the simple ray atom optics description considered so far and to properly account for the quantum nature of the atomic center-of-mass motion. The multilevel nature of the atoms is also often an essential ingredient, in particular in the design of optical traps and in laser cooling.

8.3 Dissipation

So far, our discussion of light forces has ignored the effects of spontaneous emission, except at the simple phenomenological level required to establish steady-state populations of the atomic levels. Except in the average sense invoked in the physical interpretation of the radiation pressure force, this treatment fails to account for the fact that the emission of a spontaneous photon must be accompanied by a random atomic recoil, with a recoil velocity

$$v_{\text{rec}} = \frac{\hbar k}{m}, \tag{8.24}$$

of the order of 10^{-2} m/s for alkali atoms. This recoil is oftentimes negligible, for instance when dealing with atomic samples at room temperature since thermal velocities are of the order of hundreds of meters per second. But this is not so in ultracold samples, where the random momentum kicks translate into a non-negligible heating of the atomic sample, or in atom interference or diffraction experiments, which usually require atomic beams of high monochromaticity. In such situations it is important to properly account for the impact of the random momentum fluctuations associated with spontaneous emission.

We have seen in Chap. 5 that the irreversibility of spontaneous emission makes it necessary to describe the evolution of the atomic state in terms of a master equation of the form

$$\frac{d\hat{\rho}}{dt} = -\frac{i}{\hbar}[\hat{H}, \hat{\rho}] + \hat{\mathcal{L}}[\hat{\rho}], \tag{8.25}$$

where $\hat{H}$ is the atomic Hamiltonian and the Liouvillian $\mathcal{L}[\hat{\rho}]$ accounts for irreversible processes. We now revisit this analysis to include not just the electronic degrees of freedom of the atom, but also its center-of-mass dynamics. The resulting master equation [3–5] can be derived using simple symmetry arguments and transformation properties [6].

Consider the atomic density matrix elements $\hat{\rho}_{ee}$, $\hat{\rho}_{gg}$, and $\hat{\rho}_{eg}$. As we already indicated, when the atomic center-of-mass motion is quantized they become operator-valued quantities rather than simple complex matrix elements. When atomic recoil is neglected and for atoms with upper to lower level decay the spontaneous emission contribution to their equations of motion is simply

$$\frac{\mathrm{d}\hat{\rho}_{ee}}{\mathrm{d}t}\bigg|_{\mathrm{sp}} = -\frac{\mathrm{d}\hat{\rho}_{gg}}{\mathrm{d}t}\bigg|_{\mathrm{sp}} = -\Gamma\hat{\rho}_{ee}\,, \tag{8.26}$$

$$\frac{\mathrm{d}\hat{\rho}_{eg}}{\mathrm{d}t}\bigg|_{\mathrm{sp}} = -\frac{\Gamma}{2}\hat{\rho}_{eg}\,, \tag{8.27}$$

as we have seen, the first two equations describing the familiar irreversible transfer of population from the excited to the ground state, and the third one giving the concomitant decay of the atomic coherence.

When atomic recoil is included, however, an excited atom with center-of-mass momentum $\hat{p}$ that decays to its ground state will have a shifted center-of-mass momentum $\hat{p} - \hbar\mathbf{k}$, where $\mathbf{k}$ is the momentum of the emitted photon. However, the depletion of the excited state population and the decay of the electronic coherences are then still described by Eqs. (8.26) and (8.27). This must be so, because the decay of the upper state cannot depend on the motional state of the atom, a direct consequence of Galilean invariance, and it cannot change the momentum of the excited atom, due to momentum conservation. The only modification is therefore in the ground state equation of motion. It is conveniently described by the momentum shift operator $\exp[-\mathrm{i}\mathbf{k}\cdot\hat{r}]$, which acts on the center-of-mass state vector as

$$e^{-\mathrm{i}\mathbf{k}\cdot\hat{r}}|\mathbf{p}\rangle = |\mathbf{p} - \hbar\mathbf{k}\rangle\,, \tag{8.28}$$

where $|\mathbf{p}\rangle$ is an eigenstate of the atomic center-of-mass momentum operator.

Energy conservation in the atomic rest frame requires that $k = \omega_0/c$ for the wave number of the emitted photon, so that the contribution of a spontaneous emission event in the $\mathbf{n}$ direction to the increase in population of the atomic ground state is

$$\mathrm{d}\Gamma_{\mathbf{n}}e^{-\mathrm{i}k\mathbf{n}\cdot\hat{r}}\hat{\rho}_{ee}e^{\mathrm{i}k\mathbf{n}\cdot\hat{r}}\,, \tag{8.29}$$

where the differential rate of spontaneous emission $\mathrm{d}\Gamma_{\mathbf{n}}$ along $\mathbf{n}$ is

$$\mathrm{d}\Gamma_{\mathbf{n}} = \Gamma\Phi(\mathbf{n})\mathrm{d}^2\mathbf{n}, \tag{8.30}$$

and $\Phi(\mathbf{n})\mathrm{d}^2\mathbf{n}$ is the probability of emission into the infinitesimal solid angle $\mathrm{d}^2\mathbf{n}$. For example, in the case of a linear dipole transition in free space one would have

$$\Phi(\mathbf{n}) = \frac{3}{8\pi}\left(1 - \frac{(\mathbf{n}\cdot\mathbf{d})^2}{d^2}\right)\,. \tag{8.31}$$

Integrating over all directions yields then

$$\frac{\mathrm{d}\hat{\rho}_{gg}}{\mathrm{d}t}\bigg|_{\mathrm{sp}} = \int \mathrm{d}\Gamma_{\mathbf{n}}e^{-\mathrm{i}k\mathbf{n}\cdot\hat{r}}\hat{\rho}_{ee}e^{\mathrm{i}k\mathbf{n}\cdot\hat{r}}\,, \tag{8.32}$$

and inserting Eqs. (8.26), (8.27), and (8.32) into Eq. (8.25) gives the master equation describing spontaneous emission by a freely traveling two-level atom as

$$\frac{d\hat{\rho}}{dt} = -\frac{i}{\hbar}[\hat{H}_{\text{eff}}\hat{\rho} - \hat{\rho}\hat{H}_{\text{eff}}^{\dagger}] + \Gamma \int d^2\mathbf{n}\Phi(\mathbf{n})e^{-i\mathbf{k}\mathbf{n}\cdot\hat{r}}\hat{\sigma}_-\hat{\rho}\hat{\sigma}_+ e^{i\mathbf{k}\mathbf{n}\cdot\hat{r}}. \tag{8.33}$$

As in the Monte Carlo wave functions approach of Sect. 5.4 we have introduced the effective non-Hermitian Hamiltonian

$$\hat{H}_{\text{eff}} = \hat{H}_A - i\hbar\frac{\Gamma}{2}\hat{\sigma}_+\hat{\sigma}_-. \tag{8.34}$$

The limit where photon recoil is neglected is recovered by setting $k = 0$ in the exponents of the integrand of Eq. (8.33), resulting in the master equation of Eqs. (5.120) and (5.121),

$$\frac{d\hat{\rho}}{dt} = -\frac{i}{\hbar}[\hat{H}_{\text{eff}}\hat{\rho} - \hat{\rho}\hat{H}_{\text{eff}}] + \Gamma\hat{\sigma}_-\hat{\rho}\hat{\sigma}_+. \tag{8.35}$$

The integral term in Eq. (8.33) accounts for the irreversible increase in population of the electronic ground state. It prevents a description of the atomic dynamics in terms of a simple Schrödinger equation for a state vector. If however one is only interested in the dynamics of the excited electronic state, such a description is still possible and compatible with the master equation (8.33). This can be seen from the following argument: The equation for $\hat{\rho}_{ee}$ is readily obtained from Eq. (8.33) as

$$\frac{d\hat{\rho}_{gg}}{dt} = -\frac{i}{\hbar}\left[\frac{\hat{p}^2}{2m}, \hat{\rho}_{ee}\right] - \Gamma\hat{\rho}_{ee}. \tag{8.36}$$

Writing then $\hat{\rho}_{ee} = |\phi_e(t)\rangle\langle\phi_e(t)|$, where $|\phi_e(t)\rangle$ is a ket describing the center-of-mass motion of the atom in its excited state, the equation for $\hat{\rho}_{ee}$ is immediately recognized to be compatible with the effective Schrödinger equation

$$i\hbar\frac{d|\phi_e\rangle}{dt} = \left[\frac{\hat{p}^2}{2m} + \hbar\left(\omega_0 - i\frac{\Gamma}{2}\right)\right]|\phi_e\rangle, \tag{8.37}$$

which is typically much easier to solve than the corresponding master equation, as we discussed in the analysis of the Monte Carlo quantum trajectories of Sect. 5.4. From Eq. (8.37) one finds immediately the upper state probability

$$P_e(t) = e^{-\Gamma t}\langle\phi_e(0)|\phi_e(0)\rangle, \tag{8.38}$$

with a lifetime that is independent of the center-of-mass state of the atom, as we argued should be the case.[2]

8.4 Atomic Diffraction

In the analysis of the gradient and radiation pressure forces of Sect. 8.2 we mentioned the difficulty in evaluating the center-of-mass expectation value of the operator $\nabla \left[\Omega_r(z) \exp(-i\Phi(z)) \right]$ and restricted ourselves to the ray atom optics regime where the center-of-mass atomic wave function is approximated by its center-of-mass position. This approach is particularly useful in situations where the thermal de Broglie wavelength of the atoms remains short compared to an optical wavelength, that is, in atomic samples above the so-called recoil temperature

$$T_{\text{recoil}} = \hbar^2 k^2 / 2 m k_B . \tag{8.39}$$

However, it typically fails in the subrecoil temperature regime, atomic diffraction experiments, and the description of the interaction between Bose–Einstein condensates and light. This is the regime considered in this section, which discusses a first example of *wave atom optics*, the diffraction of atomic matter waves by a light field.

In 1933, P. Kapitza and P. M. Dirac [7] predicted that an electron beam could be diffracted by a standing light field as a result of stimulated Compton scattering. However, they concluded that the experiment was not feasible, due to the lack of a suitable light source. It was not until 1965 that the experiment was finally carried out by L. S. Bartell et al. [8]. Shortly thereafter it was suggested that diffraction by light fields could occur for neutral atoms as well [9, 10]. It was noted that the effect could be significantly stronger than with electrons, as a result of the resonant enhancement of the atom–field interaction. The first experimental observations of atomic diffraction by optical gratings were carried out by D. Pritchard and his coworkers [11–13], following a series of earlier experiments on atomic deflection that lacked however the resolution required to separate various diffraction orders [14, 15].

A typical atomic diffraction experiment consists of a monoenergetic beam of atoms interacting with a standing wave light field, see Fig. 8.5. After leaving the field region, the atoms further propagate toward a screen or some other detection system. The resulting near-resonant atomic Kapitza–Dirac effect can be categorized into three major regimes, commonly named the Raman–Nath, Bragg, and Stern–Gerlach regimes. In the first two cases the wave packets of the impinging atoms

---

[2]This last statement assumes that the Doppler effect does not shift the atomic transition frequency ω_0 to a value for which the density of modes of the electromagnetic vacuum exhibits unusual features, such as might be the case near a photonic band gap with a field mode density equal to zero.

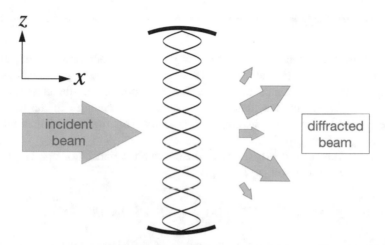

Fig. 8.5 Schematic of an atomic diffraction experiment at a standing wave light field. The atoms are typically detected in the far-field region to the right of the optical grating

are large compared to the period of the standing wave pattern. As a result, the atoms probe the full spatial structure of the potential. In contrast, the Stern–Gerlach regime is characterized by wave packets that are well localized compared to the periodicity of the optical potential.

To describe these various regimes theoretically we consider an atomic beam of high enough momentum p_x in the direction perpendicular to the direction of propagation z of the field that it can be treated classically. In contrast, its transverse momentum $\hat{p}_z$ is assumed to be comparable to or smaller than the recoil momentum $\hbar k$ and is treated quantum mechanically. Ignoring for now the effects of spontaneous emission, the Hamiltonian describing the interaction of the atoms with the standing wave is therefore, in a frame rotating at the laser frequency ω,

$$\hat{H} = \frac{\hat{p}_z^2}{2m} + \hbar\Delta |e\rangle\langle e| + \hbar\Omega_r \cos(k\hat{z}) f(t) \left(|e\rangle\langle g| + \text{h.c.}\right), \qquad (8.40)$$

where $[\hat{z}, \hat{p}_z] = i\hbar$, $\Omega_r = d\mathcal{E}/\hbar$ is the resonant Rabi frequency (1.62), and $\Delta = \omega_0 - \omega$. The function $f(t)$ accounts for the details of interaction time and strength of the dipole coupling between the atom and the laser beam. It is determined by the velocity v_x of the atoms and the laser beam profile. In the following we assume for simplicity that this profile is rectangular of width L, so that

$$f(t) = \Theta(t) - \Theta(t - L/v_x), \qquad (8.41)$$

where Θ is the Heaviside function.

8.4.1   Raman–Nath Regime

Raman–Nath diffraction refers to a regime where it is appropriate to neglect the kinetic energy term in the Hamiltonian (8.40), while still treating $\hat{z}$ as an operator. This approximation amounts to considering an atom of infinite mass. As such it accounts properly for the light-induced momentum changes in the transverse atomic motion but neglects its changes in velocity. This regime was first analyzed theoretically in Refs. [16–18].

It follows from the property (8.28) of the momentum translation operator that the action of $\cos(k\hat{z})$ on the ket $|p\rangle$ is simply[3]

$$\cos(k\hat{z})|p\rangle = \frac{1}{2}\left(e^{ikz} + e^{-ikz}\right)|p\rangle = \frac{1}{2}\left(|p + \hbar k\rangle + |p - \hbar k\rangle\right), \qquad (8.42)$$

an expression that provides a simple picture of the effect of the optical field on the atomic center-of-mass motion. Each time the atom absorbs energy from the wave propagating in the $+z$ direction, its center-of-mass momentum is increased by $\hbar k$. Conversely, each time it decays by emitting light in that same direction, its momentum is decreased by $\hbar k$. Since the situation is reversed for the wave propagating in the $-z$ direction, it follows that each atomic transition can result in a momentum kick $\pm \hbar k$. The result of successive absorption and emission events is that the atom acquires ever higher and lower momentum components, which differ from the initial momentum by integer numbers of $\hbar k$.

In order to quantify this effect it is convenient to work in the momentum representation and to express the state vector of the atom as

$$|\psi(t)\rangle = C_e(p, t)|e\rangle + C_g(p, t)|g\rangle, \qquad (8.43)$$

where $C_e(p, t)$ and $C_g(p, t)$ are the momentum representation wave functions associated with the atom in its excited and ground electronic states, respectively. Substituting $|\psi(t)\rangle$ into the Schrödinger equation yields the infinite set of coupled ordinary differential equations

$$i\hbar \frac{dC_e(p, t)}{dt} = \frac{\hbar \Omega_r}{2} \left[C_g(p + \hbar k, t) + C_g(p - \hbar k, t) \right] + \hbar \Delta \, C_e(p, t)$$

$$i\hbar \frac{dC_g(p, t)}{dt} = \frac{\hbar \Omega_r}{2} \left[C_e(p + \hbar k, t) + C_e(p - \hbar k, t) \right]. \qquad (8.44)$$

Consider for concreteness the resonant situation $\Delta = 0$ and the initial condition $|\psi(0)\rangle = C_g(p = 0, t = 0)|g\rangle$. It describes an atom in its ground state with a well-defined transverse momentum $p = 0$, corresponding to a plane wave for

---

[3]To simplify the notation we omit the subscript "z" in the z-component $\hat{p}_z$ of the momentum operator in the following unless confusion is possible.

the center-of-mass atomic wave function. This initial condition, combined with Eq. (8.44), implies that only transverse momenta that are integer numbers of $\hbar k$ can ever become populated. We can therefore expand the partial wave functions $C_e(p, t)$ and $C_g(p, t)$ as

$$C_e(p, t) - \sum_{\ell=-\infty}^{\infty} e_\ell(t)\delta(p - \ell\hbar k),$$

$$C_g(p, t) = \sum_{\ell=-\infty}^{\infty} g_\ell(t)\delta(p - \ell\hbar k), \tag{8.45}$$

with the initial condition[4] $e_\ell(0) = 0$, $g_\ell(0) = \delta_{\ell 0}$. The equations of motion (8.44) reduce then to

$$i\hbar \frac{dx_\ell}{dt} = \frac{\hbar\Omega_r}{2} [x_{\ell-1} + x_{\ell+1}], \tag{8.46}$$

where $x_\ell = e_\ell$ for ℓ odd and $x_\ell - g_m$ for ℓ even, since for our initial condition even scattering orders always correspond to ground state atoms and odd scattering orders to excited atoms. The solution of this equation is known to be in the form of ℓth-order Bessel functions of the first kind [16],

$$x_\ell(t) = i^\ell J_\ell(\Omega_r t) . \tag{8.47}$$

This gives the probability $P_\ell(t)$ for the atom to have a transverse momentum $\ell\hbar k$ as

$$P_\ell(t) = J_\ell^2(\Omega_r t) . \tag{8.48}$$

The Raman–Nath approximation is valid provided that the transverse kinetic energy of the atoms remains small compared with the interaction energy $\hbar\Omega_r$. Clearly, as more and more scattering orders are excited, this condition eventually ceases to be valid. The transverse kinetic energy corresponding to the ℓth scattering order is easily seen to be $\hbar\ell^2\omega_{\text{rec}}$, which implies that we must have $\ell^2\omega_{\text{rec}} \ll \Omega_r$. From the properties of the Bessel functions J_ℓ one can show that after an interaction time t, $2\ell_{\max}$ translational states are populated, with $\ell_{\max} \simeq 2\Omega_0 t$. This implies that the Raman–Nath approximation holds provided that $t \ll 1/\sqrt{4\Omega_r\omega_{\text{rec}}}$.

Figure 8.6 shows a numerical solution of Eq. (8.44) that also includes the kinetic energy term of the Hamiltonian (8.40) and the effects of spontaneous emission. It illustrates the linear increase in the number of scattering orders as a function of time predicted by the Raman–Nath approach and shows how this growth is

---

[4]The δ-functions should not be taken literally. What is meant instead is a series of sharply peaked and normalizable functions whose momentum width is much smaller than the recoil momentum $\hbar k$.

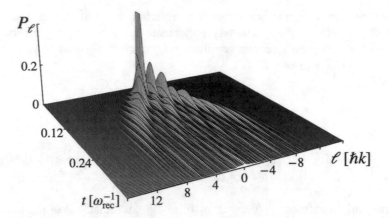

Fig. 8.6 Momentum distribution P_ℓ of an atomic beam interacting with a resonant standing wave optical field as a function of time. The units are recoil units, i.e. time is in units of ω_{rec}^{-1} and the momentum is in units of $\hbar k$. The figure results from simulations also including the kinetic energy part of the Hamiltonian and spontaneous emission at rate $\Gamma = 20\omega_{rec}$. This leads for long enough times to the washing-out of the diffraction structure

eventually stopped by the effects of the atomic kinetic energy. Physically, this saturation results from a violation of energy–momentum conservation. Specifically, because the dispersion relation of light is linear while that of atoms is quadratic, it is impossible to conserve both energy and momentum at large scattering orders, a feature reminiscent of phase mismatch in nonlinear optics. A numerical solution of the problem that accounts the kinetic energy term and spontaneous emission normally is therefore necessary in the general case, see e.g. Refs. [17, 18].

8.4.2   Bragg Regime

The Bragg regime of atomic diffraction is that regime where the effects of energy–momentum conservation are sufficiently important that the kinetic energy term in the atomic Hamiltonian cannot be ignored. As a result the number of allowed diffraction orders is severely limited when compared to the Raman–Nath regime. This is the atom optics analog of optical Bragg diffraction, where substantial diffraction only occurs if the Bragg condition is fulfilled. Atomic Bragg diffraction was first demonstrated in Ref. [13] and a more detailed theory that outlined here can be found e.g. in Refs. [18, 19].

Rather than using the momentum representation as in the analysis of Raman–Nath diffraction we treat Bragg diffraction in the coordinate representation, expending the state vector of the atom as

$$|\psi(t)\rangle = C_e(z, t)|e\rangle + C_g(z, t)|g\rangle \,, \tag{8.49}$$

where $e(z, t)$ and $g(z, t)$ are the center-of-mass wave functions corresponding to the excited and ground electronic states, respectively. The equations of motion for these wave functions are

$$i\hbar\frac{\partial C_e(z, t)}{\partial t} = -\frac{\hbar^2}{2m}\frac{\partial^2 C_e(z, t)}{\partial z^2} + \hbar\Omega_r\cos(kz)C_g(z, t) + \hbar\Delta e(z, t)$$

$$i\hbar\frac{\partial C_g(z, t)}{\partial t} = -\frac{\hbar^2}{2m}\frac{\partial^2 C_g(z, t)}{\partial z^2} + \hbar\Omega_r\cos(kz)C_e(z, t).$$

For large detunings $|\Delta| \gg (\Omega_r, \omega_{\text{rec}})$ and atoms initially in their ground electronic state, we can adiabatically eliminate the upper electronic state. The ground state wave function equation of motion reduces then to

$$i\hbar\frac{\partial C_g(z, t)}{\partial t} = -\frac{\hbar}{2m}\frac{\partial^2 C_g(z, t)}{\partial z^2} - \frac{\hbar\Omega_r^2}{\Delta}\cos^2(kz)C_g(z, t). \tag{8.50}$$

This is a Mathieu equation, whose analytical solution is not possible in general. We proceed to find an approximate solution in the regime of Bragg diffraction by first transforming it into an infinite set of ordinary differential equations via the introduction of the Fourier series expansion

$$C_g(z, t) = \sum_\ell g_\ell(t)e^{i\ell kz}, \tag{8.51}$$

where ℓ labels again the units of transverse momentum. When inserted into the Mathieu equation (8.50) this yields the coupled difference–differential equations

$$i\hbar\frac{dg_\ell(t)}{dt} = \left(\ell^2\hbar\omega_{\text{rec}} - \frac{\hbar\Omega_r^2}{2\Delta}\right)g_\ell(t) - \frac{\hbar\Omega_r^2}{4\Delta}(g_{\ell+2}(t) + g_{\ell-2}(t)) \tag{8.52}$$

or, concentrating on first-order Bragg scattering, $g_\ell(0) = \delta_{\ell,1}$,

$$i\hbar\frac{dg_1(t)}{dt} = \left(\hbar\omega_{\text{rec}} - \frac{\hbar\Omega_r^2}{2\Delta}\right)g_1(t) - \frac{\hbar\Omega_r^2}{4\Delta}(g_3(t) + g_{-1}(t)),$$

$$i\hbar\frac{dg_{-1}(t)}{dt} = \left(\hbar\omega_{\text{rec}} - \frac{\hbar\Omega_r^2}{2\Delta}\right)g_{-1}(t) - \frac{\hbar\Omega_r^2}{4\Delta}(g_1(t) + g_{-3}(t)). \tag{8.53}$$

These two equations are coupled to equations for $\ell = \pm 3$, which are in turn coupled to equations for $\ell = \pm 5$, etc. As such they belong to an infinite set of difference–differential equations. However, the energy difference between an initial state of momentum $p_i = \ell_i$ and a final state with momentum $\ell_f\hbar k$ is $\Delta E = (\ell_i + \ell_f)^2\hbar^2k^2 - \ell_i^2\hbar^2k^2$, so that conservation of energy demands that $2\ell_f = \ell_i \pm \ell_i$. In particular, for $\ell_i = 1$ as considered here, the only other energy conserving diffraction order is for $\ell_f = -1$. This permits one to truncate Eq. (8.53) at $\ell = \pm 1$. The resulting equations can be solved straightforwardly to give

$$g_1(t) = \exp\left[-\mathrm{i}\left(\omega_{\mathrm{rec}} - \Omega_r^2/2\Delta\right)t\right]\cos\left(\omega_p t\right),$$

$$g_{-1}(t) = -\mathrm{i}\exp\left[-\mathrm{i}\left(\omega_{\mathrm{rec}} - \Omega_r^2/2\Delta\right)t\right]\sin\left(\omega_p t\right),$$

where $\omega_p = \Omega_r^2/4|\Delta|$. This shows that Bragg diffraction provides a method to coherently split an atomic beam into two parts, much like a diffraction grating in optics. It is characterized by a periodic oscillation between the $\ell = 1$ and $\ell = -1$ scattering orders, an effect known in neutron diffraction as Pendellösung oscillations, and observed in atom optics experiments by P. J. Martin et al. [13].

8.4.3 Stern–Gerlach Regime

The final limiting case of atomic diffraction that we consider is the Stern–Gerlach regime, where in contrast to Raman–Nath and Bragg diffraction the atomic beams are spatially narrow compared to the period of the optical potential. We already encountered Stern–Gerlach diffraction in the discussion of the inverse Stern–Gerlach effect of Sect. 7.3, where we showed how the different number states comprising a quantized cavity field mode deflect excited and ground state atoms differently. We now revisit this situation more quantitatively for the case of a classical field, noting that with a proper change of the classical Rabi frequency Ω_r to the appropriate n-photon Rabi frequency, the same analysis would also hold for a quantized field in some number state $|n\rangle$.

Stern–Gerlach diffraction is conveniently analyzed by considering the optical forces acting on the initially well-localized wave packet. They can be determined simply by ignoring the kinetic energy term in the Hamiltonian (8.40) and locally diagonalizing the remaining potential energy term. At resonance $\Delta = 0$ this term is simply

$$\hat{H} \rightarrow \hat{H}_{\mathrm{local}}(z) = \hbar\Omega_r \cos(kz)\left(|e\rangle\langle g| + \mathrm{h.c.}\right), \qquad (8.54)$$

and its local eigenstates are the spatially dependent dressed states

$$|1\rangle f(z) = \frac{1}{\sqrt{2}}\left(|g\rangle + |e\rangle\right)f(z),$$

$$|2\rangle f(z) = \frac{1}{\sqrt{2}}\left(|g\rangle - |e\rangle\right)f(z), \qquad (8.55)$$

with local eigenvalues

$$E_1(z) = \Omega_r \cos(kz),$$

$$E_2(z) = -\Omega_r \cos(kz). \tag{8.56}$$

Consider then what happens to an atomic wave packet initially in its ground electronic state $|g\rangle$ and at location z_0,

$$|\psi(z, 0)\rangle = |g\rangle f(z_0), \tag{8.57}$$

or, in terms of the local eigenstates (8.56),

$$|g\rangle f(z) = \frac{1}{\sqrt{2}}\big[|1\rangle f(z) + |2\rangle f(z)\big]. \tag{8.58}$$

The two components of $|g\rangle f(z)$ in Eq. (8.58) are subject to equal and opposite forces

$$F_1(z) = -\frac{\mathrm{d}E_1(z)}{\mathrm{d}z} = \hbar k\, \Omega_0 \sin(kz),$$

$$F_2(z) = -\frac{\mathrm{d}E_2(z)}{\mathrm{d}z} = -\hbar k\, \Omega_0 \sin(kz), \tag{8.59}$$

which are π out of phase with each other. As a result, the mean positions $\langle z_1 \rangle$ and $\langle z_2 \rangle$ of the partial wave packets associated with the atom in the dressed states $|1\rangle$ and $|2\rangle$ are subject to the equations of motion

$$\frac{\mathrm{d}^2 \langle z_1 \rangle}{\mathrm{d}t^2} = \frac{\hbar k}{m} \Omega_r \sin(k\langle z_1 \rangle),$$

$$\frac{\mathrm{d}^2 \langle z_2 \rangle}{\mathrm{d}t^2} = -\frac{\hbar k}{m} \Omega_r \sin(k\langle z_2 \rangle), \tag{8.60}$$

with $\langle z_1(0) \rangle = \langle z_2(0) \rangle = z_0$. These are pendulum equations. They show that an atomic wave packet initially at rest is split into two parts that oscillate within a potential well with period of small oscillations $\sqrt{2\Omega_r \omega_{\mathrm{rec}}}$, with $\omega_{\mathrm{rec}} = \hbar k^2/2m$ the recoil frequency, as illustrated in Fig. 8.7. This behavior, which is the analog for atomic matter waves of the Stern–Gerlach effect for spin-1/2 particles in a magnetic field gradient, was first experimentally verified by T. Sleator and colleagues [20].

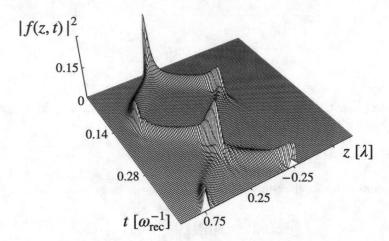

Fig. 8.7 Splitting and oscillations of the two dressed states of a two-level atom in a Stern–Gerlach experiment as a function of time. The atomic beam is characterized by an initial width (FWHM) of $\Delta x = 0.25\lambda$. Time and position are in recoil units

8.5   Spontaneous Emission

The previous section showed that in the absence of spontaneous emission, the treatment of atomic diffraction by periodic light gratings is relatively simple. For a plane wave incident atomic beam, it leads to sharply defined diffraction peaks. However, these peaks are smeared as a result of the random momentum kick imparted on the atoms by spontaneous decay. By increasing the number of spontaneous decays during the time of interaction between the atom and the optical field, the system undergoes a transition from a diffractive regime to a diffusion-dominated regime [21, 22]. It appears that in that regime the only way to obtain a good agreement with the experiments is by direct numerical solution of the master equation accounting for the random atomic recoil resulting from the spontaneous emission, see e.g. Refs. [23, 24].

More specifically, when taking spontaneous emission into account, the atomic dynamics is governed by the master equation (8.25),

$$\frac{d\hat{\rho}}{dt} = -\frac{i}{\hbar}[\hat{H}, \hat{\rho}] + \left.\frac{d\hat{\rho}}{dt}\right|_{\text{sp}} \equiv (\hat{\mathcal{L}}_H + \hat{\mathcal{L}}_D)\hat{\rho}, \tag{8.61}$$

where $\hat{H}$ is given by Eq. (8.40). Consistently with the discussion of Sect. 8.3 we express $\hat{\mathcal{L}}_D\hat{\rho}$ as

$$\hat{\mathcal{L}}_D\hat{\rho} = \frac{\Gamma}{2}\left(\hat{\sigma}_+\hat{\sigma}_-\hat{\rho} + \hat{\rho}\hat{\sigma}_+\hat{\sigma}_-\right) + \Gamma \int d^2\mathbf{n}\Phi(\mathbf{n})e^{-ik\mathbf{n}\cdot\hat{r}}\hat{\sigma}_-\hat{\rho}\hat{\sigma}_+e^{ik\mathbf{n}\cdot\hat{r}}, \tag{8.62}$$

where Eq. (8.31) gives the angular distribution of spontaneously emitted photons for a dipole-allowed transition. The master equation (8.61) can be solved numerically

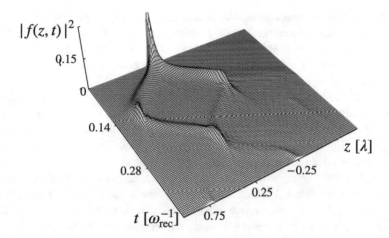

Fig. 8.8 Same as Fig. 8.7, but including the effects of spontaneous emission at rate $\Gamma = 10\omega_{rec}$

using, for example, a Lie–Trotter product formula to disentangle its formal solution as

$$\hat{\rho}(t + \Delta t) = \mathcal{T} \exp\left[\int_t^{t+\Delta t} dt (\hat{\mathcal{L}}_H + \hat{\mathcal{L}}_D)\right] \hat{\rho}(t) \simeq \exp[\hat{\mathcal{L}}_H \Delta t] \exp[\hat{\mathcal{L}}_D \Delta t] \hat{\rho}(t), \tag{8.63}$$

where $\mathcal{T}$ stands for "time-ordered" and commutators of order Δt^2 have been neglected in the exponents. Its solution permits one to study the transition from the diffractive regime to the spontaneous emission dominated diffusive regime of interaction between the atoms and the light grating. Figure 8.8 shows one such example for the case of Stern–Gerlach diffraction. This clearly demonstrates the need to avoid spontaneous emission as much as possible in atomic diffraction experiments.

8.6 Atom Interferometers

Very much like mirrors and beam splitters are the basic building blocks of optical interferometers, it is not hard to imagine that, by simple role reversal, optical gratings and mirrors can serve as basic elements of atom interferometers by coherently separating an atomic matter wave into two components that will follow different paths before being recombined at a detector. Changes in the resulting matter-wave interference pattern can then provide a signature of a perturbation on one of the arms, or a differential perturbation on both arms. Because in contrast to photons atoms are massive, they present several important advantages in a number of applications, in particular related to measurements of accelerations and gravitational forces. Two such examples are gravimetry, the measurement of gravitational acceleration, and gravity gradiometry, which uses combinations of two or more atom interferometers to study variations in the earth gravitational field, as further discussed in Problems 8.7–8.9. Gravity gradiometry is for instance an important resource for mineral exploration, where it relies on the different densities of types of rocks or liquids.

Atom interferometers were by no means the first matter-wave interferometers. Electron interferometry goes back to the 1950s [25–27] and neutron interferometry was developed in the 1960s by H. Maier-Leibnitz [28] and brought to a great degree of sophistication by H. Rauch and his coworkers [29, 30]. While these interferometers are valuable tools for probing fundamental physics, atom interferometry has a number of advantages of its own. It offers a wealth of possibilities stemming from the different internal structures of atoms, the wide range of properties that they possess, and their great variety of interactions with the environment, including other atoms, electromagnetic and gravitational fields, and perhaps even dark matter, as will be further elaborated upon in Chap. 12.

The previous section suggests several ways to exploit atomic diffraction by light fields to realize atomic beam splitters and mirrors. However, a number of potential difficulties need to be dealt with, most importantly perhaps spontaneous emission and the associated random changes in atomic momentum, as we have seen. In addition, it is usually important to limit the number of partial beams in the interferometer, ruling out the use of Raman–Nath diffraction. Bragg diffraction is more promising from this point of view, but finding the right balance between the large detunings required to avoid a significant population of the excited electronic state $|e\rangle$ and limit spontaneous emission, while at the same time allowing for a fast enough atom–field interaction, usually proves to be a challenge.[5]

For this reason we concentrate in the following on the two-photon stimulated Raman transitions approach [32, 33], a method that has proven remarkably successful in a number of situations and largely avoids the issues associated with spontaneous emission. In this approach two-photon stimulated Raman transitions are exploited to coherently split an ultracold atomic wave function. This is realized by two counter-propagating lasers of amplitudes E_1 and E_2 and frequencies $\omega_1 = ck_1$ and $\omega_2 = ck_2$, respectively, which drive a transition between two hyperfine ground states $|g\rangle$ and $|e\rangle$ via a far off-resonant intermediate level $|i\rangle$, as illustrated in Fig. 8.9.[6]

Our starting point is the Hamiltonian of a three-level system interacting with two classical fields at frequencies ω_1 and ω_2. It is a straightforward generalization of Eq. (1.51), but including the center-of-mass kinetic energy of the atom as well since

---

[5]There are however notable but relatively rare exceptions, see for instance Ref. [31], which uses a ultranarrow clock transition in ^{88}Sr to develop an atom interferometer based on single-photon transitions, a system to which we will return in Chap. 12.

[6]Although that figure may be somewhat intimidating due to the various detunings that it involves, the only complication compared to the familiar two-level situation is the need for careful bookkeeping of these detunings and the fields involved. Except for that, the analysis follows very much the lines that we are already familiar with.

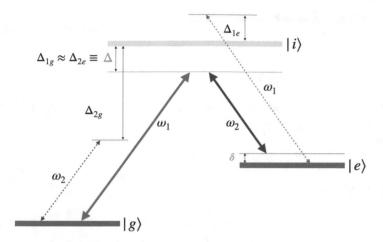

Fig. 8.9 Level scheme for an atom interferometer beam splitter or mirror element based on stimulated Raman transitions. The far off-resonant intermediate level $|i\rangle$ is drawn in light grey to indicate that it is adiabatically eliminated, leaving only the effective two-level system $\{|g\rangle, |e\rangle\}$. The two driving fields are shown in different colors for clarity, with the solid arrows showing the relevant transitions. The most important detunings are Δ and δ

we are interested in mechanical effects. This Hamiltonian reads

$$
\hat{H} = \frac{\hat{p}^2}{2m} + \hbar\omega_g|g\rangle\langle g| + \hbar\omega_e|e\rangle\langle e| + \hbar\omega_i|i\rangle\langle i| - \hbar\big[\Omega_1|i\rangle\langle g| + \Omega_2|i\rangle\langle e| + \text{h.c.}\big],
$$

$$(8.64)$$

where we have introduced the Rabi frequencies $\Omega_i = dE_i/\hbar, i = \{1, 2\}$ of the two dipole-allowed transitions $|g\rangle - |i\rangle$ and $|e\rangle - |i\rangle$, with dipole moments d taken to be equal for simplicity, and

$$
\mathbf{E}(\mathbf{r}, t) = \mathbf{E}_1 \cos(\mathbf{k}_1 \cdot \mathbf{r} - \omega_1 t + \phi_1) + \mathbf{E}_2 \cos(\mathbf{k}_2 \cdot \mathbf{r} - \omega_2 t + \phi_2), \qquad (8.65)
$$

with $\mathbf{k}_1 \approx -\mathbf{k}_2$ for counter-propagating fields.

For this geometry the field $\mathbf{E}_1(\mathbf{r}, t)$ couples the state $|g, \mathbf{p}\rangle$ to the state $|i, \mathbf{p}+\hbar\mathbf{k}_1\rangle$, and that state is coupled in turn to $|e, \mathbf{p} + \hbar(\mathbf{k}_1 + \mathbf{k}_2)\rangle \approx |e, \mathbf{p} + 2\hbar\mathbf{k}\rangle$ by the field $\mathbf{E}_2(\mathbf{r}, t)$. The atoms making a transition from $|g\rangle$ to $|e\rangle$ experience therefore a momentum recoil $\hbar(\mathbf{k}_1 + \mathbf{k}_2) \equiv \hbar\mathbf{k}_{\text{eff}} \approx 2\hbar\mathbf{k}$, and the general state of the system can therefore be decomposed into a series of uncoupled manifolds $\{|g, \mathbf{p}\rangle, |i, \mathbf{p} + \hbar\mathbf{k}\rangle, |e, \mathbf{p} + 2\hbar\mathbf{k}\rangle\}$ characterized by states of the general form

$$
|\psi_{\mathbf{p}}(t)\rangle = C_{g,\mathbf{p}}(t)|g, \mathbf{p}\rangle + C_{i,\mathbf{p}+\mathbf{k}}|i, \mathbf{p} + \hbar\mathbf{k}\rangle + C_{e,\mathbf{p}+2\mathbf{k}}|e, \mathbf{p} + 2\hbar\mathbf{k}\rangle. \qquad (8.66)
$$

We consider the limit of strongly off-resonant transitions,

$$\Delta_{1g} \equiv \omega_1 - (\omega_i - \omega_g) \gg \Omega_1 \,,$$

$$\Delta_{2e} \equiv \omega_2 - (\omega_i - \omega_e) \gg \Omega_2 \,, \tag{8.67}$$

and also assume that the two-photon transition between $|e\rangle$ and $|g\rangle$ is nearly resonant,[7] $\omega_e - \omega_g \approx \omega_1 - \omega_2$, with

$$(\omega_1 - \omega_2) - (\omega_e - \omega_g) \equiv \delta \ll \Delta_{1g} \,, \Delta_{2e} \,,$$

so that we have approximately

$$\Delta_{1g} \approx \Delta_{2e} \equiv \Delta \,, \tag{8.68}$$

as shown in Fig. 8.9.

The detuning conditions (8.67) allow us to adiabatically eliminate the intermediate state $|i\rangle$ by proceeding along lines analog to those already encountered in deriving the dispersive limit of the Jaynes–Cummings model. As was the case there, an important outcome of this elimination is the appearance of a light shift similar to Eq. (3.15). Especially in precision measurement applications of atom interferometry these shifts cannot be ignored, and neither can the Doppler shifts resulting from atomic motion. The resulting effective Hamiltonian describing the dynamics of the remaining two-state system manifold $\{|g, \mathbf{p}\rangle \,, |e, \mathbf{p} + \mathbf{k}_{\text{eff}}\rangle\}$ is therefore

$$\hat{H} = \frac{\hbar}{2} \begin{pmatrix} \frac{|\Omega_2|^2}{2\Delta} & \Omega_{\text{eff}} e^{-i(\delta_{12}+\phi_{\text{eff}})} \\ \Omega_{\text{eff}} e^{i(\delta_{12}t+\phi_{\text{eff}})} & \frac{|\Omega_1|^2}{2\Delta}, \end{pmatrix} \tag{8.69}$$

where

$$\delta_{12} = (\omega_1 - \omega_2) - \left[(\omega_e - \omega_g) + \frac{\mathbf{p} \cdot \mathbf{k}_{\text{eff}}}{m} + \frac{\hbar k_{\text{eff}}^2}{2m} \right] \,, \tag{8.70}$$

$\mathbf{k}_{\text{eff}} = \mathbf{k}_1 + \mathbf{k}_2$, and the relative phase $\phi_{\text{eff}} \equiv \phi_1 - \phi_2$ between the two fields is chosen to make the effective two-photon Rabi frequency

$$\Omega_{\text{eff}} = \frac{\Omega_1 \Omega_2^*}{2\Delta} e^{i\phi_{\text{eff}}} \tag{8.71}$$

---

[7] For convenience, the detunings are expressed in this section in the form of field frequencies minus atomic frequencies, in contrast to the convention adopted in much of this book.

real and positive. Note that in addition to the photon recoil $\hbar k_{\text{eff}}^2/2m$ the detuning δ_{12} also includes the Doppler shift $\mathbf{p} \cdot \mathbf{k}_{\text{eff}}/m$, with $\mathbf{p}$ the center-of-mass atomic momentum, as shown in Problem 8.7.

Except for the light shifts $|\Omega_i|^2/2\Delta$ appearing in the diagonal of $\hat{H}$ the situation is now formally equivalent to that of a driven two-level system of Chap. 1. For atoms in the initial state

$$|\psi(t_0)\rangle = C_{g,\mathbf{p}}(t_0)|g, \mathbf{p}\rangle + C_{e,\mathbf{p}+\mathbf{k}_{\text{eff}}}(t_0)|e, \mathbf{p} + \mathbf{k}_{\text{eff}}\rangle$$

we have therefore

$$C_{e,\mathbf{p}+\mathbf{k}_{\text{eff}}}(t_0 + \tau) = e^{-i[(\Omega_1^2+\Omega_2^2)/4\Delta+\delta_{12}]\tau/2}$$

$$\times \Big\{ C_{e,\mathbf{p}+\mathbf{k}_{\text{eff}}}(t_0)\big[\cos(\Omega\tau/2) - i\cos\theta\sin(\Omega\tau/2)\big]$$

$$-iC_{g,\mathbf{p}}(t_0)e^{-i(\delta_{12}t_0+\phi_{\text{eff}})}\sin\theta\sin(\Omega\tau/2)\Big\}$$

$$C_{g,\mathbf{p}}(t_0 + \tau) = e^{-i[(\Omega_1^2+\Omega_2^2)/4\Delta-\delta_{12}]\tau/2}$$

$$\times \Big\{ -iC_{e,\mathbf{p}+\mathbf{k}_{\text{eff}}}(t_0)e^{i(\delta_{12}t_0+\phi_{\text{eff}})}\sin\theta\sin(\Omega\tau/2)$$

$$+C_{g,\mathbf{p}}(t_0)\big[\cos(\Omega\tau/2) + i\cos\theta\sin(\Omega\tau/2)\big]\Big\}, \qquad (8.72)$$

where

$$\Omega = \sqrt{\Omega_{\text{eff}}^2 + (\delta_{12} - (|\Omega_1|^2 - |\Omega_2|^2)/4\Delta)^2},$$

$$\sin\theta = \Omega_{\text{eff}}/\Omega,$$

$$\cos\theta = -\frac{\delta_{12} - (|\Omega_1|^2 - |\Omega_2|^2)/4\Delta}{\Omega}, \qquad (8.73)$$

where we recognize the correction to δ_{12} from the light shifts in the last expression.

The expressions (8.72) are somewhat cumbersome, a consequence of the multiple detunings associated with two-photon processes in three-level systems, combined with the light shifts and effective two-photon Rabi frequencies that result from the elimination of the intermediate state $|i\rangle$. The key point, though, is that qualitatively they describe essentially the same physics as the Rabi oscillations familiar from driven two-level systems, with periodic oscillations between the two states $|g, \mathbf{p}\rangle$ and $|e, \mathbf{p} + \hbar\mathbf{k}_{\text{eff}}\rangle$, as further elaborated upon in Problem 8.7. In particular, it is possible to chose the field amplitudes and interaction times τ such that an equal superposition of the probability amplitudes of these two states is realized, thereby forming a 50/50 beam splitter for the atomic wave function. Such a pulse is called a $\pi/2$ pulse in reference to the evolution of the state vector on the Bloch sphere. At a later time T one can then apply a π pulse that acts as a mirror by exchanging $|g, \mathbf{p}\rangle$ and $|e, \mathbf{p} + \hbar\mathbf{k}_{\text{eff}}\rangle$, as sketched in Fig. 8.10. This has the effect of redirecting

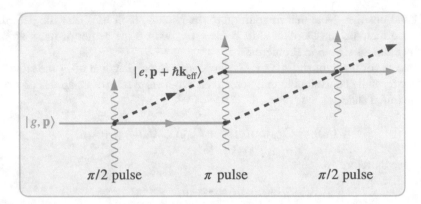

$|e, \mathbf{p} + \hbar\mathbf{k}_{\text{eff}}\rangle$

$|g, \mathbf{p}\rangle$

$\pi/2$ pulse π pulse $\pi/2$ pulse

Fig. 8.10 Diagram of a stimulated Raman atom interferometer scheme. The green solid and red dashed lines indicate paths for which the atom is in the states $|g, \mathbf{p}\rangle$ and $|e, \mathbf{p} + \hbar\mathbf{k}_{\text{eff}}\rangle$, respectively. The $\pi/2$, π, and $\pi/2$ pulses act as beam splitters (the $\pi/2$ pulses) and "mirror" (the π pulse)

the partial waves, so that they overlap again at a time $2T$, at which time a second $\pi/2$ pulse recombines them and produces a matter-wave interference pattern.

Because of their remarkable potential sensitivity, in particular when using "clock transitions" such as for instance the ultranarrow $^1S_0 - {}^3P_1$ intercombination line of ^{88}Sr, atom interferometers are a tool of choice to carry out precision measurements both in applied science and in engineering as well as in fundamental science. For example, Problem 8.9 discusses their use as gravity gradiometers. Laser cooled atoms in atomic fountains can result in transit times through the arms of the interferometer approaching one second and separations between the partial atomic waves of the order of centimeters. These, and other improvements in atom interferometers, can yield extraordinary sensitivities to minute changes in forces and positions. Chapter 12 will illustrate how these properties can be exploited to address profound questions in fundamental science.

Mechanical Gratings   Instead of optical gratings it is also possible to develop atom interferometers that use nanofabricated diffraction gratings consisting, for example, of thin, low-stress silicon nitride membranes with precisely patterned holes. The first experimental verification of atomic diffraction by a transmission mask was given by J. A. Leavitt and F. A. Bills [34] who used a self-made single slit for that purpose. To make mechanical gratings suitable for use in atom interferometers, great care must be taken to obtain a pattern with positional accuracy below a small fraction of the grating period. Their fabrication process is described in Refs. [35, 36]. In addition to their applications in matter-wave diffraction and atom interferometers, nanofabricated gratings have been used in the generation of matter-wave holograms.

Problems

Problem 8.1 Determine the gradient force $F_{gr}(x, y)$ and optical potential $U_{opt}(x, y)$ resulting from two standing waves propagating in perpendicular directions $\vec{x}$ and $\vec{y}$. Plot the potential for two values of the detuning Δ of opposite signs, with $|\delta| = \Gamma$.

Problem 8.2

(a) Solve analytically the Bragg diffraction equations of motion

$$i\hbar \frac{dg_\ell(t)}{dt} = \left(\ell^2 \hbar \omega_{rec} - \frac{\hbar \Omega_r^2}{2\Delta} \right) g_\ell(t) - \frac{\hbar \Omega_r^2}{4\Delta} (g_{\ell+2}(t) + g_{\ell-2}(t))$$

after truncating them at $\ell = \pm 1$.

(b) Keeping then terms up to $\ell = \pm 5$, solve these equations numerically to investigate, and verify the role of energy conservation in the system dynamics.

Problem 8.3 Using atomic diffraction in the Bragg regime and two standing wave fields, determine quantitatively how to design a beam splitters and "mirrors" arrangement that operates as an atom interferometer.

Problem 8.4 Consider the problem of atomic diffraction, but instead of a classical field, with the interaction of the atoms driven by a quantized field described by the standing wave $\hat{E} = \mathcal{E}_s (\hat{a} + \hat{a}^\dagger) \sin(k\hat{z})$. Extend the Hamiltonian (8.40) to handle this situation, and solve the atom diffraction problem in the Raman–Nath regime for the resonant case $\Delta = 0$ and a field in a number state $|n\rangle$, ignoring spontaneous emission into other modes.

Problem 8.5 Solve that same problem, but for an electromagnetic field consisting of two counter-propagating running waves, $\hat{E} = \mathcal{E}_r(\hat{a}_1 + \hat{a}_1^\dagger) \exp(ik\hat{z}) + \mathcal{E}_r(\hat{a}_2 + \hat{a}_2^\dagger) \exp(-ik\hat{z})$. Explain why the diffraction pattern of the atoms is different in that case. Hint: Think of the discussion of standing waves *vs.* running waves of Sect. 2.2.2 and of quantum entanglement.

Problem 8.6 Extend the analysis of the Stern–Gerlach regime of atomic diffraction to the case of a single-mode quantized field, with an atom–field interaction of the form $\hat{H} \rightarrow \hat{H}_{local}(z) = \hbar g \cos(kz) \left(\hat{a} |e\rangle \langle g| + \text{h.c.} \right)$, by extending the dressed states of Eq. (8.55) to include the state of the field, much as in the discussion of the Jaynes–Cummings problem, see Eqs. (3.7), but with spatially dependent dressed states.

Problem 8.7 (Atom Interferometers and Gradiometry) *This and the next two problems dig somewhat deeper into atom interferometry and in particular derive a result that motivates the use of atom interferometers as gradiometers, gravity gradiometers, and even gravity wave antennas.*

(a) Consider the stimulated Raman problem of Sect. 8.6, but ignoring the recoil effects associated with the absorption and emission of light, that is, ignore the kinetic energy term in the Hamiltonian (8.64), and consider the resonant case $\delta = 0$. The general state of the system is then

$$|\psi(t)\rangle = C_g(t)|g\rangle + C_e(t)|e\rangle + C_i(t)|i\rangle .$$

In the limit of large detunings Δ, eliminate adiabatically the intermediate state $|i\rangle$, determine the ac Stark shift and the effective two-photon Rabi frequency that couples then the remaining levels $|g\rangle$ and $|e\rangle$, and derive the equations of motion for the probability amplitudes $C_e(t)$ and $C_g(t)$.

(b) How are these equations modified for a moving atom whose center-of-mass momentum when it is in the ground electronic state is $\mathbf{p}$? Show that in that case the dynamics of the system is described by the effective Hamiltonian (8.69).

Problem 8.8 Determine the conditions under which the combined effects of the fields E_1 and E_2 act as (a) a $\pi/2$ and (b) a π pulse for the two states $|e\rangle$ and $|g\rangle$ in the stimulated Raman scattering situation described by the effective Hamiltonian (8.69).

Problem 8.9 We will encounter in Chap. 12 atom interferometers relying on "clock transitions," that is, transitions between a ground state $|g\rangle$ and an excited state $|e\rangle$ with extremely long spontaneous lifetimes, possibly as long as tens of seconds, see e.g. Ref. [31]. These operate on true two-state transitions rather than on effective ones, in which case the equations of motion (8.72) simplify to

$$C_{e,\mathbf{p}+\mathbf{k}}(t_0 + \tau) = e^{-i\delta\tau/2}\Big\{ C_{e,\mathbf{p}+\mathbf{k}}(t_0)\big[\cos(\Omega\tau/2) - i\cos\theta \sin(\Omega\tau/2)\big]$$

$$- iC_{g,\mathbf{p}}(t_0)e^{-i(\delta t_0 + \phi)} \sin\theta \sin(\Omega\tau/2)\Big\} ,$$

$$C_{g,\mathbf{p}}(t_0 + \tau) = e^{i\delta\tau/2}\Big\{ - iC_{e,\mathbf{p}+\mathbf{k}}(t_0)e^{i(\delta t_0 + \phi)} \sin\theta \sin(\Omega\tau/2)$$

$$+ C_{g,\mathbf{p}}(t_0)\big[\cos(\Omega\tau/2) + i\cos\theta \sin(\Omega\tau/2)\big]\Big\} ,$$

where $\Omega_r = dE_0/\hbar$ is the field Rabi frequency and $\Omega = \sqrt{|\Omega_r|^2 + \delta^2}$ the generalized Rabi frequency associated with the electric field

$$\mathbf{E}(\mathbf{r}, t) = \mathbf{E}_0 \cos(\mathbf{k} \cdot \mathbf{r} - \omega t + \phi]) ,$$

and the detuning

$$\delta = \omega - \left[(\omega_e - \omega_g) + \frac{\mathbf{p} \cdot \mathbf{k}}{m} + \frac{\hbar k^2}{2m}\right]$$

accounts for both the Doppler shift associated with atomic motion and the photon recoil frequency.

(a) Show that for an atom with initial ground and electronic state probability amplitudes $C_{g,\mathbf{p}}(t_0)$ and $C_{e,\mathbf{p}+\mathbf{k}}(t_0)$ the state of the system after a π pulse of duration τ will be

$$C_{e,\mathbf{p}+\mathbf{k}}(t_0 + \tau) = -iC_{g,\mathbf{p}}(t_0)e^{-i\delta\tau/2}e^{-i[\delta t_0 + \phi]},$$

$$C_{g,\mathbf{p}}(t_0 + \tau) = -iC_{e,\mathbf{p}+\mathbf{k}}(t_0)e^{i\delta\tau/2}e^{i[\delta t_0 + \phi]}.$$

Similarly, for a $\pi/2$ pulse or duration $\tau/2$,

$$C_{e,\mathbf{p}+\mathbf{k}}(t_0 + \tau/2) = e^{-i\delta\tau/4}\left[C_{e,\mathbf{p}+\mathbf{k}}(t_0) - iC_{g,\mathbf{p}}(t_0)e^{-i[\delta t_0 + \phi]}\right]/\sqrt{2},$$

$$C_{g,\mathbf{p}}(t_0 + \tau/2) = e^{i\delta\tau/4}\left[-iC_{e,\mathbf{p}+\mathbf{k}}(t_0)e^{i[\delta t_0 + \phi]} + C_{g,\mathbf{p}}(t_0)\right]/\sqrt{2}.$$

(b) Consider an atom interferometer with a $\pi/2$-π-$\pi/2$ configuration, with the first $\pi/2$ pulse occurring at time t_1, the π pulse at $t_2 = t_1 + T + \tau/2$, and the second $\pi/2$ pulse at $t_3 = t_1 + 2T + 3\tau/2$. Show then that for an atom initially in its ground electronic state, we have, following these three pulses,

$$C_{e,\mathbf{p}+\mathbf{k}}(t_3 + \tau/2) = -(i/2)e^{-i\delta\tau/2}e^{-i[\delta t_2 + \phi(t_2)]}(1 - e^{-i\delta\tau/2}e^{-i\Delta\phi}),$$

where

$$\Delta\phi = \phi(t_1) - 2\phi(t_2) + \phi(t_3),$$

and $\phi(t_i)$ is the phase of the light relative to the atom at time t_i, referenced to the phase at some fixed time point.

(c) If the atom is falling under the action of gravity and the $\pi/2$-π-$\pi/2$ sequence is produced by a vertically oriented laser, show that $\Delta\phi = -\mathbf{k} \cdot \mathbf{g}T^2$, where g is the acceleration of gravity. As such this system permits to measure $\mathbf{g}$, or any other acceleration, and a gravity gradiometer can be realized with the use of two interferometers that measure $\mathbf{g}$ at two elevations.

References

1. I. Ridpath, *A Comet Called Halley* (Cambridge University, Cambridge, 1985)
2. C. Cohen-Tannoudji, *Atomic Motion in Laser Light: In Fundamental Systems in Quantum Optics*, ed. by J. Dalibard, J.M. Raimond, J. Zinn-Justin (North-Holland, Amsterdam, 1992), p. 3
3. A.P. Kazantsev, G.I. Surdutovich, V.P. Yakovlev, The motion of atoms and molecules in a resonance light field. J. Phys (Paris) **42**(9), 1231–1237 (1981)
4. J. Dalibard, C. Cohen-Tannoudji, Atomic motion in laser light: Connection between semiclassical and quantum descriptions. J. Phys. B **18**, 1661 (1985)

5. C. Cohen-Tannoudji, J. Dupont-Roc, G. Grynberg, *Atom-Photon Interactions: Basic Processes and Applications* (Wiley-Interscience, New York, 1992)
6. P. Meystre, M. Wilkens, *Spontaneous emission by moving atoms: in Cavity Quantum Electrodynamics*, ed. by P. Berman (Academic Press, Boston, 1994), p. 301
7. P.L. Kapitza, P.A.M. Dirac, The reflection of electrons from standing light waves. Proc. Cambridge Phil. Soc. **29**, 297 (1933)
8. L.S. Bartell, H.B. Thompson, R.R. Roskos, Observation of stimulated Compton scattering of electrons by laser beam. Phys. Rev. Lett **14**, 851 (1965)
9. S. Altschuler, L.M. Franz, R. Braunstein, Reflection of atoms from standing light waves. Phys. Rev. Lett. **5**, 231 (1966)
10. A. Ashkin, Atomic beam deflection by resonance radiation pressure. Phys. Rev. Lett. **25**, 1321 (1970)
11. P.E. Moskowitz, P.L. Gould, S.R. Atlas, D.E. Pritchard, Diffraction of an atomic beam by standing-wave radiation. Phys. Rev. Lett. **51**, 370 (1983)
12. P.L. Gould, G.A. Ruff, D.E. Pritchard, Diffraction of atoms by light: the near-resonant Kapitza-Dirac effect. Phys. Rev. Lett. **56**, 827 (1986)
13. P.J. Martin, B.G. Oldaker, A.H. Miklich, D.E. Pritchard, Bragg scattering of atoms from a standing light wave. Phys. Rev. Lett. **60**, 515 (1988)
14. R. Schieder, H. Walther, L. Wöste, Atomic beam deflection by the light of a tunable dye laser. Opt. Commun. **5**, 337 (1972)
15. J.L. Picqué, J.L. Vialle, Atomic beam deflection and broadening by recoils due to photon absorption or emission. Opt. Commun. **5**, 402 (1972)
16. R.J. Cook, A.F. Bernhardt, Deflection of atoms by a resonant standing electromagnetic wave. Phys. Rev. A **18**, 2533 (1978)
17. E. Arimondo, A. Bambini, S. Stenholm, Quasiclassical theory of laser-induced atomic beam dispersion. Phys. Rev. A **24**, 898 (1981)
18. A.F. Bernhardt, B.W. Shore, Coherent atomic deflection by resonant standing waves. Phys. Rev. A **23**, 1290 (1981)
19. P. Meystre, E. Schumacher, E.M. Wright, Quantum Pendellösung in atom diffraction by a light grating. Ann. Phys. (Leipzig) **48**, 141 (1991)
20. T. Sleator, T. Pfau, V. Balykin, O. Carnal, J. Mlynek, Experimental demonstration of the optical Stern-Gerlach effect. Phys. Rev. Lett. **68**, 1996 (1992)
21. P.L. Gould, P.J. Martin, G.A. Ruff, R.E. Stoner, J.L. Picqué, D.E. Pritchard, Momentum transfer to atoms by a standing wave: Transition from diffraction to diffusion. Phys. Rev. A **43**, 585 (1991)
22. C. Tanguy, S. Reynaud, C. Cohen-Tannoudji, Deflection of an atomic-beam by a laser wave—transition between diffractive and diffusive regimes. J. Phys. B **17**, 4623 (1984)
23. S.M. Tan, D.F. Walls, Atomic deflection in the transition between diffractive and diffusive regimes: a numerical simulation. Phys. Rev. A **44**, 2779 (1991)
24. M. Wilkens, E. Schumacher, P. Meystre, Transition from diffraction to diffusion in the near-resonant Kapitza-Dirac effect: a numerical approach. Opt. Commun. **86**, 34 (1991)
25. L. Marton, Electron interferometer. Phys. Rev. **85**, 1057 (1952)
26. L. Marton, J.A. Simpson, J.A. Suddeth, An electron interferometer. Rev. Sci. Instrum **25**, 1099 (1954)
27. G. Mollenstedt, H. Duker, Fresnelscher Interferenzversuch mit einem Biprisma für Elektronenwellen. Naturwissenschaften **42**, 41 (1955)
28. H. Maier-Leibnitz, T. Springer, Ein interferometer für langsame neutronen. Z. Phys. **167**, 368 (1962)
29. H. Rauch, W. Treimer, U. Bonse, Test of a single-crystal neutron interferometer. Phys. Lett. A **47**, 369 (1974)
30. U. Bonse, H. Rauch, *Neutron Interferometry* (Oxford University, New York, 1979)
31. J. Rudolph, T. Wilkason, M. Nantel, H. Swan, C.M. Holland, Y. Jiang, B.E. Garber, S.P. Carman, J.S. Hogan, Large momentum transfer clock atom interferometry on the 689 nm intercombination line of strontium. Phys. Rev. Lett. **124**, 083604 (2020)

32. M. Kasevich, S. Chu, Atomic interferometry using stimulated Raman transitions. Phys. Rev. Lett. **67**, 181 (1991)
33. B. Young, M. Kasevich, S. Chu, in *Atom Interferometry*, ed. by P. Berman (Academic Press, San Diego, 1997), pp. 363–406
34. J.A. Leavitt, F.A. Bills, Single-slit diffraction pattern of a thermal atomic potassium beam. Am. J. Phys. **37**, 905 (1969)
35. D.W. Keith, R.J. Soave, M.J. Rooks, Free-standing gratings and lenses for atom optics. J. Vac. Sci. Technol. B **9**, 2846 (1991)
36. C.R. Ekstrom, D.W. Keith, D.E. Pritchard, Atom optics using microfabricated structures. Appl. Phys. B **54**, 369 (1992)

Chapter 9
Laser Cooling

This chapter discusses how the mechanical effects of light can be exploited
to cool atomic samples. We focus first on Doppler cooling, Sisyphus cooling,
and subrecoil cooling, three techniques that allow the cooling of free space
atomic samples to temperatures increasingly close to absolute zero. We
then turn to cavity cooling, which relies on the spatial dependence of
the field Rabi frequency in optical resonators, before concluding with the
sideband cooling of trapped ions, an important method for applications in
quantum information and quantum metrology and that can also be extended
to mesoscopic and macroscopic mechanical oscillators.

We understand intuitively that shining light on an object tends to heat it up, but what
is perhaps less evident is that light can also be used to cool objects to remarkably
low temperatures. Laser cooling of atoms and ions is a well-established area of
AMO physics, and similar techniques are now exploited to cool mesoscopic and
macroscopic objects as well, a topic to which we will return in Chap. 11. In this
chapter, we concentrate largely on atomic cooling and discuss a series of methods
that permit to reach increasingly low temperatures, to the point where the atoms are
so cold and their thermal de Broglie wavelength so large and that their individual
center-of-mass wave packets start to overlap with those of neighboring atoms. This
can result in the onset of phase transitions to many-body states of the atomic sample
as a whole, more famously perhaps to the Bose–Einstein condensation of bosonic
atoms which will be the topic of Chap. 10.

9.1 Doppler Cooling

The ray atom optics description of the effect of light forces on atomic motion
introduced in Sect. 8.2 is sufficient to understand the simplest form of laser cooling,
Doppler cooling, which was proposed and demonstrated in the 1970s [1, 2]. All that
is required is to generalize the form of the radiation pressure and gradient forces of

© The Author(s), under exclusive license to Springer Nature Switzerland AG 2021
P. Meystre, *Quantum Optics*, Graduate Texts in Physics,
https://doi.org/10.1007/978-3-030-76183-7_9

Eqs. (8.17) and (8.18) to account for the situation of moving atoms rather than atoms at rest. This generalization will allow us to see how a properly detuned light beam can slow down atoms and thereby cool an atomic sample. Since as we shall see Doppler cooling does not typically permit to approach the recoil limit $v = \hbar k/m$ of atomic velocities, it is sufficient for now to describe the atomic motion classically.

Consider specifically an atom moving at constant velocity $\mathbf{v}_0$ in the plane monochromatic running wave

$$\mathbf{E}(\mathbf{r}, t) = \vec{\epsilon}\, \mathcal{E} \cos(\omega t - \mathbf{k} \cdot \mathbf{r})\,. \tag{9.1}$$

Assuming that it is initially at the origin $\mathbf{r} = 0$, the field $\mathbf{E}(\mathbf{r}, t)$ at its position at time t will be

$$\mathbf{E}(\mathbf{r}, t) = \vec{\epsilon}\, \mathcal{E} \cos(\omega t - \mathbf{k} \cdot \mathbf{v}_0 t) = \vec{\epsilon}\, \mathcal{E} \cos[(\omega - \omega_d)t], \tag{9.2}$$

where we recognize $\omega_d = \mathbf{k} \cdot \mathbf{v}_0$ as a Doppler frequency shift. For moving atoms, then, the atom–field detuning Δ appearing in Eqs. (8.20) and (8.21) becomes

$$\Delta \to \Delta_d = \omega_0 - (\omega - \omega_d) = \Delta + \omega_d\,. \tag{9.3}$$

We recall from Sect. 8.2 that the gradient force acting on an atom vanishes for plane running waves, since they do not exhibit any field amplitude gradient. We are therefore left with the radiation pressure force F_{rp} only, and for an atom moving along the axis of the running wave, Eq. (8.20) yields

$$F_{\text{rp}}(v_0) = \frac{1}{2}\hbar k \Gamma \left(\frac{\Omega_r^2/2}{\Delta_d^2 + (\Gamma/2)^2 + \Omega_r^2/2} \right). \tag{9.4}$$

For small velocities, we can expand this expression to lowest order in v_0 as

$$F_{\text{rp}}(v_0) \simeq F_{\text{rp}}(v_0 = 0) - \eta v_0\,, \tag{9.5}$$

where

$$\eta = \hbar k^2 \Gamma \left[\frac{\Delta\, \Omega_r^2/2}{[\Delta^2 + (\Gamma/2)^2 + \Omega_r^2/2]^2} \right]. \tag{9.6}$$

The first term in Eq. (9.5) is a constant, and the second term acts as the detuning-dependent friction force illustrated in Fig. 9.1. The interpretation of this force is quite simple: because of the Doppler effect, the atom sees a field propagating toward it as having a higher frequency than a field propagating in the same direction. Hence, an optical field that is red-detuned from the atomic frequency for an atom at rest ($\Delta > 0$) will appear closer to resonance in the first case and further from resonance in the second case. As a result of this imbalance, atoms moving toward the light field

Fig. 9.1 Frictional
coefficient η, in arbitrary
units, as a function of the
detuning $\Delta = \omega_0 - \omega$
between the atomic transition
and field frequencies, for
$\Omega_r = \Gamma/4$. Detuning in units
of Γ

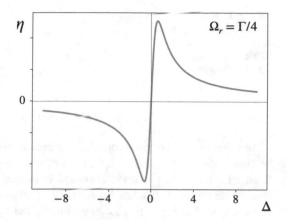

are decelerated more than atoms moving away from it are accelerated. Cooling can therefore be achieved by combining two counter-propagating red-detuned fields. In case these fields are weak enough that the two running waves treated as independent, we have immediately that the net force acting on an atom is

$$F_{\text{Doppler}} \simeq -2\eta v_0 . \tag{9.7}$$

The situation is more complicated in the case of strong laser fields, since the two counter-propagating waves can no longer be assumed to act independently. Under these conditions, the atomic dynamics can become quite complicated, and one can even reach a situation where damping requires the laser to be blue-detuned. A complete theory of the strong-field regime of Doppler cooling is given in Ref. [3], which also discusses in detail the theoretical limit to the temperature that can be achieved by this method. Here, we give a simple heuristic argument to evaluate this limit, based on the balance between the cooling effect of the friction force (9.7) and the heating effect of spontaneous emission.

Doppler Cooling Limit   We saw in Sect. 8.3 that each spontaneous emission event is accompanied by a momentum kick $\hbar\mathbf{k}$ of the atom in some random direction. This results in a random momentum diffusion, that is, in the heating of the atomic center-of-mass motion. In steady state, the number of random momentum kicks in the time interval δt is given by

$$\delta n = \Gamma \rho_{ee} \delta t, \tag{9.8}$$

and the resulting momentum spread of the atom is

$$\langle \delta p^2 \rangle = \hbar^2 k^2 \Gamma \rho_{ee} \delta t = \frac{\hbar^2 k^2 \Gamma}{2} \left(\frac{s}{1+s} \right) \delta t , \tag{9.9}$$

where we have used the steady-state inversion (1.87) to evaluate ρ_{ee}.

Table 9.1 Tabulated values
of the Doppler temperature
T_{Doppler} and the recoil
temperature T_{recoil} for
selected elements

Atom	T_{Doppler} (μK)	T_{recoil} (μK)
^{1}H	2389	1285
^{7}Li	142.11	6.06
^{23}Na	240.18	2.40
^{85}Rb	143.41	0.37
^{133}Cs	124.39	0.20

In addition to the randomness of spontaneous emission, a second source of heating is the randomness of light absorption, a result of the fact that the direction from which a given photon is absorbed is uncorrelated with the others. It can be shown that this leads to a heating roughly equal to that associated with spontaneous emission, so that Eq. (9.9) should be multiplied by a factor of 2. Doppler cooling reaches a steady state when the decrease in momentum due to the cooling force (9.7) balances this diffusion,

$$\left(\frac{\delta p^2}{\delta t}\right)_{\text{cooling}} = \left(\frac{\delta p^2}{\delta t}\right)_{\text{diffusion}}, \tag{9.10}$$

that is, with $\delta p^2 = 2p\delta p$, for

$$-2\eta\frac{p^2}{m} = \hbar^2 k^2 \Gamma \left(\frac{s}{1+s}\right). \tag{9.11}$$

From Eq. (9.6), one finds readily that the maximum friction is obtained for $\Delta = \Gamma/2$, which yields the lowest equilibrium temperature T_{Doppler}. In the weak-field limit, this gives

$$k_B T_{\text{Doppler}} = \frac{\delta p^2}{m} = \hbar\Gamma/4, \tag{9.12}$$

where k_B is Boltzmann's constant. A more precise analysis [3] finds

$$T_{\text{Doppler}} \simeq \hbar\Gamma/2k_B, \tag{9.13}$$

the so-called Doppler limit of laser cooling. This limit is of the order of 240 μK for sodium and is listed in Table 9.1 for several other elements.

It was long believed that this was the fundamental limit of laser cooling, but multilevel effects ignored in the current discussion, and to which we now turn, show that this is far from being the case.[1]

---

[1] W. D. Phillips is fond of saying that "there is no such thing as a two-level atom, and Sodium is not one of them."

9.2 Sisyphus Cooling

At the Doppler temperature limit T_{Doppler}, the atomic thermal de Broglie wavelength

$$\Lambda_{\text{dB}} \equiv \frac{2\pi\hbar}{\sqrt{2mk_B T}} \tag{9.14}$$

is still very small, of the order of $10\,\text{nm}$ or so, the precise value depending on the atom under consideration. Since Λ_{dB} is a rough measure of the spatial extent of the atomic wave packet, this indicates that above or near the Doppler temperature limit the center-of-mass wave packet is at least one or two orders of magnitude smaller than an optical wavelength. Hence, its dynamics is still well described in the ray atom optics approximation. One way to reach the wave optics regime is to filter the velocity distribution of the atoms, as this has been the case in many atomic diffraction experiments. Another route consists in further cooling the sample below the Doppler limit. Polarization gradient cooling, or weak-field *Sisyphus cooling*, is an efficient laser cooling mechanism that achieves this goal. It was first observed by P. Lett et al. [4] and explained by J. Dalibard and C. Cohen-Tannoudji [5].

In its simplest form, polarization gradient cooling results from the optical force associated with two optical potentials out of phase with each other and acting on an atom with two degenerate ground states. Under appropriate conditions, the atom preferentially jumps from one to the other ground state when it approaches a maximum of the potential associated with the first ground state and hence a minimum of the other. As a result, it mostly moves "uphill," much like Sisyphus of the Greek legend, and loses its kinetic energy in the process. The underlying mechanism is an "optical pumping" process whereby the atom undergoes a transition to an excited electronic state followed by spontaneous emission. This section briefly reviews the main characteristics of this cooling method in the simple case of a multilevel atom with a $J_g = 1/2 \leftrightarrow J_e = 3/2$ atomic transition between its ground and excited states.

In one dimension, one possible way to generate an appropriate optical potential is to use two counter-propagating light fields in a so-called lin$\perp$lin geometry,

$$\mathbf{E}(z, t) = \frac{1}{2}\left[\vec{\epsilon}_x \mathcal{E}_0 e^{i(kz-\omega t)} - i\vec{\epsilon}_y \mathcal{E}_0 e^{-i(kz+\omega t)} + \text{c.c.}\right]. \tag{9.15}$$

Both fields are traveling waves of amplitude $\mathcal{E}_0$, wave number k, and frequency ω, but they have perpendicular linear polarizations $\vec{\epsilon}_x$ and $\vec{\epsilon}_y$ along the $+x$ and $+y$ directions, respectively, hence the denomination lin$\perp$lin. The phases of the two traveling waves are chosen such that the total field has a σ_- circular polarization at the locations $z = \pm n\lambda/2$ and a σ_+ polarization at $z = \lambda/4 \pm n\lambda/2$, where n is an integer. The field (9.15) may then be reexpressed as

$$\mathbf{E}(z, t) = \frac{1}{2}\left[\vec{\epsilon}(z)\mathcal{E}e^{-i\omega t} + \text{c.c.}\right], \tag{9.16}$$

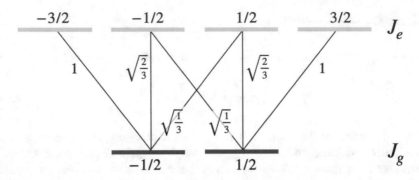

Fig. 9.2 Level scheme, with the Clebsch–Gordan coefficients indicated, for polarization gradient (Sisyphus) cooling on a $J_g = 1/2 \leftrightarrow J_e = 3/2$ transition. The $|e\rangle$ level manifold is shown in light gray to emphasize the fact that these levels are only virtually populated by far off-resonant transitions and adiabatically eliminated, so that the optical fields and spontaneous emission effectively couple only the two levels of the ground state manifold

where $\mathcal{E} = \sqrt{2}\mathcal{E}_0$ and $\vec{\epsilon}(z)$ is the position-dependent polarization vector

$$\vec{\epsilon}(z) = \vec{\epsilon}_- \cos(kz) - i\vec{\epsilon}_+ \sin(kz) \tag{9.17}$$

with

$$\vec{\epsilon}_\pm = \pm\frac{1}{\sqrt{2}}\left(\vec{\epsilon}_x \mp i\vec{\epsilon}_y\right). \tag{9.18}$$

As we shall see, this position-dependent polarization is essential in achieving Sisyphus cooling.

In contrast to the two-level atoms considered in much of this book, the optical field interacts now with the $J_g = 1/2 \leftrightarrow J_e = 3/2$ transition of the multilevel atomic system sketched in Fig. 9.2. We indicate the ground state manifold with the subscript "g" and the excited state manifold with "e," so that the corresponding atomic Hamiltonian reads

$$\hat{H} = \frac{\hat{p}_z^2}{2m} + \hbar\omega_g \hat{P}_g + \hbar\omega_e \hat{P}_e, \tag{9.19}$$

where we have introduced the projection operators $\hat{P}_g$ and $\hat{P}_e$ for the ground and excited manifolds as

$$\hat{P}_g = \sum_{m_g} |g_{m_g}\rangle\langle g_{m_g}| \ , \ \ m_g = \left\{-\tfrac{1}{2}, \tfrac{1}{2}\right\},$$

$$\hat{P}_e = \sum_{m_e} |e_{m_e}\rangle\langle e_{m_e}| \ , \ \ m_e = \left\{-\tfrac{3}{2}, -\tfrac{1}{2}, \tfrac{1}{2}, \tfrac{3}{2}\right\}. \tag{9.20}$$

The electric dipole interaction $\hat{H}_{AF}$ between the atom and the laser field (9.16) must be modified from its two-level atom expression to include the various levels involved and the relative strengths of their transitions, given by appropriate Clebsch–Gordan coefficients. For the $J_g = 1/2 \leftrightarrow J_e = 3/2$ transition at hand, it takes the form [3, 5]

$$
\begin{aligned}
\hat{V} &= -\hat{d}^{\,\dagger} \cdot \mathbf{E}(\hat{z}, t) e^{-i\omega t} + \text{h.c.} \\
&= -\frac{i}{2}\hbar\Omega_r \sin(k\hat{z}) \left[|e_{3/2}\rangle\langle g_{1/2}| + \frac{1}{\sqrt{3}}|e_{1/2}\rangle\langle g_{-1/2}| \right] e^{-i\omega t} \\
&\quad + \frac{1}{2}\hbar\Omega_r \cos(k\hat{z}) \left[|e_{-3/2}\rangle\langle g_{-1/2}| + \frac{1}{\sqrt{3}}|e_{-1/2}\rangle\langle g_{1/2}| \right] e^{-i\omega t} + \text{h.c.},
\end{aligned}
\tag{9.21}
$$

which displays explicitly the spatial dependence of the various transitions involved, a critical element of Sisyphus cooling. The reduced atomic raising operator $\hat{d}^+$ and its adjoint $\hat{d}^-$ are defined in terms of the Clebsch–Gordan coefficients for the allowed transitions as

$$
\begin{aligned}
\vec{\epsilon}_q \cdot \hat{d}^+ |m_g\rangle &= \langle J_g 1 m_g q | J_e m_e\rangle |m_e\rangle \,, \\
\vec{\epsilon}_q \cdot \hat{d}^+ |m_e\rangle &= 0 \,,
\end{aligned}
\tag{9.22}
$$

with $q = \pm$ gives the polarization of the laser field. Their values for a $J_g = 1/2 \leftrightarrow J_e = 3/2$ transition are shown in Fig. 9.2. We have also introduced the Rabi frequency

$$
\Omega_r = -\langle e_{3/2}|\hat{d} \cdot \vec{\epsilon}_+|g_{1/2}\rangle \frac{\mathcal{E}}{\hbar} \,.
\tag{9.23}
$$

In weak-field polarization gradient cooling, one always considers situations where the laser field is detuned far to the red of the atomic transition. In that case, it is possible to adiabatically eliminate the upper electronic states. The evolution of the remaining two magnetic ground states is then governed by the effective atomic Hamiltonian

$$
\hat{H}_{\text{eff}} = \frac{\hat{p}^2}{2m} - \hbar\Delta s_0 \hat{\Lambda}^{\dagger}(z)\hat{\Lambda}(z) \,,
\tag{9.24}
$$

where the operator $\hat{\Lambda}(z)$ is

$$
\hat{\Lambda}(z) = \vec{\epsilon}(\hat{z}) \cdot \hat{d}^+ = \left[\vec{\epsilon}_- \cos(k\hat{z}) - i\vec{\epsilon}_+ \sin(k\hat{z})\right] \cdot \hat{d}^+,
\tag{9.25}
$$

and

$$
s_0 = \frac{1}{4}\left(\frac{\Omega_r^2}{\Delta^2 + (\Gamma/2)^2} \right)
\tag{9.26}
$$

is the individual saturation parameter corresponding to each of the two counter-propagating waves.

The Hamiltonian (9.24) is the sum of the atomic kinetic energy and an effective optical potential

$$\hat{V}_{\text{opt}} = -\hbar \Delta s_0 \hat{\Lambda}^\dagger(\hat{z}) \hat{\Lambda}(\hat{z}) \tag{9.27}$$

that is local in the z-coordinate and diagonal in the ground state basis $|g_{\pm 1/2}\rangle$. Problem 9.2, see also Ref. [5], shows that

$$\hat{V}_{\text{opt}}|g_{1/2}\rangle = -\hbar \Delta s_0 \left[1 - \frac{2}{3}\cos^2(k\hat{z})\right]|g_{1/2}\rangle \equiv U_+(z)|g_{1/2}\rangle ,$$

$$\hat{V}_{\text{opt}}|g_{-1/2}\rangle = -\hbar \Delta s_0 \left[1 - \frac{2}{3}\sin^2(k\hat{z})\right]|g_{-1/2}\rangle \equiv U_-(z)|g_{-1/2}\rangle . \tag{9.28}$$

This demonstrates that the two ground state sublevels $|g_{1/2}\rangle$ and $|g_{-1/2}\rangle$ are subject to optical potentials, or light shifts, resulting from both the σ_+ and σ_- standing waves appearing in Eq. (9.22). They are illustrated in Fig. 9.3, which shows that the minima of the light shift $\hat{U}_-(z)$ repeat periodically at $z = n\lambda/2$, while those of $\hat{U}_+(z)$ occur at $z = \lambda/4 + n\lambda/2$. These are the locations where the laser field exhibits either pure σ_+ or pure σ_- polarization, respectively.

If the atomic dynamics were only driven by the optical field $\mathbf{E}(z, t)$, then atoms in the ground states $|g_{1/2}\rangle$ and $|g_{-1/2}\rangle$ would simply move in the periodic potentials $U_+(z)$ and $U_-(z)$, respectively, without undergoing any electronic transition. In the ray atom optics regime, this motion is similar to that of a ball being alternatively

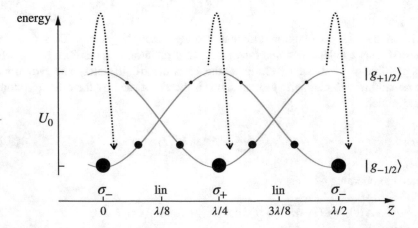

Fig. 9.3 Light shifts of the magnetic sublevels $|g_{\pm 1/2}\rangle$ as a function of position. The dotted arrows indicate the virtual transitions to the excited state manifold, followed by spontaneous transitions back from one to the other of the two magnetic ground states. The size of the dots is proportional to the steady-state populations of these levels for an atom at rest at position z, for $\Delta > 0$

accelerated and decelerated as it moves down and up the successive potential "hills" of a corrugated roof. In the wave atom optics regime, the situation is a bit more subtle, as the periodic potential acts as a diffraction grating for the atoms, as we saw in Sect. 8.4. However, neither in the ray optics nor in the wave optics regime would the interaction of the atoms with these potentials alone lead to cooling: what is still missing is a mechanism through which the atomic center-of-mass energy can be dissipated. In this case, this mechanism is spontaneous emission, which we have ignored so far. Specifically, the effective ground state Hamiltonian (9.24) accounts only for virtual transitions from one of the magnetic ground states to the excited state manifold, followed by *stimulated* transitions back to the other ground state.

The omission of spontaneous emission can be corrected along the same lines as in the discussion of Sect. 8.3, that is, by describing the atomic dynamics in terms of a master equation for the ground state manifold that includes both a non-Hermitian effective Hamiltonian and a "jump" Liouvillian. Following the adiabatic elimination of the excited state manifold, it takes the form [3]

$$\frac{d\hat{\rho}}{dt} = \frac{1}{i\hbar}\left(\hat{H}_{\text{eff}}\hat{\rho} - \hat{\rho}\hat{H}_{\text{eff}}^{\dagger}\right) + \hat{\mathcal{L}}[\hat{\rho}], \tag{9.29}$$

where

$$\hat{H}_{\text{eff}} = \frac{\hat{p}^2}{2m} - \hbar s_0\left(\Delta + \frac{i\Gamma}{2}\right)\hat{\Lambda}^{\dagger}(\hat{z})\hat{\Lambda}(\hat{z}), \tag{9.30}$$

$$\hat{\mathcal{L}}[\hat{\rho}] = \Gamma s_0 \sum_{q=0,\pm}(\vec{\epsilon}_q{}^{\star}\cdot\hat{d}^-)\hat{\Lambda}(\hat{z})\hat{\rho}\hat{\Lambda}^{\dagger}(\hat{z})(\vec{\epsilon}_q\cdot\hat{d}^+), \tag{9.31}$$

and the vectors $\vec{\epsilon}_q$ give the polarization of the spontaneously emitted photon, with $q = 0$ corresponding to linear polarization. Importantly, the master equation (9.29) is valid only to zeroth order in the parameter

$$\eta = \frac{kv_0}{\Gamma}, \tag{9.32}$$

where v_0 is the atomic velocity. As such, it does not include the effects of Doppler cooling and hence would apply to a precooled atomic sample. Furthermore, it does not account for the random atomic recoil associated with spontaneous emission that we already encountered in Sect. 8.3, a point to which we will return when discussing the Sisyphus cooling limit at the end of this section.

Note that the Liouvillian $\hat{\mathcal{L}}[\hat{\rho}]$ involves the laser field through the operators $\hat{\Lambda}(\hat{z})$ and $\hat{\Lambda}^{\dagger}(\hat{z})$. This is because after eliminating the excited state manifold, the effects of spontaneous emission appear in the ground state only, and for these levels, the Liouvillian must account for the combined effects of the virtual absorption of a laser photon, then, followed by spontaneous emission. This contribution to the atomic evolution is key to the cooling process, as it is the mechanism that couples the two degenerate ground states $|g_{1/2}\rangle$ and $|g_{-1/2}\rangle$. This is easily seen intuitively, because the laser field can, for example, induce a transition from the ground state $|g_{-1/2}\rangle$ to the excited state $|e_{1/2}\rangle$, from which the atom can then spontaneously decay to the ground state $|g_{1/2}\rangle$, as indicated by the dotted arrows in Fig. 9.3.

The master equation (9.29) permits to compute the dynamics of the populations $P_{\pm 1/2}$ of the two sublevels of the ground state manifold resulting from this optical pumping mechanism. Treating the atomic center-of-mass motion classically, its evolution is governed by the rate equations

$$\frac{dP_{1/2}(z,t)}{dt} = -\frac{2\Gamma'}{9}\cos^2(kz)P_{1/2}(z,t) + \frac{2\Gamma'}{9}\sin^2(kz)P_{-1/2}(z,t),$$

$$\frac{dP_{-1/2}(z,t)}{dt} = -\frac{dP_{1/2}}{dt}(z,t), \tag{9.33}$$

where we have introduced the scaled decay rate $\Gamma' = s_0\Gamma$.

The first of these equations shows that optical pumping out of $|g_{1/2}\rangle$ into $|g_{-1/2}\rangle$ is largest at the positions $z = n\lambda/2$ of σ_- polarization, that is, at the locations of maxima of the optical potential $U_+(z)$, from which the atom undergoes a transition to a minimum of $U_-(z)$. As a result, then, the atom has a tendency to always move uphill, resulting in the cooling of its center-of-mass motion.

Sisyphus Cooling Limit Returning for a moment to the full internal atomic dynamics, the interaction potential (9.21) readily allows one to extend the gradient force (8.18) to the multilevel system considered here to give

$$F_{gr} = \frac{\hbar k\Omega_r}{2}\cos(kz)\left[\langle g_{1/2}|\hat{\rho}|e_{3/2}\rangle e^{-i\omega t} + \frac{1}{\sqrt{3}}\langle g_{-1/2}|\hat{\rho}|e_{1/2}\rangle e^{-i\omega t} + \text{c.c.}\right]$$

$$+ \frac{\hbar k\Omega_r}{2}\sin(kz)\left[\langle g_{-1/2}|\hat{\rho}|e_{-3/2}\rangle e^{-i\omega t} + \frac{1}{\sqrt{3}}\langle g_{1/2}|\hat{\rho}|e_{-1/2}\rangle e^{-i\omega t} + \text{c.c.}\right]$$

along the z-axis. In the far off-resonant situation considered here and for velocities low enough that $kv \ll \Gamma$, the upper electronic states can be adiabatically eliminated, and this equation reduces approximately to

$$F_{gr} = kU_0 \sin 2kz \left[P_{1/2}(z,t) - P_{-1/2}(z,t)\right], \tag{9.34}$$

where

$$U_0 = \frac{2}{3}\hbar\Delta\left(\frac{\Omega_r^2}{\Delta^2 + (\Gamma/2)^2}\right) \simeq \frac{2}{3}\left(\frac{\hbar\Omega_r^2}{\Delta}\right) \tag{9.35}$$

and $|U_0|$ is the depth of the optical potential. The reactive force is therefore proportional to the difference in populations of the two ground state sublevels, as would be intuitively expected. From Eq. (9.33), we note that the characteristic rate governing optical pumping between these states is

$$\frac{1}{\tau_P} \equiv \frac{2\Gamma'}{9} = \frac{2s_0\Gamma}{9}. \tag{9.36}$$

For low intensities, $s_0 \ll 1$, it can be orders of magnitude smaller than the spontaneous emission rate Γ. This is why polarization gradient cooling can lead to a temperature orders of magnitude lower than Doppler cooling. Indeed, one could intuitively argue that the final temperature that can be achieved by this method is roughly given by the depth $\hbar|\Delta|s_0$ of the periodic optical potentials. In the case of large detunings $|\Delta| \gg \Gamma$, this gives

$$k_B T \simeq \frac{\hbar\Omega_r^2}{|\Delta|}. \tag{9.37}$$

This result might convey the impression that arbitrarily low temperatures can be achieved simply by reducing the Rabi frequency Ω_r. This, however, is not correct: in finding the expression (9.37), we assumed that the energy lost by the atom while climbing a potential well is large compared with the kinetic energy change in a fluorescence cycle due to the atomic recoil, an assumption implicit in the neglect of spontaneous recoil in the master equation (9.29). This condition clearly ceases to hold when $\hbar\Omega_r^2/|\Delta| \simeq \hbar^2 k^2/2m$. Consequently, the true limit of Sisyphus cooling turns out to be a few recoil temperatures.

9.3 Subrecoil Cooling

In both Doppler and Sisyphus cooling, the final cooling temperature is a result of a balance between the cooling force and spontaneous heating. Hence, it appears that to reach temperatures below the recoil limit, one needs to find a mechanism that leaves the atoms in a final state that is completely immune to spontaneous emission. Such *dark states* are central to cooling via velocity-selective coherent population trapping (VSCPT) [6], the cooling mechanism to which we now turn.

For concreteness, we consider a $J_g = 1 \leftrightarrow J_e = 1$ atomic transition driven by two counter-propagating laser fields of orthogonal polarizations, see Fig. 9.4. The positive frequency part of the combined laser fields is then

$$\mathbf{E}(z, t) = \frac{1}{2}\mathcal{E}\left[\vec{\epsilon}_+ e^{ikz} + \vec{\epsilon}_- e^{-ikz}\right]. \tag{9.38}$$

A key element of this level scheme is the vanishing electric dipole matrix element between the levels $|e_0\rangle$ and $|g_0\rangle$, so that it can thought of as consisting of a "V"-type and a "Λ"-type three-level systems that are not coupled to each other by an electric field of the form (9.38). In addition, spontaneous emission can optically

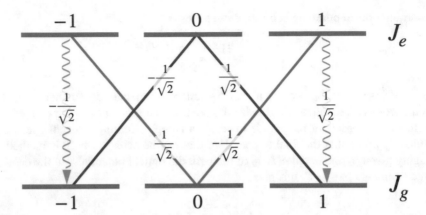

Fig. 9.4 Schematics of a $J_g = 1 \leftrightarrow J_e = 1$ atomic transition, showing the Λ (red) and V (blue) subsystems, and the levels coupled by spontaneous emission, with the non-zero Clebsch–Gordan coefficients indicated

pump population from the "V" into the "Λ" system, but not the other way round, as illustrated in Fig. 9.4.

For atoms initially in the ground state manifold, it is therefore possible to ignore the levels associated with the "V"-subsystem and restrict the analysis to the "Λ" system only, for which the laser–atom interaction takes the form

$$\hat{V} = -\frac{\hbar\Omega_r}{2\sqrt{2}} \left(|e_0\rangle\langle g_1| e^{-ikz} - |e_0\rangle\langle g_{-1}| e^{ikz} \right) e^{-i\omega t} + \text{h.c.}, \tag{9.39}$$

where $\Omega_r = d\mathcal{E}/\hbar$ is the Rabi frequency. The minus sign in front of the second contribution to $\hat{H}$ follows from the relative signs of the Clebsch–Gordan coefficients associated with a $J_g = 1 \leftrightarrow J_e = 1$ electric dipole transition and indicated in Fig. 9.4.

In the momentum representation, the interaction Hamiltonian (9.39) becomes

$$\hat{V} = \frac{\hbar\Omega_r}{2\sqrt{2}} \int dp \, (|e_0, p\rangle\langle g_1, p + \hbar k| - |e_0, p\rangle\langle g_{-1}, p - \hbar k|) \, e^{-i\omega t} + \text{h.c.}, \tag{9.40}$$

where we have used the integral representation of the translation operator

$$e^{\pm ikz} = \int dp |p\rangle\langle p \mp \hbar k|. \tag{9.41}$$

From Eq. (9.40), we have that

$$\hat{V}|e_0, p\rangle = -\frac{\hbar\Omega_r}{2\sqrt{2}} \left(|g_1, p + \hbar k\rangle - |g_{-1}, p - \hbar k\rangle \right) e^{-i\omega t}, \tag{9.42}$$

that is, the laser field couples the excited electronic state $|e_0\rangle$ of the atom with momentum p to a coherent superposition of the two electronic ground states $|g_1, p + \hbar k\rangle$ and $|g_{-1}, p - \hbar k\rangle$ with momenta differing by $2\hbar k$. All such dipole-coupled triplets form closed manifolds $\mathcal{M}(p)$ that we label with the value of the excited state momentum,

$$\mathcal{M}(p) \equiv \{|e_0, p\rangle, |g_1, p + \hbar k\rangle, |g_{-1}, p - \hbar k\rangle\}. \tag{9.43}$$

This suggests introducing the new basis set

$$|\psi_c(p)\rangle = \frac{1}{\sqrt{2}} (|g_1, p + \hbar k\rangle - |g_{-1}, p - \hbar k\rangle)$$

$$|\psi_{nc}(p)\rangle = \frac{1}{\sqrt{2}} (|g_1, p + \hbar k\rangle + |g_{-1}, p - \hbar k\rangle)$$

$$|\psi_e(p)\rangle = |e_0, p\rangle, \tag{9.44}$$

for which one readily finds

$$\hat{V}|\psi_c(p)\rangle = -\frac{\hbar \Omega_r}{2} e^{-i\omega t} |\psi_e(p)\rangle \tag{9.45a}$$

$$\hat{V}|\psi_e(p)\rangle = -\frac{\hbar \Omega_r}{2} e^{i\omega t} |\psi_c(p)\rangle \tag{9.45b}$$

$$\hat{V}|\psi_{nc}(p)\rangle = 0. \tag{9.45c}$$

Importantly, though, two additional mechanisms need to be considered in the description of subrecoil cooling limit, and both of them couple the various manifolds $\mathcal{M}(p)$. They are of as always spontaneous emission, which produces an incoherent coupling between them resulting from atomic recoil kicks randomly distributed between $-\hbar k$ and $\hbar k$ and, in addition, a coherent motional coupling between the levels $|\psi_c(p)\rangle$ and $|\psi_{nc}(p)\rangle$ due to the kinetic energy part of the atomic Hamiltonian. Both are essential in understanding the subrecoil cooling limit.

Dark States Equation (9.45c) shows that the state $|\psi_{nc}(p)\rangle$ is not coupled to any other state by the laser field. Hence, if spontaneous emission resulted in a transition from the excited state $|e(p)\rangle$ to $|\psi_{nc}(p)\rangle$, it would appear that the atom would remain trapped in that state forever. This, however, is not correct, because we still need to account for the effects of the kinetic energy $\hat{p}^2/2m$ on the atomic dynamics. While the states $|e_0, p\rangle$, $|g_1, p+\hbar k\rangle$ and $|g_{-1}, p-\hbar k\rangle$ are eigenstates of the kinetic energy operator, such is not the case for the superpositions $|\psi_c(p)\rangle$ and $|\psi_{nc}(p)\rangle$. This is what leads to the advertised *motional coupling* of $|\psi_c(p)\rangle$ and $|\psi_{nc}(p)\rangle$, with matrix elements

$$\langle \psi_c(p)| \frac{\hat{p}^2}{2m} |\psi_{nc}(p)\rangle = \frac{\hbar k\, p}{m}. \tag{9.46}$$

As a result, atoms can therefore escape from the state $|\psi_{nc}(p)\rangle$ except, however, for $p = 0$: the state $|\psi_{nc}(0)\rangle$ is a perfect trap, a dark state uncoupled to any other state of the system, be it via laser-induced dipole transitions, spontaneous emission, or motional coupling.

This observation leads to a simple qualitative explanation of atomic cooling via velocity-selective coherent population trapping. We discuss the one-dimensional situation for simplicity, but an extension to three dimensions is straightforward. Consider an atom initially in the manifold $\mathcal{M}(p)$ and spontaneously emitting a photon of momentum $\hbar k_0$, with $k_0 = \omega_0/c$, in some random direction. Denoting its component along the z-axis $\hbar k_{0,z} = u$, this results in a transition from the state $|e_0, p\rangle$ to some linear superposition of the states $|g_1, p - u\rangle$ and $|g_{-1}, p + u\rangle$. From Eq. (9.43), it is apparent that the first of these states belongs to the manifold $\mathcal{M}(p - u - \hbar k)$ and the second to $\mathcal{M}(p + u + \hbar k)$. Because $-\hbar k \leq u \leq \hbar k$, spontaneous emission results therefore in a redistribution of states in the manifold $\mathcal{M}(p)$ into the manifolds $\mathcal{M}(p')$, with

$$p - 2\hbar k \leq p' \leq p + 2\hbar k, \tag{9.47}$$

where we have assumed that $k_0 \simeq k$.

In addition to this recoil effect, spontaneous emission is the cause of the finite linewidth Γ of the upper states $|\psi_e(p)\rangle$ and $\Gamma' \simeq \Omega_r^2/\Gamma$ of the ground states $|\psi_c(p)\rangle$—this latter value, which results from the laser coupling between the ground and excited electronic states, being valid at resonance and in the weak-field limit $\Omega_r \ll \Gamma$. Also, the states $|\psi_{nc}(p)\rangle$ acquire a linewidth Γ'', a result of their motional coupling to the states $|\psi_c(p)\rangle$, which are in turn optically coupled to the excited states $|\psi_e(p)\rangle$. For $kp/M \ll \Gamma'$, one finds

$$\Gamma'' = 4\left(\frac{kp}{M}\right)^2 \frac{\Gamma}{\Omega_r^2}, \tag{9.48}$$

which can be interpreted as the probability per unit time that the atom leaves the state $|\psi_{nc}(p)\rangle$.

Because the only state with an infinite lifetime is $|\psi_{nc}(0)\rangle$, subsequent fluorescence cycles eventually lead to an accumulation of atoms in that state. Hence, the final momentum distribution of the atoms will be given (in one dimension) by two arbitrarily narrow peaks centered at $p = \pm\hbar k$. A detailed theory of VSCPT cooling can be found in Refs. [3, 7].

9.4 Cavity Cooling

We have seen in Chap. 3.4 that cavity environments provide significant additional flexibility compared to free space situations for the control of atom–field interactions. It is therefore not too surprising that these added features can also be exploited to cool atoms. This section discusses a method of cavity cooling [8, 9] that is somewhat reminiscent of Sisyphus cooling, except that it relies now on dressed states of the atom–cavity system. To provide an intuitive understanding of this approach, we first discuss a fully classical model that describes the cavity cooling of the Lorentz atom of Sect. 1.2 before turning to a full, albeit simplified description where both the atom and the intracavity field are quantized.

Classical Oscillator We consider a single-mode optical cavity supporting a standing wave field mode of frequency ω_c and wave vector K, driven by a classical field of amplitude $\mathcal{E}_{in}$ and frequency ω. After a straightforward modification of Eq. (5.147) for classical fields, see Eq. (5.158) and Problem 9.3, the equation of motion for the slowly varying amplitude $\mathcal{E}(t)$ of the intracavity field is

$$E(z, t) = \mathcal{E}(t)e^{-i\omega t} \cos(Kz) + \text{c.c.} \tag{9.49}$$

given by

$$\frac{d\mathcal{E}(t)}{dt} + (\kappa/2 - \Delta_c)\mathcal{E}(t) = \frac{i\omega}{2\epsilon_0}\mathcal{P}(t) + \sqrt{\kappa}\mathcal{E}_{in}, \tag{9.50}$$

where $\mathcal{P}$ is the slowly varying polarization, $\kappa/2$ is the amplitude decay rate of the resonator of length L and $\Delta_c = \omega - \omega_c$. The intracavity field interacts with a harmonically bound electron of frequency ω_0 described as a Lorentz atom. Its position $x(t)$ relative to the "atomic nucleus" located at position z_a undergoes damped oscillations governed by Eq. (1.20),

$$\left(\frac{d^2}{dt^2} + 2\gamma\frac{d}{dt} + \omega_0^2\right)x(t) = -\frac{e}{m}E(z_a, t). \tag{9.51}$$

In the slowly varying amplitude and phase approximation, and with Eqs. (9.49) and (1.50), the Maxwell wave equation simplifies to

$$\frac{d\mathcal{E}(t)}{dt} + (\kappa/2 - i\Delta_c)\mathcal{E}(t) = [-\Gamma(z_a) - i\mathcal{U}(z_a)]\mathcal{E}(t) + \sqrt{\kappa}\mathcal{E}_{in}, \tag{9.52}$$

where $\Delta = \omega_0 - \omega$,

$$\Gamma(z_a) = \left(\frac{e^2}{4m\epsilon_0 V}\right)\frac{\gamma}{\Delta^2 + \gamma^2}\cos^2(Kz_a) \tag{9.53}$$

is a spatially dependent decay rate resulting from the scattering of light from the dipole oscillator, and

$$\mathcal{U}(z_a) = -\left(\frac{e^2}{4m\epsilon_0 V}\right)\frac{\Delta}{\Delta^2 + \gamma^2}\cos^2(Kz_a) \tag{9.54}$$

is the associated spatially dependent frequency shift of the cavity mode frequency. The cavity volume V was introduced to convert $\mathcal{P}(t)$ into a proper polarization density.

For convenience, we now scale the intracavity field to

$$\mathcal{E} = \sqrt{\frac{\hbar\omega}{\epsilon_0 V}}\alpha, \qquad (9.55)$$

keeping in mind that this is a somewhat misleading step due to the appearance of $\hbar$ in a completely classical theory. The motivation for this scaling is merely to allow for an easy generalization to a quantum description, where α will then be interpreted as the square root of the mean number of intracavity photons. The coupled self-consistent equations that describe the coupled evolution of the field and the center of mass of the classical "atom" read then

$$\frac{d\alpha}{dt} = [-\kappa/2 - \Gamma(z_a) - i\Delta_c - i\mathcal{U}(z_a)]\alpha + \eta, \qquad (9.56a)$$

$$\frac{dp_a}{dt} = -|\alpha|^2 \frac{d}{dz_a}\mathcal{U}(z_a), \qquad (9.56b)$$

$$\frac{dz_a}{dt} = \frac{p_a}{m}, \qquad (9.56c)$$

where η accounts for the coupling of α to the driving field $\mathcal{E}_{in}$ and p_a is the particle momentum. Equation (9.56b) shows that the force $-|\alpha|^2(d\mathcal{U}(z_a)/dz_a)$ acting on the classical oscillator is essentially the gradient force introduced in Sect. 8.2, except that as it now acts on a classical oscillator in a cavity instead of on a two-state system.

The cooling of the oscillator results from the periodic dependence of the intracavity field on the atomic position. Equation (9.56a) shows that it is driven both by $\mathcal{U}(z_a)$, which changes the mode frequency locally, and by $\Gamma(z_a)$, which is due to the spatially dependent spontaneous light scattering by the dipole, resulting in a spatially dependent cavity response time.

More specifically, for empty cavities, the maximum transmission and intracavity intensity occurs when it is driven by a resonant field, but the situation is different with an atom inside the resonator: The atom–field interaction results in a shift in the cavity resonance frequency $\mathcal{U}(z_a)$ that depends on the atomic position. In particular, the strong coupling for an atom at an antinode of the cavity mode may shift the pump into resonance, thereby increasing the intracavity field. However, the field is also subject to a position-dependent damping $\kappa/2 + \Gamma(z_a)$ that results in a time delayed response, and for appropriate conditions, field maxima can therefore occur after the particle reaches a location where the cavity mode function—in the present example, $\sin Kz$ with $K = n\pi/L$, n integer—imposes a minimum to $\mathcal{U}(z)$. As a consequence, the particle will climb up the optical potential hills at times when they

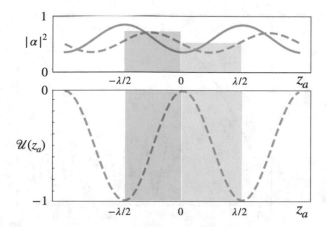

Fig. 9.5 Schematic of the classical picture of cavity cooling. The lower figure shows $\mathcal{U}(z_a)$ (negative in this example), which governs the position-dependent change in the frequency of the cavity field due to its interaction with the particle. Its zeros are the zeros of the cavity mode function $\sin(Kz)$. The upper figure shows both the intracavity field $|\alpha|^2_{\text{static}}(z_a)$ for a dipole at rest (solid line) and its value $|\alpha|^2(z_a)$ (dashed line) for a dipole moving at constant velocity, as a function of the dipole position z_a. The red shaded area shows a region where the particle is moving uphill against a field $|\alpha|^2(z_a)$. That field is larger than in the green area, where the particle moves downhill against a weaker field $|\alpha|^2(z_a)$, resulting in its cooling. Arbitrary units

are higher than when it moves downhill. This results in the cooling of its motion, as illustrated in Figs. 9.5 and 9.6. This mechanism, whose origin is a properly time-delayed gradient force, is reminiscent of Sisyphus cooling, as will also become evident from a simplified quantum description in terms of dressed states to which we now turn.

Simplified Quantum Description A simple quantum interpretation of cavity cooling follows directly from the Jaynes–Cummings model, with the only differences that the two-level atom is now allowed to propagate along the cavity axis and that the system is driven by an external field, see Eq. (5.147) or its approximate form (5.158) for a classical driving field.

Since the vacuum Rabi frequency is now spatially dependent, the dressed states (3.3) of the (non-driven) Jaynes–Cummings model are spatially dependent as well,

$$|1, n\rangle = \sin\theta_n(z)|e, n\rangle + \cos\theta_n(z)|g, n + 1\rangle$$
$$|2, n\rangle = \cos\theta_n(z)|e, n\rangle - \sin\theta_n(z)|g, n + 1\rangle, \tag{9.57}$$

with

$$\tan[2\theta_n(z_a)] = \frac{-2g(z_a)\sqrt{n + 1}}{\Delta} \tag{9.58}$$

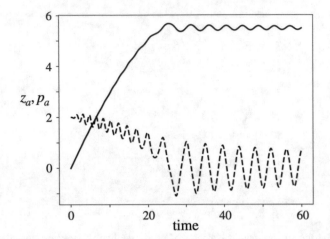

Fig. 9.6 Solid line: evolution of the particle position z_a (solid line) and momentum p_a (dashed line) during the cavity cooling of a classical oscillator of mass m and frequency ω_0. The position z_a is in units of the wavelength λ, the momentum p_a is in units of $\hbar k$, and time is in units of κ. In this example, $\Gamma(z_a = 0) = 0.16$, $\mathcal{U}(z_a = 0) = 0.45$, $\Delta = -2$, $\Delta_c = 0$, and $\eta = 1.0$. For long enough times, the oscillator becomes trapped in one potential well of the optical potential

and corresponding eigenenergies

$$E_{1n}(z_a) = \hbar(n + 1/2)\omega + \hbar\Omega_n(z_a),$$
$$E_{2n}(z_a) = \hbar(n + 1/2)\omega - \hbar\Omega_n(z_a),\qquad(9.59)$$

where

$$\Omega_n(z_a) = \frac{1}{2}\sqrt{\Delta^2 + 4g^2(z_a)(n + 1)}.\qquad(9.60)$$

In order to identify the basic physics underlying the cavity cooling mechanism, it is sufficient to consider the limit of weak excitation, where the atoms are mostly in the ground state. For large negative detunings, the dressed state $|1, n\rangle$ of the undriven system approaches $|g, n + 1\rangle$, with eigenenergy $E_{1,n} \approx \hbar\omega(n + 1) - \hbar\omega_0/2$ near the nodes of the intracavity field, see Eqs. (7.43). Similarly, $|2, n\rangle \rightarrow |e, n\rangle$, with energy $E_{2n} \approx \hbar\omega n + \hbar\omega_0/2$. In particular, the energy difference between the states $E_{1,0}$ and $E_{2,0}$ becomes

$$E_{1,0} - E_{2,0} \approx -\hbar\Delta,\qquad(9.61)$$

as illustrated in Fig. 9.7. At these locations, the energy difference between the state $|1, 0\rangle$ and the ground state $|g, 0\rangle$ is readily found to be $\hbar\omega$, that is, this transition is resonant with the driving field. It follows that at these locations, the atom will be preferentially pumped by that field from $|g, 0\rangle$ to the state $|1, 0\rangle$. From the point on,

Fig. 9.7 Schematic of the cavity cooling mechanism, illustrating the Sisyphus cooling mechanism whereby for negative detunings $\Delta = \omega_0 - \omega$, the atoms are pumped to the dressed state $|1, 0\rangle$ near the nodes of the field and then move uphill on the spatially dependent potential associated with that state before decaying spontaneously back to the ground state $|g, 0\rangle$. Cooling is achieved by the atom being forced to always move uphill and thereby losing its kinetic energy

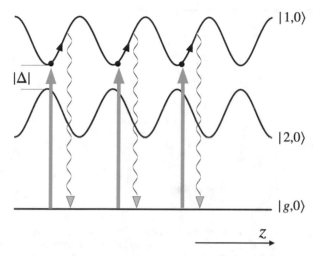

a slow atom will adiabatically follow that state. As it moves further away from the node of the field, it is forced to climb up the potential $E_{1,0}(z)$, loosing kinetic energy in the process, until spontaneous emission brings it back to the state $|g, 0\rangle$. The excitation followed by spontaneous emission sequence is then repeated, much like in Sisyphus cooling. It will eventually cease when the atom no longer has sufficient kinetic energy to climb up the potential hills and is trapped in one of the wells.

Reference [9] discusses additional conditions under which this type of cavity cooling can also be achieved, in particular by a careful choice of cavity detuning $\Delta_c = \omega - \omega_c$ and positive detunings $\Delta = \omega_0 - \omega$. This situation is the topic of Problem 9.6.

9.5 Sideband Cooling

So far, we have considered the optical cooling of neutral atoms in atomic vapors or beams. We now turn to the case of trapped particles, more specifically trapped ions. These are of considerable interest for quantum information applications, where the capability to manipulate the internal state of strings of trapped ions cooled to their motional ground state provides a promising approach to quantum state engineering and quantum information processing.

Ion Traps Ions can be trapped by electromagnetic restoring forces, either in combinations of static electric and magnetic fields as in Penning traps [10] or in time-dependent electric fields as in Paul traps [11], see Fig. 9.8. In both cases, the resulting potential can be approximated as a quadratic potential, so that the motion of an ion near the trap center can be characterized by three frequencies. In the Penning trap, the axial motion is at a frequency ν_{ax}, while the radial motion is the sum of a harmonic cyclotron motion of cyclotron frequency ν_{cyc} and a repulsive

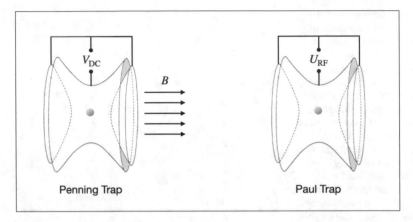

Fig. 9.8 The Paul trap and the Penning trap are geometrically similar quadrupole traps. The Paul trap exploits an intermediate frequency AC voltage between a hyperbolic ring and hyperbolic end electrodes, while the Penning trap has a repulsive DC voltage on the end electrodes and a strong magnetic field in the axial direction

magnetron motion at frequency ν_{mag} due to the fact that the electric field forces the ion away from the trap center, with $\nu_{\text{cyc}} \gg \nu_{\text{ax}} \gg \nu_{\text{mag}}$. Similarly, the ion motion near the center of a Paul trap is well described by a harmonic motion in three dimensions characterized by three frequencies ν_x, ν_y, and ν_z. In the following, we restrict ourselves for simplicity to one dimension. The generalization to three dimensions is straightforward.

Resolved Sideband Cooling We consider a single two-level ion of mass m trapped in a one-dimensional harmonic trap of frequency ν and interacting with a classical field $E(z, t) = E_0 \vec{\epsilon} \cos(\omega t - kz)$. This system is described by adding the harmonic center-of-mass motion of the ion to the Hamiltonian (1.61). With the results of Sect. 2.1, this gives readily

$$
\hat{H} = \frac{\hat{p}^2}{2m} + \frac{1}{2}m\nu^2\hat{z}^2 + \hbar\Delta|e\rangle\langle e| + \frac{\hbar\Omega_r}{2}\left[\hat{\sigma}_+ e^{ik\hat{z}} + \text{h.c.}\right]
$$

$$
= \hbar\nu\hat{b}^\dagger\hat{b} + \hbar\Delta|e\rangle\langle e| + \frac{\hbar\Omega_r}{2}\left[|e\rangle\langle g|e^{ik\hat{z}} + \text{h.c.}\right], \tag{9.62}
$$

where the Hamiltonian $\hbar\nu\hat{b}^\dagger\hat{b}$, with eigenstates $|n\rangle$, accounts for the harmonic motion of the trapped ion. The operators $\hat{b}$ and $\hat{b}^\dagger$, with $[\hat{b}, \hat{b}^\dagger] = 1$, are annihilation and creation operators of vibrational quanta or phonons.

As before, the optical field-induced center-of-mass recoil of the ion is described by the translation operator (8.28) or its integral form (9.41)

$$
e^{\pm ik\hat{z}} = \int \mathrm{d}p|p\rangle\langle p \mp \hbar k|. \tag{9.63}
$$

On the basis $\{|n\rangle\}$ of vibrational Fock states, this operator is associated with transitions $|n\rangle \to |n'\rangle$, with transition matrix elements

$$F_{n\to n'} = \langle n'|e^{\pm ik\hat{z}}|n\rangle . \tag{9.64}$$

It follows that a state $|g, n\rangle$ of the ion is in general coupled to a set of states $\{|e, n'\rangle\}$, with transition frequencies

$$\omega_{n,n'} = \omega_0 + (n' - n)\nu . \tag{9.65}$$

The linewidth of the excited electronic state $|e\rangle$ is given by its spontaneous decay rate Γ, and for large $\Gamma \gg \nu$, there will be a number of accessible states $|e, m\rangle$ whose transition frequency (9.65) falls within it. This should be contrasted with the opposite limit $\Gamma \ll \nu$, where the various frequencies $\omega_{n,n'}$ will be resolved. This is the so-called *resolved sideband* limit of sideband cooling. In that regime, the laser can be tuned to a specific sideband and selectively drive transitions between bands.

Consider then the transition matrix elements (9.64), and reexpress the position operator $\hat{z}$ of the ion in terms of the annihilation and creation operators $\hat{b}$ and $\hat{b}^\dagger$ as

$$\hat{z} = z_{\text{zpf}}(\hat{b} + \hat{b}^\dagger), \tag{9.66}$$

where z_{zpf} is the zero-point position

$$z_{\text{zpf}} = \sqrt{\frac{\hbar}{2m\nu}} . \tag{9.67}$$

Further introducing the *Lamb–Dicke parameter*

$$\eta = kz_{\text{zpf}} , \tag{9.68}$$

with $\eta^2 = \omega_{\text{rec}}/\nu$, Eq. (9.64) becomes

$$\begin{aligned} F_{n\to n'} &= \langle n'|e^{i\eta(\hat{b}+\hat{b}^\dagger)}|n\rangle \\ &= \langle n'|1 + i\eta(\hat{b} + \hat{b}^\dagger) + O(\eta^2)|n\rangle . \end{aligned} \tag{9.69}$$

If the trap frequency ν is much larger than the recoil frequency ω_{rec}, then $\eta \ll 1$. In that case Eq. (9.69) shows that transitions where n' differs from n by more than one are largely suppressed, and it is sufficient to consider ground to excited transitions involving the levels

$$|g, n\rangle \to |e, n \pm 1\rangle , \tag{9.70}$$

as illustrated in Fig. 9.9.

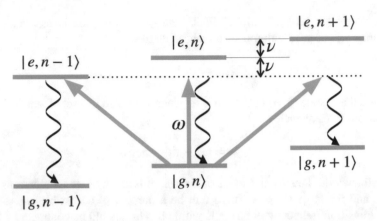

Fig. 9.9 Schematics of the resolved sideband cooling of trapped ion, with v the trap frequency and ω the frequency of the cooling laser. The straight green arrows indicate the laser-induced transitions $|g, n\rangle \leftrightarrow |e, n'\rangle$, with $n' = n$ or $n \pm 1$ in the Lamb–Dicke limit, and the wavy arrows indicate spontaneous emission between the states $|e, n\rangle$ and $|g, n\rangle$

Motional cooling will be achieved if transitions from the state $|n\rangle$ to $|n - 1\rangle$ are favored over transitions from $|n\rangle$ to $|n + 1\rangle$. This is because from a given initial state $|e, n - 1\rangle$, spontaneous emission will on the average result in the decay $|e, n - 1\rangle \to |g, n - 1\rangle$, so that following subsequent transitions, the ion will eventually approach the $|n = 0\rangle$ state.

From the analysis of the rate equation approximation of Chap. 1, we recall that the transition rate between two levels coupled by a field with Rabi frequency Ω_r and with atom–field detuning Δ is given by Eq. (1.85),

$$R = \frac{\Omega_r^2}{\Gamma} \left(\frac{(\Gamma/2)^2}{\Delta^2 + (\Gamma/2)^2} \right). \tag{9.71}$$

With Eqs. (9.69) and (9.65), this result becomes, when considering the Lamb–Dicke limit of the motional transition matrix elements,

$$R_{n \to n-1} = n\eta^2 \frac{\Omega_r^2}{\Gamma} \left(\frac{(\Gamma/2)^2}{\Delta_-^2 + (\Gamma/2)^2} \right) \equiv n A_- \tag{9.72}$$

$$R_{n \to n+1} = (n + 1)\eta^2 \frac{\Omega_r^2}{\Gamma} \left(\frac{(\Gamma/2)^2}{\Delta_+^2 + (\Gamma/2)^2} \right) \equiv (n + 1) A_+, \tag{9.73}$$

with

$$\Delta_\pm = (\omega_0 \pm v) - \omega. \tag{9.74}$$

The rate A_- is maximally favored over A_+ for $\Delta_- = 0$, or $\omega = \omega_0 - \nu$, and hence $|\Delta_+| = 2\nu$, that is, for a laser detuned to the red of the transition frequency ω_0 by the trap frequency ν, see Fig. 9.9. From a detailed balance argument along the lines of Sect. 5.2, the rate of change of the probability p_n to be in the motional state $|n\rangle$ is

$$\frac{dp_n}{dt} = -nA_- p_n + (n+1)A_- p_{n+1} - (n+1)A_+ p_n + nA_+ p_{n-1}, \qquad (9.75)$$

see in particular the discussion leading to Eq. (5.63). This results in the steady-state solution

$$p_n = \frac{A_+}{A_-} p_{n-1}, \qquad (9.76)$$

which gives, for a field resonant with the blue-shifted transition $|g, n\rangle \to |e, n-1\rangle$ so that $\Delta_- = 0$ and $\Delta_+ = \nu$ and $\Gamma \ll \nu$,

$$p_n \approx \left(\frac{\Gamma}{2\nu}\right)^2 p_{n-1} \qquad (9.77)$$

corresponding to a thermal distribution with $\bar{n} = (\Gamma/2\nu)^2 \ll 1$.

In addition to being of considerable importance in precision measurements and in the implementation of trapped ion-based quantum information systems, sideband cooling also finds significant applications in quantum optomechanics, where the same idea is exploited to cool mesoscopic mechanical systems to their ground state of vibration. This will be discussed in Chap. 11.

9.6 Evaporative Cooling

While laser cooling has found a number of applications in quantum optics and optomechanics, it turns out not to always be sufficient to achieve the combined phase space densities and low temperatures required to investigate the many-body dynamics of quantum gases, such as e.g. Bose–Einstein condensation. In such cases laser cooling can be complemented by other techniques, the most common being evaporative cooling. For the sake of completeness, we briefly discuss elementary aspects of this approach of interest for quantum optics applications.

In contrast to the laser cooling methods discussed so far, evaporative cooling is quite familiar in everyday life. It is the mechanism that leads to the cooling of a cup of coffee and that is also used as a very effective, low-tech, and inexpensive way to cool houses in hot and dry climates. As time goes on, coffee cools down as the most energetic (warmest) molecules escape from the cup and the remaining ones rethermalize at a lower temperature.

The evaporative cooling of atomic samples works in much the same way, except that the coffee cup is replaced by an atomic trap, and, more importantly, the atomic densities and temperatures involved are much lower. It was first proposed by Hess [12], and its first experimental demonstration was accomplished in spin-polarized atomic hydrogen [13]. Extensive reviews of evaporative cooling can be found in Refs. [14] and [15].

To achieve efficient cooling, the high energy tail of the thermal distribution of atoms must be constantly repopulated by collisions so that an equilibrium distribution can be maintained and the cooling process sustained. Hence, an essential condition for evaporative cooling is that the lifetime of the sample to be cooled must be long compared with the collisional thermalization time. The problem, however, is that at the low densities typical of atom optics experiments, the collisions are rare, with a concomitant long thermalization time. Indeed, typical collisional thermalization times in Bose–Einstein condensation experiments are of the order of seconds.

An efficient way to improve evaporative cooling in atomic vapors is the technique of *radio frequency-induced evaporation*, first proposed by D. Pritchard et al. [16] and by T. W. Hijmans et al. [17]. The basic idea beyond this approach is quite simple: consider a magnetic trap with trapping potential of the form

$$V(r) = m_F g \mu_B [B(r) - B_0],\tag{9.78}$$

where m_F is the magnetic quantum number of the trapped magnetic sublevel, g is the Landé g-factor, μ_B is the Bohr magneton, B_0 is the bias field of the trap and $B(r)$ is the radially dependent trap magnetic field, and apply a radio frequency field at frequency ω_{rf} that can flip the atomic magnetic spin, hence leaving the atoms in a state expelled from the trap. The resonance condition for such transitions is

$$\hbar \omega_{\text{rf}} = |g| \mu_B B_{\text{res}}(r),\tag{9.79}$$

which, combined with Eq. (9.78), shows that only atoms with an energy

$$E > |m_F||g|\mu_B [B_{\text{res}}(r) - B_0] = \hbar |m_F|(\omega_{\text{rf}} - \omega_0),\tag{9.80}$$

where ω_0 is the frequency associated with the bias field B_0, will escape the trap, or evaporate. Slowly varying ω_{rf} in a way appropriate for the details of the trap under consideration allows one to achieve sustained evaporation without having to reduce the potential depth of the actual magnetic trap, making it easier to cool the sample in an efficient fashion.

To discuss evaporative cooling in a slightly more quantitative way, consider a sample of atoms of initial total energy E_T and average energy per atom $\langle E \rangle = E_T/N$, and assume that the average energy of the evaporated atoms is $(1 + \epsilon)\langle E \rangle$. The energy removed from the sample by dN particles is therefore $(1 + \epsilon)\langle E \rangle dN$, and the change in average energy per atom is

$$\langle E \rangle - d\langle E \rangle = \frac{E_T - (1 + \epsilon)\langle E \rangle dN}{N - dN}, \tag{9.81}$$

so that to the lowest order

$$\frac{d\langle E \rangle}{\langle E \rangle} = \epsilon \frac{dN}{N} \tag{9.82}$$

or

$$\frac{\langle E \rangle}{\langle E_0 \rangle} = \left(\frac{N}{N_0} \right)^\epsilon, \tag{9.83}$$

where $\langle E_0 \rangle$ and N_0 are initial values. In this simple model, the average energy of the particles, and hence their temperature, decreases as a power of the number of particles. While this is an oversimplified picture, which assumes that ϵ is independent of N and ignores the detailed role of the thermalizing collisions and of possible external influences such as radio frequency fields, it does give a rough idea of the way evaporative cooling works.

Problems

Problem 9.1 Evaluate the recoil temperature, recoil velocity, recoil temperature, and Doppler temperature, as well as the acceleration associated with the force $F = \hbar k \Gamma$ for the D2 ($3^2 S_{1/2} \rightarrow 3^2 P_{3/2}$) line of a sodium atom (mass $= 3.82 \cdot 10^{-26}$ kg, $\omega = 2\pi \times 508.8$ THz, $\Gamma = 62 \cdot 10^6 \text{s}^{-1}$). Evaluate also its de Broglie wavelength at both the Doppler temperature and the recoil temperature.

Problem 9.2 Derive the optical potentials (9.28)

$$\hat{V}_{\text{opt}}|+\tfrac{1}{2}\rangle = -\hbar \Delta s_0 \left[1 - \tfrac{2}{3} \cos^2(k\hat{z}) \right] |+\tfrac{1}{2}\rangle \equiv \hat{U}_+(\hat{z})|+\tfrac{1}{2}\rangle ,$$

$$\hat{V}_{\text{opt}}|-\tfrac{1}{2}\rangle = -\hbar \Delta s_0 \left[1 - \tfrac{2}{3} \sin^2(k\hat{z}) \right] |-\tfrac{1}{2}\rangle \equiv \hat{U}_-(\hat{z})|-\tfrac{1}{2}\rangle ,$$

which account for the stimulated dynamics of the ground state manifold in Sisyphus cooling by adiabatically eliminating the evolution of the excited manifold.

Problem 9.3 Derive equation (9.50)

$$\frac{d\mathcal{E}(t)}{dt} + (\kappa/2 - \Delta_c)\mathcal{E}(t) = \frac{i\omega}{2\epsilon_0} \mathcal{P}(t) + \sqrt{\kappa} \mathcal{E}_{\text{in}} ,$$

for the intracavity field in case it is driven by an external field $\mathcal{E}_{\text{in}}$.

Problem 9.4 Derive Eq. (9.52) and Eqs. (9.29)–(9.31) by following an approach that parallels the analysis leading to Eq. (5.115) and adiabatically eliminating the upper state manifold.

Problem 9.5 Derive Eq. (9.50) by extending the analysis of the slowly varying approximation of Sect. 1.2 to the case of a cavity with damping rate κ and mode frequency ω_c driven by an external field of frequency ω. Derive also Eqs. (9.56a)–(9.56c) by extending Eqs. (1.26) and (1.27) to the case of a Lorentz atom inside the resonator.

Problem 9.6 Show that a cavity cooling scheme similar to the one discussed in Sect. 9.4 can be achieved by a proper choice of cavity detuning $\Delta_c = \omega - \omega_c$ and positive atom–field detunings $\Delta = \omega_0 - \omega$. In that case, you should find that the atom needs to be pumped into the minima of the lower, rather than the upper dressed state.

Problem 9.7

(a) Derive the rate equations that govern the sideband cooling of a single ion confined to a one-dimensional trap in the Lamb–Dicke limit.
(b) Solve these equations numerically for $\eta = 0.1$, $\omega_r^2/\Gamma = 1$, $\Delta_- = 0$ and parameters Γ, $(\omega - \omega_o)$ and ν such that the resolved sideband regime is satisfied. Assume in the numerical solution that the ion motion is initially in thermal equilibrium with a mean phonon number $\bar{n} = 3$.
(c) Verify that the steady state reached by the ion satisfies the detailed balance condition (9.77).

References

1. T.W. Hänsch, A. Schawlow, Cooling of gases by laser radiation. Opt. Commun. **13**, 68 (1975)
2. D.J. Wineland, W.M. Itano, Laser cooling of atoms. Phys. Rev. A **20**, 1521 (1979)
3. C. Cohen-Tannoudji, *Atomic Motion in Laser Light: In Fundamental Systems in Quantum Optics*, ed. by J. Dalibard, J.M. Raimond, J. Zinn-Justin (North-Holland, Amsterdam, 1992), p. 3
4. P.D. Lett, R. Watts, C. Westbrook, W. Phillips, P. Gould, H. Metcalf, Observation of atoms laser-cooled below the Doppler limit. Phys. Rev. Lett. **61**, 169 (1988)
5. J. Dalibard, C. Cohen-Tannoudji, Laser cooling below the Doppler limit by polarization gradients: simple theoretical models. J. Opt. Soc. Am. B **6**, 2023 (1989)
6. A. Aspect, E. Arimondo, R. Kaiser, N. Vansteenkiste, C. Cohen-Tannoudji, Laser cooling below the one-photon recoil by velocity-selective coherent population trapping. Phys. Rev. Lett. **61**, 826 (1988)
7. A. Aspect, E. Arimondo, R. Kaiser, N. Vansteenkiste, C. Cohen-Tannoudji, Laser cooling below the one-photon recoil energy by velocity-selective coherent population trapping: theoretical analysis. J. Opt. Soc. B **6**, 2112 (1989)
8. P. Horak, G. Hechenblaikner, K.M. Gheri, H. Stecher, H. Ritsch, Cavity-induced atom cooling in the strong coupling regime. Phys. Rev. Lett. **79**, 4974 (1997)
9. G. Hechenblaikner, M. Gangl, P. Horak, H. Ritsch, Cooling an atom in a weakly driven high-Q cavity. Phys. Rev. A **58**, 3030 (1998)

10. L.S. Brown, G. Gabrielse, Geonium theory: Physics of a single electron or ion in a Penning trap. Rev. Mod. Phys. **58**, 233 (1986)
11. W. Paul, Electromagnetic traps for charged and neutral particles. Rev. Mod. Phys. **62**, 531 (1990)
12. H.H. Hess, Evaporative cooling of magnetically trapped and compressed spin-polarized hydrogen. Phys. Rev. B **34**, 3476 (1986)
13. N. Masuhara, J.M. Doyle, J.C. Sandberg, D. Kleppner, T.J. Greytak, H.F. Hess, G.P. Kochanski, Evaporative cooling of spin-polarized hydrogen. Phys. Rev. Lett. **61**, 935 (1988)
14. W. Ketterle, N.J. van Druten, Evaporative cooling of trapped atoms, in *Advances in Atomic, Molecular and Optical Physics*, vol. 37, ed. by B. Bederson, H. Walther (Academic Press, San Diego, 1996), p. 181
15. J.T.M. Walraven, Atomic hydrogen in magnetostatic traps, in *Quantum Dynamics of Simple Systems*, ed. by G.L. Oppo, S.M. Barnett (Institute of Physics Publication, London, 1996), p. 315
16. D.E. Pritchard, K. Helmerson, A.G. Martin, Atom traps, in *Atomic Physics*, vol. 11, ed. by S. Haroche, J. C. Gay, G. Grynberg (World Scientific, Singapore, 1989), p. 179
17. T.W. Hijmans, O.J. Luiten, I.D. Setija, J.T.M. Walraven, Optical cooling of atomic hydrogen in a magnetic trap. J. Opt. Soc. Am. B **6**, 2235 (1989)

Chapter 10
Bose–Einstein Condensation

Bosonic atomic vapors cooled to extremely low temperatures may undergo transitions to Bose–Einstein condensates (BEC). After a brief review of this effect in non-interacting free space systems this chapter discusses new features resulting from trapped samples. We then turn to an introduction to Schrödinger field quantization and apply this formalism to the derivation of the Gross–Pitaevskii equation and the description of Bogoliubov quasi-particles. The final part of this chapter focuses on optical lattices, a type of optical traps that permits to develop powerful bridges with condensed matter physics, for example in realizations of the Bose–Hubbard model. We conclude with a brief overview of atom microscopes, which allow for a direct characterization, atom by atom, of ultracold atomic samples on lattices.

The experimental realization of quantum degenerate atomic systems, in particular of atomic Bose–Einstein condensation (BEC), is arguably one of the most important achievements resulting from the availability of laser cooling and advances in evaporative cooling techniques. In addition to opening a new area of investigation in AMO physics and in quantum and atom optics, this breakthrough also provides the opportunity to build exciting bridges with other areas of physics, most importantly perhaps with condensed matter physics and field theory.

The theoretical prediction of Bose–Einstein condensation goes back to the realization by A. Einstein [1, 2], following the original ideas by S. N. Bose [3], that below a critical temperature T_c a gas of non-interacting bosons can develop a macroscopic population of its lowest energy state. Although it had been predicted for decades, the first atomic BEC was successfully realized only in 1995 by E. Cornell et al. [4], who cooled a gas of rubidium atoms to $1.7 \cdot 10^{-7}$ K above absolute zero, and shortly thereafter by W. Ketterle et al. [5], who created a BEC of sodium atoms. The requisite breakthroughs to achieve this goal were the ability to produce low enough temperatures, combined with trapping techniques that permit to confine atoms with a sufficient phase space density without the need for material containers.

The first experimental demonstrations of atomic Bose–Einstein condensation were rapidly followed by a broad spectrum of advances in the theoretical and experimental study of ultracold atomic gases. They offer exciting new avenues to investigate a wide range of topics at the boundary between AMO science, condensed matter physics, and field theory and are driven to a significant extent by the ability to develop *quantum simulators* that permit the quantitative study of strongly correlated bosonic or fermionic systems, quantum phase transitions, topological phases of matter, and much more.

The theory of BEC is well documented in most books on statistical mechanics [6–8], although surprisingly, there are still difficulties being addressed, especially relative to the fluctuations in particle numbers predicted by different statistical ensembles, see e.g. Refs. [9–12]. However, these issues seem to be restricted to the case of non-interacting gases and we will not dwell on them here.

Although our focus in this chapter will be on trapped, low-density, and weakly interacting alkali bosonic gases, we begin with a brief review of the phenomenology of BEC in free, non-interacting systems, concentrating on the mean occupation of the various states of the particles. We then discuss in Sect. 10.2 the new features resulting from trapped samples before turning to a more formal development of Schrödinger field quantization, a formalism of considerable use in the discussion of the more elaborate aspects of BEC. The last part of this chapter focuses on optical lattices, an important type of optical traps that permit to build cold atom quantum simulators with controlled inter-particle interactions of many-body condensed matter systems, the Bose–Hubbard model being an example. We also discuss the use of "atomic microscopes" as unique tools for the manipulation and readout atom by atom of these simulators.

10.1 Phenomenology

We consider a Bose gas of free, non-interacting particles described by the Hamiltonian

$$\hat{H} = \sum_i \hat{H}_i = \sum_i \frac{\hat{p}_i^2}{2m} \tag{10.1}$$

or equivalently[1]

$$\hat{\mathcal{H}} = \sum_j E_j \hat{c}_j^\dagger \hat{c}_j \,, \tag{10.2}$$

---

[1] $\hat{\mathcal{H}}$ is the so-called second-quantized form of $\hat{H}$. Both Hamiltonians describe the same ensemble of N non-interacting particles of energies E_i and they are fully equivalent, as will be shown in some detail in Sect. 10.3. Here we briefly "jump the gun" and use $\hat{\mathcal{H}}$ without formal justification so as to exploit the intuition that we have gained from quantized optical fields, whose Hamiltonian is formally identical.

where $\hat{c}_j$ and $\hat{c}_j^\dagger$ are annihilation and creation operators that satisfy the boson commutation relations $[\hat{c}_j, \hat{c}_k^\dagger] = \delta_{jk}$, that annihilate, respectively, create a particle of energy $E_j = \hat{p}_j^2/2m$, with $\hat{n}_j = \hat{c}_j^\dagger \hat{c}_j$ the number operator for these particles.

As shown in Problem 10.1, this system is characterized by the thermal equilibrium density operator (compare also to Eq. (2.85))

$$\hat{\rho} = \frac{1}{Z} \prod_j e^{-\beta(E_j - \mu)\hat{c}_j^\dagger \hat{c}_j} ,$$ (10.3)

where $\beta = 1/k_B T$, Z is the partition function, and the energies E_j are those of free particles of mass m. Finally μ is the chemical potential of the system, determined in the case of a canonical ensemble from the normalization condition $N = \sum n_j$, with N the number of particles in the system.

The associated Bose–Einstein distribution for the mean occupation number of the level of energy E_j is

$$\langle n_j \rangle = \frac{1}{\exp[\beta(E_j - \mu)] - 1} .$$ (10.4)

As a consequence, for the non-interacting particles considered here μ must satisfy the condition

$$\mu < 0 ,$$ (10.5)

a direct consequence of the form of the distribution (10.4) and the fact that the number of particles in a given state must remain positive.

For simplicity, we adopt periodic boundary conditions for a cubic box of volume $V = L^3$. The particle center-of-mass eigenfunctions are then

$$\varphi_n(x, y, z) = \frac{1}{\sqrt{V}} \exp\left[\frac{i}{\hbar}(p_x x + p_y y + p_z z) \right],$$ (10.6)

where

$$p_x = \frac{2\pi n_x}{L}\hbar , \qquad p_y = \frac{2\pi n_y}{L}\hbar , \qquad p_z = \frac{2\pi n_z}{L}\hbar ,$$ (10.7)

and n_i are integer numbers, and the corresponding energy eigenvalues are

$$E_j = \frac{1}{2m} \left(p_x^2 + p_y^2 + p_z^2 \right) \equiv \frac{p^2}{2m} .$$ (10.8)

For a large enough box, the allowed momenta become quasicontinuous, and the number of eigenstates in a momentum space volume $d^3 p$ is $V d^3 p/(2\pi\hbar)^3$. The number of states with energy below some given value E is therefore

$$
\mathcal{N}(E) = \frac{V}{(2\pi\hbar)^3} \int_0^{2\pi} d\phi \int_0^{\pi} \sin\theta \, d\theta \int_0^{\sqrt{2mE}} p^2 dp
$$

$$
= \frac{4\pi}{3} \frac{V}{(2\pi\hbar)^3} (2mE)^{3/2} , \tag{10.9}
$$

with corresponding density of states

$$
\mathcal{D}(E) \equiv \frac{d\mathcal{N}(E)}{dE} = \left(\frac{V}{\sqrt{2}\pi^2\hbar^3}\right) m^{3/2} \sqrt{E} . \tag{10.10}
$$

In the continuum limit the total number of particles is therefore

$$
N = \langle n_0 \rangle + \int_0^{\infty} dE \, \mathcal{D}(E) \frac{1}{\exp[\beta(E-\mu)] - 1} , \tag{10.11}
$$

where we have used Eq. (10.4) and the ground state population n_0 was added "by hand," a step required since $\mathcal{D}(0) = 0$.

Critical Temperature As previously mentioned, the chemical potential μ can be determined from the number of particles in the system. To achieve this goal, Eqs. (10.11) and (10.5) must be solved self-consistently, a task performed numerically in general. However, we can gain further insight into the physics involved by introducing the *fugacity*

$$
z \equiv e^{\beta\mu} \tag{10.12}
$$

with

$$
0 \le z < 1 , \tag{10.13}
$$

as follows directly from Eq. (10.5).

With the density of states (10.10), Eq. (10.11) then becomes, neglecting for now the ground state population $\langle n_0 \rangle$,

$$
4\pi^2 \left(\frac{\hbar^2\beta}{2m}\right)^{3/2} \frac{N}{V} = \int_0^{\infty} dx \frac{x^{1/2}}{e^x/z - 1} , \tag{10.14}
$$

where we have introduced the dimensionless energy $x \equiv \beta E$, and the quantity to be determined is now the fugacity rather than the chemical potential. The solution of Eq. (10.14) is obtained by intersecting a horizontal line at a value equal to its

Fig. 10.1 The function $I(z)$ of Eq. (10.15)

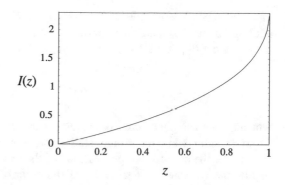

left-hand side with the function

$$I(z) \equiv \int_0^\infty dx \, \frac{x^{1/2}}{e^x/z - 1}, \tag{10.15}$$

which is plotted in Fig. 10.1. This procedure does yield a unique solution as long as $z < 1$, that is, as long as

$$4\pi^2 \left(\frac{\hbar^2\beta}{2m}\right)^{3/2} \frac{N}{V} < I(z=1) \simeq 2.315\ldots, \tag{10.16}$$

However, there appears to be no solution otherwise. For a fixed density N/V, the condition (10.16) implies that the sample temperature must be larger than a *critical Bose temperature* T_c given by

$$k_B T_c \simeq 6.632 \left(\frac{\hbar^2}{2m}\right) \left(\frac{N}{V}\right)^{2/3}. \tag{10.17}$$

This is a very disturbing result, since it should be possible in principle both to have as many particles as we wish in the sample, and also to cool it to an arbitrarily low temperature. So something seems to be missing in our analysis. What could it be?

Ground State Population The answer to this question is immediately apparent if we return for a minute to the approximation that leads to Eq. (10.14): The absence of a solution below T_c simply results from the fact that the ground state population $\langle n_0 \rangle$ was ignored in that expression. From the Bose–Einstein distribution (10.4) it is given by

$$\langle n_0 \rangle = \frac{1}{e^{-\beta\mu} - 1} = \frac{z}{1 - z}, \tag{10.18}$$

and it becomes macroscopic as the fugacity approaches its limiting value of one, or equivalently as the chemical potential approaches zero from below. Specifically, we then have $\exp(-\beta\mu) \simeq 1 - \beta\mu = 1 + 1/\langle n_0 \rangle$, or

$$\langle n_0 \rangle \simeq -\frac{1}{\beta\mu}, \tag{10.19}$$

which is increasingly macroscopic quantity as $\beta\mu \to 0$.

Importantly, while the ground state population becomes macroscopic below T_c, such is not the case for the populations of the excited states. This is easily seen to be the case for states of energy E_j of the order of or larger than $k_B T$, since from Eq. (10.4) one has immediately

$$\langle n_j \rangle = \frac{1}{\exp[\beta(E_j - \mu)] - 1} < \frac{1}{e - 1} < 1, \tag{10.20}$$

and $\langle n_j \rangle \ll \langle n_0 \rangle$. The proof is somewhat more delicate for states of energy E_j separated from the ground state energy by much less than $k_B T$. Because $\mu \simeq 0$, we have that $\beta(E_j - \mu) \ll 1$, and hence

$$\langle n_j \rangle \simeq \frac{1}{\beta(E_j - \mu)} = \frac{1}{\beta E_j + 1/\langle n_0 \rangle}, \tag{10.21}$$

where we have used Eq. (10.19) to obtain the last equality. Consider specifically the first excited state, of energy $E_1 = 2\pi^2\hbar^2/MV^{2/3}$. Below T_c, and with Eq. (10.17), we have

$$E_1/k_B T > E_1/k_B T_c \simeq 1/N^{2/3} \tag{10.22}$$

so that for a total particle number sufficiently large that the inequality $\beta E_1 \ll |\beta\mu|$ holds

$$\frac{\langle n_1 \rangle}{\langle n_0 \rangle} \simeq \frac{|\mu|}{E_1} \simeq \frac{1}{\langle n_0 \rangle/N^{2/3} + 1}. \tag{10.23}$$

The explicit temperature dependence of the ground state population $\langle n_0 \rangle$ for temperatures below the critical temperature T_c can be obtained from Eq. (10.11), with the substitution $\mu \approx -1/\beta\langle n_0 \rangle$ from Eq. (10.19) and noting that below T_c this value is so close to zero that we can safely set $\mu = 0$ in the integral. In other words, for $T < T_c$ the chemical potential remains essentially zero, and $\langle n_0 \rangle$ adjusts itself to satisfy Eq. (10.11). Carrying out the integration as before yields, with the definition (10.17) of T_c,

$$\langle n_0 \rangle = N\left[1 - \left(\frac{T}{T_c}\right)^{3/2}\right]. \tag{10.24}$$

Fig. 10.2 Free space condensate fraction—ground state condensate fraction $\langle n_0 \rangle$ normalized to the total number of particles N—as a function of the normalized temperature T/T_c

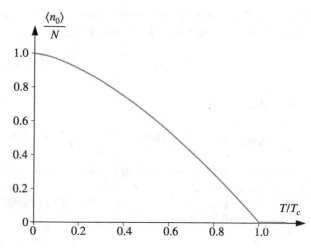

Hence, the ground state population grows continuously from 0 to N as T decreases from T_c to $T = 0$, as illustrated in Fig. 10.2. For $T \ll T_c$ we have therefore that $\langle n_0 \rangle \simeq N$, so that for $N \gg 1$

$$\frac{\langle n_1 \rangle}{\langle n_0 \rangle} \simeq \frac{1}{N^{1/3}} \ll 1 . \tag{10.25}$$

This confirms that while the ground state is macroscopically populated the individual populations of all excited levels are negligible by comparison. This is the phenomenon of Bose–Einstein condensation.

10.2   BEC in Traps

The experimental realization of atomic Bose–Einstein condensation always involves samples trapped in either magnetic or optical traps. However, the analysis of the previous section, while initially depending explicitly on the density of states $\mathcal{D}(\omega)$, was then specialized to a free space geometry. In this section we show that tailoring the environment of ultracold atomic vapors can change this behavior dramatically and that confined geometries can significantly modify Bose–Einstein condensation as compared to its free space behavior or even lead to its absence. This state of affairs is somewhat reminiscent of Chap. 7, where we showed how the radiative properties of atoms can be profoundly influenced by their environment, in that case the density of modes of the electromagnetic field.

We consider for concreteness the situation where the trapping potential of the atoms is a three-dimensional harmonic oscillator

$$V_{\text{trap}}(\mathbf{r}) = \frac{m}{2} \left(\omega_1^2 x_1^2 + \omega_2^2 x_2^2 + \omega_3^2 x_3^2 \right), \tag{10.26}$$

with energy levels

$$E_{n_1,n_2,n_3} = \left(n_1 + \tfrac{1}{2} \right) \hbar\omega_1 + \left(n_2 + \tfrac{1}{2} \right) \hbar\omega_2 + \left(n_3 + \tfrac{1}{2} \right) \hbar\omega_3, \tag{10.27}$$

where n_i are integers larger than or equal to zero. As before, the density of states is obtained by first evaluating the number of states $\mathcal{N}(E)$ whose energy is below some value E in the continuum limit. Noting that the surfaces of equal energy are planes given by Eq. (10.27) it is readily found to be

$$\mathcal{N}(E) = \frac{E^3}{6\hbar^3 \omega_1 \omega_2 \omega_3} \tag{10.28}$$

and the associated density of states is

$$\mathcal{D}(E) \equiv \frac{d\mathcal{N}(E)}{dE} = \frac{E^2}{2\hbar^3 \omega_1 \omega_2 \omega_3}. \tag{10.29}$$

It is also useful [13] to introduce a more general density of states

$$\mathcal{D}(E) = C_\alpha E^{\alpha-1}, \tag{10.30}$$

where $\alpha = 3/2$ for the free space situation of Sect. 10.1, $\alpha = 3$ for a three-dimensional harmonic trap, and $\alpha = 1$ for a gas of free particles in two dimensions. Here C_α is a constant depending on the geometry at hand.

We saw in the preceding section that the critical temperature T_c corresponds to the point where the chemical potential μ becomes equal to zero and $\langle n_0 \rangle$ starts to build up macroscopically. At precisely that temperature, we still have $\langle n_0 \rangle \simeq 0$. Setting $\mu = 0$ with the density of states (10.30) in Eq. (10.11) with $\mu = 0$ gives for the total number of particles at that temperature

$$N = C_\alpha \int_0^\infty dE \, \frac{E^{\alpha-1}}{e^{E/k_B T_c} - 1} = C_\alpha (k_B T_c)^\alpha \int_0^\infty dx \, \frac{x^{\alpha-1}}{e^x - 1}$$

$$= C_\alpha (k_B T_c)^\alpha \Gamma(\alpha) \zeta(\alpha), \tag{10.31}$$

where $x = E/k_B T_c$,

$$\Gamma(\alpha) = \int_0^\infty dx \, x^{\alpha-1} e^{-x}$$

Table 10.1 Tabulated values
of the functions $\Gamma(\alpha)$ and
$\zeta(\alpha)$

α	$\Gamma(\alpha)$	$\zeta(\alpha)$
1	1	∞
1.5	0.8862	2.612
2	1	1.645
2.5	1.3293	1.341
3	2	1.202
3.5	3.326	1.127
4	6	1.082

is the Γ function, and

$$\zeta(\alpha) = \frac{1}{\Gamma(\alpha)} \int_0^\infty dx\, \frac{x^{\alpha-1}}{e^x - 1}$$

is the Riemann ζ-function. Some numerical values of these functions are given in
Table 10.1.

From Eq. (10.31) we have that

$$k_B T_c = \left(\frac{N}{C_\alpha \Gamma(\alpha)\zeta(\alpha)}\right)^{1/\alpha}. \tag{10.32}$$

For a 3-dimensional harmonic trap we have $\alpha = 3$ and $C_\alpha = 1/2\hbar^3 \omega_1 \omega_2 \omega_3$, and
the critical temperature becomes

$$k_B T_c \simeq 0.94\hbar\, (\omega_1 \omega_2 \omega_3 N)^{1/3}, \tag{10.33}$$

compared with the free space result (10.17), in which case T_c scales as $N^{2/3}$.

The condensate fraction can be obtained as in the three-dimensional free space
case of Sect. 10.1 by reintroducing the ground state population, so that

$$\langle n_0 \rangle(T) = N - C_\alpha \int_0^\infty dE\, E^{\alpha-1} \frac{1}{e^{E/k_B T} - 1}. \tag{10.34}$$

For $\alpha = 3$ and with the definition (10.33) this gives

$$\langle n_0 \rangle(T) = N \left[1 - \left(\frac{T}{T_c}\right)^3\right], \tag{10.35}$$

as illustrated in Figs. 10.3 and 10.4. In contrast to the free space situation of
Eq. (10.24), $\langle n_0 \rangle$ now scales as T^3 rather than $T^{3/2}$.

Importantly, since $\zeta(\alpha)$ diverges for the two-dimensional case $\alpha = 1$, we observe
that in that case, Bose–Einstein condensation can occur only at $T = 0$, a dramatic

Fig. 10.3 Condensate fraction $\langle n_0 \rangle / N$ in a three-dimensional harmonic trap as a function of the normalized temperature T/T_c. The dashed curve shows that same fraction in free space, for comparison (Remember however that T_c takes different values on free space and in a trap, see Eqs. (10.17) and (10.33))

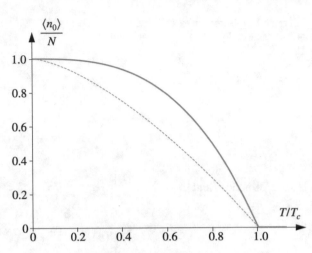

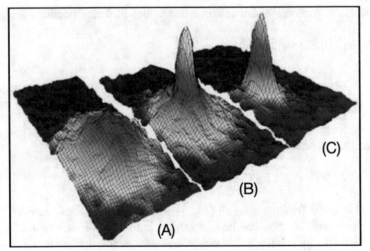

Fig. 10.4 Iconic picture of the first atomic BEC experimental demonstration at JILA. The false-color 3-D images display the velocity distribution of the cloud (**a**) just before and (**b**) just after the appearance of the condensate; and (**c**) after further evaporation has left a sample of nearly pure condensate. The circular pattern of the non-condensed fraction is indicative of an isotropic velocity distribution consistent with thermal equilibrium. In contrast, the sharp condensate fraction is elliptical, indicative that it is a highly nonthermal distribution (Credit M. Matthews, JILA. Courtesy of NIST/JILA/CU-Boulder)

demonstration of the impact of the system's density of states on its behavior. This shows in a dramatic fashion the dependence on dimensionality and density of states of Bose–Einstein condensation, a situation reminiscent of the changes in the radiative properties of atoms in tailored electromagnetic environments of cavity QED.

10.3 Schrödinger Field Quantization

We have seen in earlier chapters how the quantization of the electromagnetic field is a central element in the study of quantum optical systems. A similar tool to theoretically describe Bose–Einstein condensates is provided by the formalism of Schrödinger field quantization, or second quantization, to which we turn in this section. Note however that while the quantization of the electromagnetic field leads to the appearance of new physics, for instance spontaneous emission, such is not the case when second quantization is narrowly applied just to the nonrelativistic problems that we address in this book. In this limited context it is merely a very convenient and powerful way to describe the many-body physics of identical particles satisfying either bosonic or fermionic statistics.

We introduce second quantization by considering first a system of N non-interacting particles that evolves according to the Schrödinger equation for the N-particle Hamiltonian

$$\hat{H}_N = \sum_{i=1}^{N} \hat{H}_i \, , \tag{10.36}$$

where

$$\hat{H}_i \equiv -\frac{\hbar^2}{2m} \nabla_i^2 + \hat{V}(\mathbf{r}_i) \tag{10.37}$$

is the Hamiltonian of particle i. Its N-particle wave function $\phi(\mathbf{r}_1, \ldots, \mathbf{r}_N, t)$ is either symmetric or antisymmetric under particle exchange, depending on the particles being bosons or fermions. Second quantization recasts this same problem in a field-theoretical formalism by introducing a Fock space and particle creation and annihilation operators for these particles, see for instance Refs. [14–16]. This allows us to easily account for systems where the total number of particles is not conserved, and offers a powerful tool to treat many-body effects. This formalism also permits to account for the quantum statistics of massive particles, fermions, or bosons, in a simple way. Last but not least, it has the considerable merit of providing significant additional insight into the problem and elegant calculation techniques and also lends itself to a number of powerful approximation methods.

Field Operators It is beyond the scope of this chapter to give a rigorous treatment of second quantization starting from a canonical quantization approach. Rather, we *postulate* that the Schrödinger wave function $\psi(\mathbf{r})$ becomes an operator[2] $\hat{\Psi}(\mathbf{r})$ that for bosonic particles satisfies the commutation relations

$$[\hat{\Psi}(\mathbf{r}), \hat{\Psi}^\dagger(\mathbf{r}')] = \delta(\mathbf{r} - \mathbf{r}'),$$

$$[\hat{\Psi}(\mathbf{r}), \hat{\Psi}(\mathbf{r}')] = 0, \tag{10.38}$$

and for fermionic particles the anticommutation relations

$$[\hat{\Psi}(\mathbf{r}), \hat{\Psi}^\dagger(\mathbf{r}')]_+ = \delta(\mathbf{r} - \mathbf{r}'),$$

$$[\hat{\Psi}(\mathbf{r}), \hat{\Psi}(\mathbf{r}')]_+ = 0, \tag{10.39}$$

where $[\ldots]_+$ is an anticommutator, $[\hat{A}, \hat{B}]_+ = \hat{A}\hat{B} + \hat{B}\hat{A}$. As we shall see, $\hat{\Psi}(\mathbf{r})$ may be interpreted as an operator annihilating a particle at position $\mathbf{r}$, and $\hat{\Psi}^\dagger(\mathbf{r})$ creates a particle at location $\mathbf{r}$.

The second-quantized Hamiltonian corresponding to the N-particle Hamiltonian $\hat{H}_N = \sum_i \left[-\hbar^2 \nabla_i^2/2m + \hat{V}(\mathbf{r}_i) \right]$ is

$$\hat{\mathcal{H}} = \int d^3 r \, \hat{\Psi}^\dagger(\mathbf{r}) \left(-\frac{\hbar^2}{2m} \nabla^2 + \hat{V}(\mathbf{r}) \right) \hat{\Psi}(\mathbf{r}). \tag{10.40}$$

Physically, one can understand this Hamiltonian in the following way: the field operator $\hat{\Psi}(\mathbf{r})$ picks a particle at location $\mathbf{r}$, the Hamiltonian evolution $\hat{H} = -\hbar^2 \nabla^2/2m + \hat{V}(\mathbf{r})$ is applied to this particle, and the operator $\hat{\Psi}^\dagger(\mathbf{r})$ puts it back into place. Finally, the integral in Eq. (10.40) guarantees that all particles in the ensemble are subject to this treatment.

With the Hamiltonian $\hat{\mathcal{H}}$ the Heisenberg equation of motion for the field operator $\hat{\Psi}(\mathbf{r}, t)$ is, for the case of bosons,

$$i\hbar \frac{d\hat{\Psi}(\mathbf{r}, t)}{dt} = [\hat{\Psi}(\mathbf{r}, t), \hat{\mathcal{H}}]$$

$$= \int d^3 r' \left[\hat{\Psi}(\mathbf{r}, t), \hat{\Psi}^\dagger(\mathbf{r}', t) \left(-\frac{\hbar^2}{2m} \nabla^2 + \hat{V}(\mathbf{r}') \right) \hat{\Psi}(\mathbf{r}', t) \right]$$

$$= \left(-\frac{\hbar^2}{2m} \nabla^2 + \hat{V}(\mathbf{r}) \right) \hat{\Psi}(\mathbf{r}, t), \tag{10.41}$$

where we have used the identity $[\hat{A}, \hat{B}\hat{C}] = \hat{B}[\hat{A}, \hat{C}] + [\hat{A}, \hat{B}]\hat{C}$ and the commutation relations (10.38). Problem 10.3 shows that the same result holds for fermions, despite the different (anti)commutation relations. Hence, the Heisenberg equation

---

[2]Hence, the moniker second quantization.

of motion for the Schrödinger field operator $\hat{\Psi}(\mathbf{r}, t)$ has the same form as the Schrödinger equation for the wave function $\psi(\mathbf{r}, t)$ in usual quantum mechanics. In this sense we can think of the operator $\hat{\Psi}(\mathbf{r}, t)$ as the quantized form of the single-particle (or very loosely speaking "classical") Schrödinger field $\psi(\mathbf{r}, t)$, in much the same way that we promoted the classical electric field $\mathbf{E}(\mathbf{r}, t)$ to a quantized field operator, $\mathbf{E}(\mathbf{r}, t) \rightarrow \hat{E}(\mathbf{r}, t)$, with $\hat{E}(\mathbf{r}, t)$ obeying Maxwell's equations just like $\mathbf{E}(\mathbf{r}, t)$.

N-Particle State In terms of the Schrödinger field creation operator $\hat{\Psi}^{\dagger}(\mathbf{r})$ an N-particle state takes the general form

$$|\phi_N\rangle = \frac{1}{\sqrt{N!}} \int d^3r_1 \ldots d^3r_N \phi_N(\mathbf{r}_1, \ldots, \mathbf{r}_N, t) \hat{\Psi}^{\dagger}(\mathbf{r}_N) \ldots \hat{\Psi}^{\dagger}(\mathbf{r}_1)|0\rangle,$$

(10.42)

where $|0\rangle$ is the vacuum state (absence of particles),

$$\hat{\Psi}(\mathbf{r})|0\rangle = 0,$$

(10.43)

and $\phi_N(\mathbf{r}_1, \ldots \mathbf{r}_N, t)$ is an N-particle wave function.

Introducing the number operator

$$\hat{N} \equiv \int d^3r \, \hat{\Psi}^{\dagger}(\mathbf{r}) \hat{\Psi}(\mathbf{r})$$

(10.44)

it is easily shown that

$$\hat{N}|\phi_N\rangle = N|\phi_N\rangle$$

(10.45)

by using the commutator identity

$$[\hat{A}, \hat{B}_N \ldots \hat{B}_1] = \sum_{i=1}^{N} \hat{B}_N \ldots \hat{B}_{i+1}[\hat{A}, \hat{B}_i]\hat{B}_{i-1} \ldots \hat{B}_1,$$

which follows directly from $[\hat{A}, \hat{B}_2\hat{B}_1] = [\hat{A}, \hat{B}_2]\hat{B}_1 + \hat{B}_2[\hat{A}, \hat{B}_1]$, see Problems 10.4 and 10.5. Similarly we have

$$\hat{\Psi}(\mathbf{r})|\phi_N\rangle = \frac{N}{\sqrt{N!}} \int d^3r_1 \ldots d^3r_{N-1} \phi_N(\mathbf{r}_1, \ldots \mathbf{r}_{N-1}, \mathbf{r}) \hat{\Psi}^{\dagger}(\mathbf{r}_{N-1}) \ldots \hat{\Psi}^{\dagger}(\mathbf{r}_1)|0\rangle,$$

(10.46)

which confirms that the operator $\hat{\Psi}(\mathbf{r})$ annihilates a particle at position $\mathbf{r}$.

The equivalence of the formalism of second quantization with the conventional quantum mechanical description of the N-body problem in terms of the sum of N

individual Hamiltonians of the form (10.37) is easily demonstrated by applying the many-body Hamiltonian (10.40) to the N-particle state $|\phi_N\rangle$, showing that

$$\hat{\mathcal{H}}|\phi_N\rangle = \frac{1}{\sqrt{N!}} \int d^3r \, \hat{\Psi}^\dagger(\mathbf{r})\hat{H}(\mathbf{r})\hat{\Psi}(\mathbf{r})$$

$$\times \int d^3r_1 \ldots d^3r_N \phi_N(\mathbf{r}_1, \ldots \mathbf{r}_N)\hat{\Psi}^\dagger(\mathbf{r}_N) \ldots \hat{\Psi}^\dagger(\mathbf{r}_1)|0\rangle$$

$$= \frac{1}{\sqrt{N!}} \sum_{i=1}^{N} \int d^3r_1 \ldots d^3r_N \, \hat{\Psi}^\dagger(\mathbf{r}_N) \ldots \hat{\Psi}^\dagger(\mathbf{r}_1)|0\rangle$$

$$\times \hat{H}_i \phi_N(\mathbf{r}_1 \ldots \mathbf{r}_N), \tag{10.47}$$

where $\hat{H}_i = -\nabla_i^2/2m + \hat{V}(\mathbf{r}_i)$. That is, the multiparticle evolution governed by the Schrödinger equation of motion

$$i\hbar \frac{d}{dt}\phi_N(\mathbf{r}_1, \ldots, \mathbf{r}_N, t) = \hat{H}_N \phi_N(\mathbf{r}_1, \ldots, \mathbf{r}_N, t) \tag{10.48}$$

with $\hat{H}_N = \sum_i \hat{H}_i$ is equivalent to the evolution given by the Fock space Schrödinger equation

$$i\hbar \frac{d|\phi_N\rangle}{dt} = \hat{\mathcal{H}}|\phi_N\rangle, \tag{10.49}$$

a direct consequence of the fact that the Schrödinger field operator evolution is governed by the single-particle Hamiltonian $\hat{H}$, see Eq. (10.41).

Continuity Equation It is easily shown that the Schrödinger field operator $\hat{\Psi}(\mathbf{r})$ satisfies a conservation of probability law. Introducing the particle density operator

$$\hat{n}(\mathbf{r}, t) = \hat{\Psi}^\dagger(\mathbf{r}, t)\hat{\Psi}(\mathbf{r}, t)$$

yields the continuity equation

$$\frac{d\hat{n}(\mathbf{r}, t)}{dt} = \frac{d\hat{\Psi}^\dagger(\mathbf{r}, t)}{dt}\hat{\Psi}(\mathbf{r}, t) + \hat{\Psi}^\dagger(\mathbf{r}, t)\frac{d\hat{\Psi}(\mathbf{r}, t)}{dt}$$

$$= \frac{-i\hbar}{2m}\left[\left(\nabla^2\hat{\Psi}^\dagger(\mathbf{r}, t)\right)\hat{\Psi}(\mathbf{r}, t) - \hat{\Psi}^\dagger(\mathbf{r}, t)\nabla^2\hat{\Psi}(\mathbf{r}, t)\right]$$

$$= -\nabla \cdot \hat{\mathbf{j}}, \tag{10.50}$$

where we have introduced the probability current

$$\hat{\mathbf{j}} \equiv \left(\frac{-i\hbar}{2m}\right)\left[\hat{\Psi}^\dagger(\mathbf{r}, t)\nabla\hat{\Psi}(\mathbf{r}, t) - \left(\nabla\hat{\Psi}^\dagger(\mathbf{r}, t)\right)\hat{\Psi}(\mathbf{r}, t)\right]. \tag{10.51}$$

Mode Expansion Just as it was useful to introduce a mode expansion of the electric field $\hat{E}(\mathbf{r}, t)$, we now introduce a mode expansion of the Schrödinger field operator $\hat{\Psi}(\mathbf{r}, t)$. To illustrate this procedure, we expand $\hat{\Psi}(\mathbf{r}, t)$ on a complete set of orthonormal eigenfunctions $\varphi_n(\mathbf{r})$ of the time-independent Schrödinger equation

$$\left(-\frac{\hbar^2}{2m}\nabla^2 + \hat{V}(\mathbf{r})\right)\varphi_n(\mathbf{r}) = E_n\varphi_n(\mathbf{r})$$

as

$$\hat{\Psi}(\mathbf{r}, t) = \sum_n \varphi_n(\mathbf{r})\hat{c}_n(t), \tag{10.52}$$

where

$$\int d^3r\,\varphi_n^\star(\mathbf{r})\varphi_m(\mathbf{r}) - \delta_{nm} \tag{10.53}$$

and the label n stands for a complete set of quantum numbers necessary to characterize that mode. For example, in the case of atoms, it could be their internal state and center-of-mass momentum. The $\hat{c}_n$ and $\hat{c}_n^\dagger$ operators will soon be interpreted as annihilation and creation operators for a particle in mode n, much as was the case for the electromagnetic field.

Inserting the expression (10.52) and its Hermitian conjugate into the second-quantized Hamiltonian (10.40) gives then

$$\hat{\mathcal{H}} = \sum_n E_n\hat{c}_n^\dagger\hat{c}_n, \tag{10.54}$$

where we assume a discrete energy spectrum for simplicity.[3] With Eq. (10.53) we can furthermore express the operators $\hat{c}_n$ from Eq. (10.52) as

$$\hat{c}_n(t) = \int d^3r\,\varphi_n^\star(\mathbf{r})\hat{\Psi}(\mathbf{r}, t). \tag{10.55}$$

For a bosonic field with commutation relations (10.38) this gives

$$[\hat{c}_n, \hat{c}_m^\dagger] = \delta_{nm},$$
$$[\hat{c}_n, \hat{c}_m] = 0, \tag{10.56}$$

---

[3] We recognize this expression as the Hamiltonian for N non-interacting bosons of Eq. (10.2).

a familiar result that we recognize from the quantization of the electromagnetic field, while for fermionic particles one finds

$$[\hat{c}_n, \hat{c}_m^\dagger]_+ = \delta_{nm},$$

$$[\hat{c}_n, \hat{c}_m]_+ = 0. \tag{10.57}$$

Furthermore, combining Eqs. (10.52) and (10.56) yields the useful commutation relation

$$[\hat{\Psi}(\mathbf{r}), \hat{c}_n^\dagger] = \varphi_n(\mathbf{r}). \tag{10.58}$$

The second-quantized Hamiltonian (10.54), together with the bosonic commutation relations (10.56), shows that we have mapped the description of the system of N non-interacting bosons to a set of modes of energies E_n. We can therefore interpret $\hat{c}_n$ as the annihilation operator and $\hat{c}_n^\dagger$ as the creation operator for a particle in mode n, with

$$\hat{c}_n|N_n\rangle = \sqrt{N_n}|N_n - 1\rangle,$$

$$\hat{c}_n^\dagger|N_n\rangle = \sqrt{N_n + 1}|N_n + 1\rangle,$$

$$\hat{c}_n^\dagger\hat{c}_n|N_n\rangle = N_n|N_n\rangle, \tag{10.59}$$

where N_n is the number of particles in mode n. The total number of particles in the system $\hat{N} = \int d^3r\, \hat{\Psi}^\dagger(\mathbf{r})\hat{\Psi}(\mathbf{r})$ of Eq. (10.44) is given by the sum of the occupations $\hat{N}_n$ of the individual modes as

$$\hat{N} = \sum_n \hat{N}_n = \sum_n \hat{c}_n^\dagger\hat{c}_n \tag{10.60}$$

and is clearly a constant of motion for the Hamiltonian (10.54). Likewise, the individual populations of all modes of the matter field are also constants of motion, which is not surprising since φ_n are eigenstates of the system. But obviously this latter property ceases to hold as soon as interactions are permitted, e.g. in the presence of a light field or collisions. Much like was the case for optical fields, matter-wave fields can be in a variety of pure or mixed states, such as for example number states, thermal states, coherent states, or squeezed states, and quantum entanglement between various modes is of course also possible.

Coupling to Optical Fields In case a many-particle system is coupled to a single-mode electromagnetic field via the electric dipole interaction, the second-quantized Hamiltonian that describes their interaction is bilinear in the optical field creation and annihilation operators, but quadratic in matter field creation and annihilation operators, a direct consequence of the conservation of the total number of particles

N. An example of such an interaction is

$$\hat{V} = \hbar g_{nm} \left(\hat{a} \hat{c}_n^\dagger \hat{c}_m + \text{h.c.} \right),$$

with photon absorption resulting in the atom being "annihilated" from mode m and "created" in a mode n characterized in general by a different internal state and center-of-mass momentum.

Two-Body Collisions In the framework of second quantization, two-body collisions characterized by an interaction energy $\hat{V}(\mathbf{r}_i, \mathbf{r}_j)$ are described by the Hamiltonian

$$\hat{V} = \frac{1}{2} \int d^3 r d^3 r' \hat{\Psi}^\dagger(\mathbf{r}) \hat{\Psi}^\dagger(\mathbf{r}') \hat{V}(\mathbf{r} - \mathbf{r}') \hat{\Psi}(\mathbf{r}') \hat{\Psi}(\mathbf{r}), \tag{10.61}$$

where the operator $\hat{\Psi}(\mathbf{r}')\hat{\Psi}(\mathbf{r})$ picks two particles at locations $\mathbf{r}$ and $\mathbf{r}'$, to which the two-body Hamiltonian $\hat{V}(\mathbf{r}, \mathbf{r}')$ is applied before they are put back into place. The application of $\hat{V}$ on the state vector $|\phi_N\rangle$ gives

$$\hat{V}|\phi_N\rangle = \tag{10.62}$$

$$\frac{1}{\sqrt{N!}} \sum_{i=1}^{N} \sum_{j>i} \int d^3 r_1 \ldots d^3 r_N \, \hat{V}(\mathbf{r}_i - \mathbf{r}_j) \phi_N(\mathbf{r}_1, \ldots, \mathbf{r}_N) \hat{\Psi}^\dagger(\mathbf{r}_N) \ldots \hat{\Psi}^\dagger(\mathbf{r}_1)|0\rangle.$$

Fermions vs. Bosons As is the case for the simple harmonic oscillator or the electromagnetic field, nothing prevents one from populating a mode n with any number of bosons. This is not the case for fermions, however, as is directly apparent from the anticommutation relation $[\hat{c}_n, \hat{c}_m]_+ = 0$, which for $m = n$ yields

$$\hat{c}_m \hat{c}_m = \hat{c}_m^\dagger \hat{c}_m^\dagger = 0.$$

This implies that it is not possible to populate a single mode with more than one particle, and that its ground state $|0\rangle$ is reached once that single particle has been removed. This property is further evidenced by the fact that the number operator $\hat{N}_n$ and its square $\hat{N}_n^2$ are easily shown to be equal,

$$\hat{N}_n^2 = \hat{N}_n$$

so that the population of a given mode must be either zero or one. Since in addition

$$\hat{N}_n \hat{c}_n^\dagger |0\rangle = \hat{c}_n^\dagger |0\rangle \tag{10.63}$$

$\hat{c}_n^\dagger|0\rangle$ is an eigenstate of mode n with eigenvalue 1. This is nothing but a statement of the Pauli exclusion principle, expressed in the formalism of second quantization in terms of anticommutator relations.

10.3.1   The Hartree Approximation

With this brief introduction to the powerful formalism of second quantization at hand, let us now return to Bose–Einstein condensation. We saw in Sects. 10.1 and 10.2 that for temperatures much below the critical temperature T_c condensates are characterized by the fact that nearly all atoms are in the ground state, and hence are described by the same wave function. Under these circumstances it seems reasonable to factorize the N-particle wave function $\phi_N(\mathbf{r}_1, \dots \mathbf{r}_N)$ as

$$\phi_N(\mathbf{r}_1, \dots \mathbf{r}_N) = \prod_{\mathbf{r}=\mathbf{r}_1}^{\mathbf{r}_N} \varphi_N(\mathbf{r}), \qquad (10.64)$$

so that the N-particle state (10.42) becomes

$$|\phi_N\rangle = \frac{1}{\sqrt{N!}} \int d^3r_1 \dots d^3r_N \, [\varphi_N(\mathbf{r}_1) \dots \varphi_N(\mathbf{r}_N)] \, \hat{\Psi}^\dagger(\mathbf{r}_N) \dots \hat{\Psi}^\dagger(\mathbf{r}_1)|0\rangle, \qquad (10.65)$$

where the *effective single-particle states* $\varphi_N(\mathbf{r})$ are assumed to be normalized. In the so-called time-dependent Hartree mean-field approximation, or Hartree approximation in short, the equations of motion for these wave functions, called *Hartree wave functions*, are determined from the Hartree variational principle [17, 18]

$$\frac{\delta}{\delta \varphi_N^\star(\mathbf{r})} \left[\langle \phi_N | i\hbar \frac{\partial}{\partial t} - \hat{\mathcal{H}} | \phi_N \rangle \right] = 0. \qquad (10.66)$$

For a many-body Hamiltonian $\hat{\mathcal{H}}$ including in addition to the single-particle Hamiltonian $\hat{\mathcal{H}}_0$ a two-body interaction of the form (10.61) Problem 10.7 shows that it yields the nonlinear equation

$$i\hbar \frac{\partial \varphi_N(\mathbf{r})}{\partial t} = \hat{H}_0 \varphi_N(\mathbf{r}) + (N-1) \int d^3r' \hat{V}(\mathbf{r}, \mathbf{r}') \varphi_N^\star(\mathbf{r}) \varphi_N^2(\mathbf{r}'). \qquad (10.67)$$

The factor $(N-1)$ appearing in this expression results from the fact that the two-body Hamiltonian $\hat{V}$ involves two creation operators on the left of V, see Eq. (10.61). This leads to $N(N-1)$ equivalent terms, while the single-particle Hamiltonian, which involves only one annihilation operator on the left of $\hat{H}_0$, leads to N equivalent terms.

Many situations involve *multicomponent* Schrödinger fields rather than the scalar fields considered so far, one example of particular interest in quantum optics being situations where in addition to the center-of-mass motion the internal degrees of freedom of the atoms play an important role. In that case, Eqs. (10.40) and (10.61) must account for the complete set of quantum numbers required to fully specify the problem. Traditionally, this is done by lumping all such quantum numbers into just one symbol, often denoted as a number, as already mentioned in the discussion of the mode expansion (10.52). For example, the Schrödinger field operator $\hat{\Psi}(1)$ could stand for an electronic ground state field $\hat{\Psi}_g(\mathbf{r})$ and $\hat{\Psi}(2)$ for an excited state $\hat{\Psi}_e(\mathbf{r})$. For the case of two-body interactions the many-body Hamiltonian of the system takes then the form

$$\hat{\mathcal{H}} = \int d1 d2 \, \hat{\Psi}^\dagger(1)\langle 1|\hat{H}_0|2\rangle\hat{\Psi}(2)$$

$$+ \frac{1}{2}\int d1 d2 d3 d4 \, \hat{\Psi}^\dagger(1)\hat{\Psi}^\dagger(2)\langle 1,2|\hat{V}|3,4\rangle\hat{\Psi}(3)\hat{\Psi}(4) , \tag{10.68}$$

where we have accounted for the fact that neither the single-particle Hamiltonian $\hat{H}_0$ nor the two-body interaction $\hat{V}$ needs to be diagonal in the basis chosen to describe the system. The Hartree ansatz (10.64) reads then

$$f_N(1,\ldots,N) = \prod_{\ell=1}^{N} \varphi_N(\ell) \tag{10.69}$$

and the nonlinear Hartree equation of motion (10.67) becomes

$$i\hbar\frac{\partial\varphi_N(\ell)}{\partial t} = \int d2\langle\ell|\hat{H}_0|2\rangle\varphi_N(2)$$

$$+ (N-1)\int d2 d3 d4\langle\ell,2|\hat{V}|3,4\rangle\varphi_N^\star(2)\varphi_N(3)\varphi_N(4) . \tag{10.70}$$

Equations (10.67) and (10.70) illustrate the important point that despite the fact that the Hartree ansatz factorizes the N-particle wave function into a product of N Hartree wave functions, these are not equivalent to the single-particle wave functions of a non-interacting system: their dynamics are not governed solely by the single-particle Hamiltonian $\hat{H}_0$, but also by a nonlinear contribution resulting from the mean-field energy of the $N - 1$-particles surrounding a given atom, a contribution to the dynamics that can become completely dominant.

Gross–Pitaevskii Equation Equations of motion such as Eqs. (10.67) and (10.70) are called *nonlinear Schrödinger equations*. In particular, for a scalar Schrödinger field and a local potential of the form

$$\hat{V}(\mathbf{r},\mathbf{r}') = \hat{V}_0\,\delta(\mathbf{r}-\mathbf{r}'). \tag{10.71}$$

Eq. (10.70) reduces to the so-called Gross–Pitaevskii equation

$$i\hbar\frac{\partial\varphi_N(\mathbf{r})}{\partial t} = \hat{H}_0\varphi_N(\mathbf{r}) + (N-1)\hat{V}_0|\varphi_N(\mathbf{r})|^2\varphi_N(\mathbf{r}).$$ (10.72)

Such local potentials are of particular relevance in the study of Bose–Einstein condensates, as their dynamics are typically influenced mostly by very low energy two-body collisions between ground state atoms that can be described in the so-called s-wave scattering limit.

s-Wave Scattering The standard way to describe the scattering of low energy particles from a spherically symmetric potential is a partial wave expansion where one proceeds by decomposing the outgoing wave in terms of its angular momentum components, and considering only a few of these partial waves. For the extremely slow moving atoms in Bose condensates only the first of these, the s-wave, is important. Physically, this is because at such low energies the de Broglie wavelength of the atoms is so long that they cannot resolve the short-range structure of the potential with which they interact. In this s-wave scattering limit two-body collisions can be described to an excellent approximation by the two-body pseudo-potential [19]

$$\hat{V}(\mathbf{r}) = \frac{4\pi\hbar^2 a}{m}\delta(\mathbf{r}),$$ (10.73)

where a is the s-wave scattering length, in terms of which the total scattering cross-section is $\sigma = 4\pi a^2$. The scattering length is positive if the interaction is repulsive, and negative if the interaction is attractive.

Expanding the potential (10.61) on a momentum basis and with the pseudo-potential (10.73) we have then

$$\hat{V} = \frac{1}{2}\int d\mathbf{p}_1 d\mathbf{p}_2 d\mathbf{p}_3 d\mathbf{p}_4 \langle \mathbf{p}_3, \mathbf{p}_4|V|\mathbf{p}_1, \mathbf{p}_2\rangle \hat{c}_{\mathbf{p}_3}^\dagger \hat{c}_{\mathbf{p}_4}^\dagger \hat{c}_{\mathbf{p}_1} \hat{c}_{\mathbf{p}_2},$$ (10.74)

where $\hat{c}_{\mathbf{p}}^\dagger$ and $\hat{c}_{\mathbf{p}}$ are creation and annihilation operators for a particle of momentum $\mathbf{p}$ and

$$\langle \mathbf{p}_3, \mathbf{p}_4|V|\mathbf{p}_1, \mathbf{p}_2\rangle = \frac{1}{V}\int d^3 r V(\mathbf{r})e^{-i\mathbf{p}\cdot\mathbf{r}/\hbar} = \frac{4\pi\hbar^2 a}{mV}.$$ (10.75)

Furthermore, conservation of momentum requires that

$$\mathbf{p} = \mathbf{p}_4 - \mathbf{p}_2 = -(\mathbf{p}_3 - \mathbf{p}_1)$$ (10.76)

so that $\hat{\mathcal{V}}$ simplifies finally to

$$\hat{\mathcal{V}} = \frac{2\pi\hbar^2 a}{mV} \int d\mathbf{p}_1 d\mathbf{p}_2 d\mathbf{p} \, \hat{c}^\dagger_{\mathbf{p}_2+\mathbf{p}} \hat{c}^\dagger_{\mathbf{p}_1-\mathbf{p}} \hat{c}_{\mathbf{p}_2} \hat{c}_{\mathbf{p}_1} \,. \tag{10.77}$$

Nonlinear Schrödinger equations, of which the Gross–Pitaevskii equation is an important example, are ubiquitous in many fields of physics, including nonlinear optics [20–22], where they describe the propagation of light in media exhibiting a cubic nonlinearity, or Kerr media. They have been studied in considerable detail and are known to lead to optical effects such as four-wave mixing, self-focusing, and defocusing, the existence of solitons, etc. It is therefore not surprising that many of the concepts first developed in optics can readily be extended to ultracold bosonic samples, in particular to dilute atomic condensates, opening up the fields of nonlinear and quantum atom optics [23–27]. For example Eq. (10.72) shows explicitly that in the s-wave scattering limit, two-body collisions are the matter-wave equivalent of a local Kerr medium with instantaneous response.

10.3.2 Quasiparticles

The standard way to determine the response of a system to small perturbations is through its linearized dynamics about some equilibrium value. The Bogoliubov approach [14, 28–30], which is particularly appropriate in the description of the response of condensates to small perturbations, proceeds by decomposing the matter-wave field into a classical mean value about which fluctuations are treated quantum mechanically. Its main outcome is the determination of the spectrum of low-lying excitations, often referred to as *quasiparticle excitations*.

To illustrate how this works we consider the free space situation described by the many-body Hamiltonian

$$\hat{\mathcal{H}} = \sum_{\mathbf{k}} T_{\mathbf{k}} \hat{c}^\dagger_{\mathbf{k}} \hat{c}_{\mathbf{k}} + \frac{U_0}{2V} \sum_{\mathbf{k},\mathbf{k}',\mathbf{q}} \hat{c}^\dagger_{\mathbf{k}+\mathbf{q}} \hat{c}^\dagger_{\mathbf{k}'-\mathbf{q}} \hat{c}_{\mathbf{k}'} \hat{c}_{\mathbf{k}} \,, \tag{10.78}$$

where

$$U_0 = \frac{4\pi\hbar^2 a}{m} \,, \tag{10.79}$$

see Eq. (10.77), and

$$T_{\mathbf{k}} = \frac{\hbar^2 k^2}{2m} \tag{10.80}$$

is the kinetic energy of particles with momentum $\mathbf{k}$. For convenience we have returned from integrals to discrete sums in the expression of $\hat{\mathcal{H}}$.

We then assume that the mode $k = 0$ of the system is macroscopically populated with $n_0 \gg 1$ atoms and can be treated classically, while all other modes are only microscopically populated, and neglect all terms in the two-body Hamiltonian that are not at least proportional to n_0. This is an appropriate approximation for a Bose–Einstein condensate well below the critical temperature T_c, as we have seen. This results in the approximate Hamiltonian

$$\hat{\mathcal{H}} \simeq \frac{n_0^2 U_0}{2V} + \sum_{\mathbf{k} \neq 0} T_{\mathbf{k}} \hat{c}_{\mathbf{k}}^\dagger \hat{c}_{\mathbf{k}} + \frac{n_0 U_0}{V} \sum_{\mathbf{k} \neq 0} \left(\hat{c}_{\mathbf{k}}^\dagger \hat{c}_{\mathbf{k}} + \hat{c}_{-\mathbf{k}}^\dagger \hat{c}_{-\mathbf{k}} + \frac{1}{2} \hat{c}_{\mathbf{k}}^\dagger \hat{c}_{-\mathbf{k}}^\dagger + \frac{1}{2} \hat{c}_{\mathbf{k}} \hat{c}_{-\mathbf{k}} \right) .$$

$$(10.81)$$

Instead of using the population n_0 of the macroscopically populated mode, it is convenient to reexpress this Hamiltonian in terms of the total number of atoms N,

$$N = n_0 + \frac{1}{2} \sum_{\mathbf{k} \neq 0} \left(\hat{c}_{\mathbf{k}}^\dagger \hat{c}_{\mathbf{k}} + \hat{c}_{-\mathbf{k}}^\dagger \hat{c}_{-\mathbf{k}} \right) , \tag{10.82}$$

where the factor of 1/2 results from the fact that each mode is counted twice in this expression. This changes Eq. (10.81) to

$$\hat{\mathcal{H}} \simeq \frac{N^2 U_0}{2V} + \frac{1}{2} \sum_{\mathbf{k} \neq 0} \left[\left(T_{\mathbf{k}} + \frac{N U_0}{V} \right) \left(\hat{c}_{\mathbf{k}}^\dagger \hat{c}_{\mathbf{k}} + \hat{c}_{-\mathbf{k}}^\dagger \hat{c}_{-\mathbf{k}} \right) \right.$$

$$\left. + \frac{N U_0}{V} \left(\hat{c}_{\mathbf{k}}^\dagger \hat{c}_{-\mathbf{k}}^\dagger + \hat{c}_{\mathbf{k}} \hat{c}_{-\mathbf{k}} \right) \right] .$$

$$(10.83)$$

This Hamiltonian can be diagonalized exactly by the Bogoliubov transformation

$$\hat{c}_{\mathbf{k}} = u_{\mathbf{k}} \hat{\alpha}_{\mathbf{k}} - v_{\mathbf{k}} \hat{\beta}_{\mathbf{k}}^\dagger ,$$

$$\hat{c}_{-\mathbf{k}} = u_{\mathbf{k}} \hat{\beta}_{\mathbf{k}} - v_{\mathbf{k}} \hat{\alpha}_{\mathbf{k}}^\dagger , \tag{10.84}$$

with the new operators $\hat{\alpha}_{\mathbf{k}}$ and $\hat{\beta}_{\mathbf{k}}$ required to satisfy the bosonic commutation relations

$$[\hat{\alpha}_{\mathbf{k}}, \hat{\alpha}_{\mathbf{k}'}^\dagger] = \delta_{\mathbf{k}, \mathbf{k}'} ,$$

$$[\hat{\beta}_{\mathbf{k}}, \hat{\beta}_{\mathbf{k}'}^\dagger] = \delta_{\mathbf{k}, \mathbf{k}'} . \tag{10.85}$$

This gives, with Eqs. (10.84),

$$u_{\mathbf{k}}^2 - v_{\mathbf{k}}^2 = 1 , \tag{10.86}$$

a condition that is automatically satisfied if the u_k and v_k are parametrized as

$$u_k = \cosh \zeta_k \,,$$

$$v_k = \sinh \zeta_k \,. \tag{10.87}$$

Inserting the expressions (10.84) into the approximate Hamiltonian (10.83) yields

$$
\hat{\mathcal{H}} = \sum_k \left\{ 2v_k^2 \left(T_k + \frac{NU_0}{V} \right) - 2u_k v_k \frac{NU_0}{V} \right. \tag{10.88}
$$

$$
+ \left[\left(T_k + \frac{NU_0}{V} \right)(u_k^2 + v_k^2) - 2u_k v_k \frac{NU_0}{V} \right] (\hat{\alpha}_k^\dagger \hat{\alpha}_k + \hat{\beta}_k^\dagger \hat{\beta}_k)
$$

$$
+ \left. \left[\frac{NU_0}{V}(u_k^2 + v_k^2) - 2u_k v_k \left(T_k + \frac{NU_0}{V} \right) \right] (\hat{\alpha}_k^\dagger \hat{\beta}_k^\dagger + \hat{\alpha}_k \hat{\beta}_k) \right\} \,.
$$

The constants u_k and v_k are then chosen so that the terms proportional to $\hat{\alpha}_k^\dagger \hat{\beta}_k^\dagger + \hat{\alpha}_k \hat{\beta}_k$ vanish, that is, so that

$$
\left[\frac{NU_0}{V}(u_k^2 + v_k^2) - 2u_k v_k \left(T_k + \frac{NU_0}{V} \right) \right] = 0 \,. \tag{10.89}
$$

Substituting Eqs. (10.87) into these equations gives

$$
\tanh(2\zeta_k) = \frac{NU_0/V}{T_k + NU_0/V} \,. \tag{10.90}
$$

Rewriting u_k^2, v_k^2, and $u_k v_k$ in terms of this expression finally yields the diagonalized form of the Hamiltonian (10.83). This is the topic of Problem 10.8, which shows that these steps result in its explicit form

$$
\hat{\mathcal{H}} = \sum_k E_k \hat{\alpha}_k^\dagger \hat{\alpha}_k \,, \tag{10.91}
$$

with

$$
E_k = \sqrt{T_k^2 + 2T_k NU_0/V} \,. \tag{10.92}
$$

This Hamiltonian describes the elementary excitations, or Bogoliubov spectrum, of the condensate.

For large momenta, the eigenenergies E_k associated with these excitations are roughly the same as those of free particles,

$$
E_k \simeq T_k \,, \tag{10.93}
$$

but for small $\mathbf{k}$, the effect of the mean-field energy NU_0/V is to replace the quadratic dispersion relation of the quasiparticles by a linear dispersion relation

$$E_{\mathbf{k}} \simeq c_s \hbar k, \qquad (10.94)$$

where we have introduced the *Bogoliubov velocity*

$$c_s = \sqrt{NU_0/mV}. \qquad (10.95)$$

The characteristic wave number k_ζ at which the two contributions to the quasiparticle energy $E_{\mathbf{k}}$ become comparable defines the *healing length*

$$\zeta \equiv \frac{1}{k_\zeta} = \frac{\hbar}{\sqrt{mNU_0/V}}, \qquad (10.96)$$

which plays an important role in characterizing the spatial properties of Bose–Einstein condensates.

The Bogoliubov linearization procedure can easily be generalized to more complicated situations, such as multicomponent condensates [28]. In the spirit of quantum and nonlinear optics, it allows one to understand many of their properties in terms of wave mixing phenomena [27], much like a linear stability analysis yields a simple understanding of many nonlinear optical phenomena in terms of pump–probe arguments.

10.4   Ultracold Atoms on Optical Lattices

The first experimental demonstrations of atomic Bose–Einstein condensation were followed by a wealth of advances in the theoretical and experimental study of ultracold atomic gases. In this context, the availability of optical lattices has played a particularly important role, as they permit to address a number of questions at the boundary between AMO physics, quantum optics, and condensed matter physics. In particular they lead to the experimental realization of *quantum simulators* aimed at the study of strongly correlated bosonic or fermionic systems, quantum phase transitions, novel topological phases of matter, and much more.

Optical lattices are formed by interfering two or more laser beams to realize periodic light structures whose geometry, depth, and dimensionality can be easily controlled. The gradient force of Sect. 8.2 can then be exploited to trap atoms in these structures. For example, a standing wave interference pattern can generate an array of optical traps in one dimension, so that ultracold atoms can either be trapped in the individual wells or made to tunnel between adjacent wells. By interfering more optical beams one can also generate arrays of one-dimensional potential tubes, realizing effectively a series of one-dimensional systems, while three orthogonal

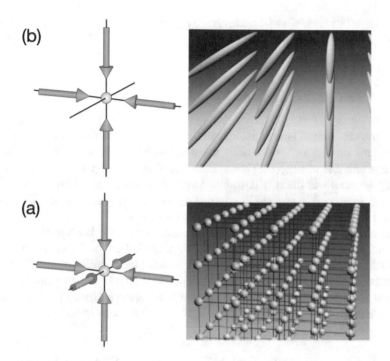

Fig. 10.5 Optical lattices formed from (**a**) two and (**b**) three interfering standing wave optical fields (From Ref. [31])

standing waves can create a three-dimensional "cubic crystal" of tightly confining harmonic oscillator potentials, as sketched in Fig. 10.5.

It is also possible to create more exotic lattices, for instance by using bichromatic optical potentials to produce Kagome lattices beset by geometric frustration [32]. Another development has been the use of synthetic dimensions [33, 34], the idea being to exploit the spin degree of freedom of atoms in situations where the different spin states behave essentially like a spatial dimension. For instance, an atom with ℓ distinct spin states and trapped in a one-dimensional optical lattice would appear to move in a two-dimensional optical lattice strip of width ℓ.

Experiments in this variety of optical lattices open the door to a new regime of ultracold atomic physics that cannot be described theoretically in terms of weakly interacting gases. The resulting possibilities to investigate the many-body physics of strongly correlated systems are particularly attractive, both theoretically and experimentally, in that in contrast to the condensed matter physics systems that they simulate the various parameters characterizing them can be controlled almost at will.

At the same time, ultracold atoms trapped in optical lattices are also revolutionizing atomic clocks. In particular, a clock formed by trapping thousands of fermionic strontium atoms in a three-dimensional optical lattice, and operating in the so-called Mott-insulating regime that will be discussed in the next section, has achieved a relative precision of 2.5×10^{-19}, or better than 1 s over the entire age of the Universe [35–37]. We will return briefly to these extraordinary clocks and some of their potential applications in Chap. 12.

10.4.1   The Bose–Hubbard Model

To illustrate the remarkable capability of ultracold atoms on optical lattices to serve as quantum simulators we consider the example of the Bose–Hubbard model of an interacting gas of bosons. This system has a long history in condensed matter physics, where it was initially introduced to study granular superconductors—cubic grains of superconductor weakly coupled by Josephson junctions. It also captures the main features of a superfluid to insulator transition.

When adapted to the situation at hand the Bose–Hubbard Hamiltonian describes a system of ultracold atoms trapped on an optical lattice, but with some hopping amplitude J between lattice sites that will delocalize them. In addition, if two atoms are on the same site, they will also feel a repulsion. This is the simplest model that contains all important aspects of the competition between kinetic energy and two-body interactions in a lattice of ultracold bosons. The corresponding Hamiltonian is [38]

$$\hat{H} = -J \sum_{<i,j>} \hat{c}_i^\dagger \hat{c}_j + \frac{U}{2} \sum_i \hat{n}_i (\hat{n}_i - 1) - \mu \sum_i \hat{n}_i \, , \tag{10.97}$$

where $\hat{c}_i^\dagger$ and $\hat{c}_i$ are boson creation and annihilation operators for atoms on site i and $\hat{n}_i = \hat{c}_i^\dagger \hat{c}_i$. The first term in $\hat{H}$ describes the tunneling of particles between neighboring lattice sites, with the $< i, j >$ summation index indicating that the sum is limited to nearest neighbors. The second term accounts for on-site atomic two-body collisions, with $U > 0$ for repulsive collisions, and μ is the chemical potential.

For a sufficiently deep optical lattice potential of the form

$$V(x, y, z) = V_0 \left[\sin^2(kx) + \sin^2(ky) + \sin^2(kz) \right] \tag{10.98}$$

with $k = 2\pi/\lambda$ and ultracold atoms the confining potential of a single site can be approximated by a harmonic potential with trapping frequencies of the order of [38]

$$\nu = (\hbar k^2 / 2\pi m) \sqrt{V_0 / \hbar \omega_{\rm rec}} \, .$$

Superfluid State It is immediately apparent from the Hamiltonian (10.97) that if tunneling dominates, $J \gg U$, the dynamics is dominated by atomic hopping between lattice sites and will result in the atoms being completely delocalized. The ground state energy of the system will be minimized if the single-particle wave functions of the individual atoms are spread over the full lattice. That is, the system will evolve toward a superfluid with many-particle wave function

$$|\Psi_{SF}\rangle \propto \left(\sum_{i=1}^{M} \hat{c}_i^\dagger\right)^N |0\rangle, \tag{10.99}$$

where M is the number of lattice sites and N the number of atoms.

Mott Insulator In the other extreme situation where tunneling is negligible and intrasite interactions dominate, the site occupation numbers $\hat{n}_i$ become constants of motion, so that the ground state of the system will consist of localized Wannier atomic wave functions with a fixed number n of atoms per site. This is a so-called Mott insulator, with many-particle wave function

$$|\Psi_{MI}\rangle \propto \prod_{i=1}^{M} (\hat{c}_i^\dagger)^n |0\rangle \tag{10.100}$$

and with no phase coherence.

To understand this behavior more quantitatively, we consider a single lattice site i in the limit where the various sites are decoupled, $J = 0$. For that site, the energy of a state with n-particles is

$$E = \left\langle n \left| \frac{U}{2} \hat{n}_i (\hat{n}_i - 1) - \mu \hat{n}_i \right| n \right\rangle = \frac{U}{2} n(n-1) - \mu n. \tag{10.101}$$

It is equal to $E = 0$ for $n = 0$, $E = -\nu$ for $n = 1$, $E = -n\mu + \frac{U}{2}n(n-1)$ for n, etc. For a given value of U the energy of the n-particle state becomes lower than that of the $(n-1)$-particle state for $(n-1)U = \mu$, so that the ground state of an isolated lattice site, the red line in Fig. 10.6, will have n atoms in the interval

$$(n-1)U < \mu < nU. \tag{10.102}$$

Mean-Field Theory Small departures from the isolated lattice sites results for $J = 0$ can be obtained by invoking a mean-field approximation, where the effects of the neighboring lattice sites of a given site are treated in an average fashion by introducing their "mean fields" $\psi_j = \langle c_j \rangle$, with the further assumption that that mean field is the same for all sites, $\psi_i = \psi_j = \psi$. With the further assumption $(\hat{c}_i - \psi_i)(\hat{c}_j^\dagger - \psi_j^*) \approx 0$, this allows one to introduce the dynamical mean-field decoupling

$$\hat{c}_i \hat{c}_j^\dagger \approx \psi_i \hat{c}_j^\dagger + \psi_j^* \hat{c}_i - \psi_i \psi_j^* = \psi \hat{c}_j^\dagger + \psi^* \hat{c}_i - |\psi|^2 \tag{10.103}$$

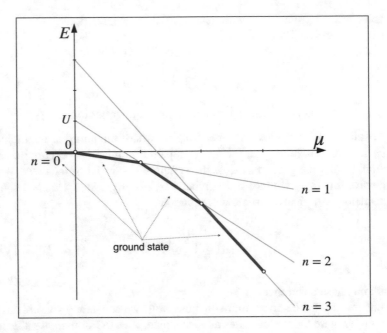

Fig. 10.6 Illustration of the level crossings (circles) between states with different integer site fillings n for $\mu = nU$. Away from these points these states are separated by an energy gap, so that they are stable against small changes in the Hamiltonian, such as a small amount of tunneling. The red line shows the ground state energy, with corresponding number of particles per site indicated, as a function of μ

which also implies that $\langle \hat{c}_i^\dagger \hat{c}_j \rangle \approx \langle \hat{c}_i^\dagger \rangle \langle \hat{c}_j \rangle$, an approximation sometimes referred to as the random phase approximation. The Hamiltonian (10.97) reduces then to the effective mean-field Hamiltonian

$$\hat{H}_{\mathrm{MF}} = \sum_i \hat{H}_i \tag{10.104}$$

with

$$\hat{H}_i = \frac{U}{2}\hat{n}_i(\hat{n}_i - 1) - \mu\hat{n}_i - J'(\psi^*\hat{c}_i + \psi\hat{c}_i^\dagger) + J'|\psi|^2, \tag{10.105}$$

where $J' = zJ$ and the coordination number z is the number of nearest neighbors of the site i. The zeroth-order energy of the i-th lattice occupied with m atoms is therefore

$$E_n^{(0)} = -\mu n + \frac{U}{2}n(n - 1) + J'|\psi|^2, \tag{10.106}$$

and the second-order correction to that energy resulting from the coupling to neighboring sites is

$$E_n^{(2)} = J'^2 |\psi|^2 \sum_{n \neq m} \frac{|\langle m | \hat{c}_i + \hat{c}_i^\dagger | n \rangle|^2}{E_n^{(0)} - E_m^{(0)}} = -J'^2 |\psi|^2 \left[\frac{U + \mu}{(\mu - Un)(U(n-1) - \mu)} \right].$$

(10.107)

If tunneling between nearest neighbor sites results in a decrease in energy of the system, the outcome will be a transition from the Mott insulator phase to a Bose condensate phase. In the mean-field description considered here the ground state energy of the system $E_n = E_n^{(0)} + E_n^{(2)}$ is

$$E_n = \frac{U}{2} n(n-1) - \mu n + J' |\psi|^2 \left[1 - J' \frac{U + \mu}{(\mu - Un)(U(n-1) - \mu)} \right],$$

(10.108)

so that the boundary between the Mott insulator phase with occupation n and the superfluid phase is given by

$$E(\psi) = J' |\psi|^2 \left[1 - J' \frac{U + \mu}{(\mu - Un)(U(n-1) - \mu)} \right] = 0.$$

(10.109)

The domains of Mott insulator, or Mott shells, and superfluidity resulting from this condition are plotted in Fig. 10.7.

Insulator to Superfluid Transition   The Mott insulator to superfluid transition in an ultracold atomic gas was first experimentally demonstrated by M. Greiner et al. [38]. A magnetically trapped condensate was transferred into an optical lattice potential, and after raising the lattice potential the condensate was distributed over

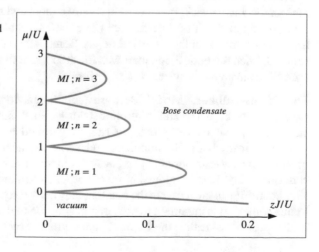

Fig. 10.7 Mean-field phase diagram of the Bose–Hubbard model showing the Mott-insulating (MI) phases, or Mott shells, with commensurate occupations n, the BEC phase, and vacuum, as a function of μ/U and zJ/U

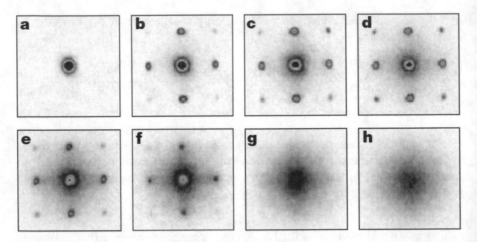

Fig. 10.8 Absorption images of multiple matter-wave interference patterns. These were obtained after suddenly releasing the atoms from an optical lattice potential with different potential depths V_0. The values of V_0 were: (**a**) 0 E_{rec}; (**b**) 3 E_{rec}; (**c**) 7 E_{rec}; (**d**) 10 E_{rec}; (**e**)13 E_{rec}; (**f**) 14 E_{rec}; (**g**)16 E_{rec}; and (**h**) 20 E_{rec} (From Ref. [38])

more than 150,000 lattice sites, with an average atom number of up to 2.5 atoms per lattice site at the center of the lattice. The lattice potential was then suddenly switched off, resulting in the atomic wave functions expanding freely and interfering with each other. In the superfluid regime all atoms are delocalized over the entire lattice with equal relative phases between different lattice sites, resulting in the high-contrast three-dimensional interference pattern expected for a periodic array of phase coherent matter-wave sources, see Fig. 10.8. The atoms could also re-enter the Mott insulator phase by increasing the lattice potential depth. In this regime the interference pattern changed markedly, as the higher localization of the atoms at individual lattice sites resulted in a decrease of the interference maxima. At the same time an incoherent background of atoms gained more and more strength, until above some potential depth no interference pattern was visible at all. Phase coherence was completely lost at that potential depth. Remarkably, though, it could be rapidly restored when the optical potential was lowered again to a value where the ground state of the many-body system is a superfluid.

The Fermi–Hubbard Model Because their interactions and geometry can relatively easily be tuned and controlled with external fields, systems of ultracold atoms on optical lattices provide a rich playground to design and investigate a variety of many-body Hamiltonians and serve as quantum simulators, with the potential to answer important open questions in material science. While this chapter concentrated on bosonic atoms, ultracold atomic fermions are of course at least as significant, due in great part to the fact that they obey the same statistics as electrons. In this context, it is important to note that in addition to the Bose–Hubbard model, it is also possible to realize the Fermi–Hubbard model by loading ultracold fermions in optical lattices [39, 40]. In particular, the Fermi–Hubbard model is believed to capture the essential physics of high-temperature superconductors and of other quantum materials. The quantum simulations that can now be realized in optical lattices promise therefore to have a large impact on the understanding of the physics of these systems.

Quantum Gas Microscopes   While experiments on quantum gases rely typically on measuring ensemble properties of the system, there would also considerable merit in probing it at the single atom level, as this would permit to determine directly particle–particle correlation functions. However, because strongly correlated atomic systems such as realized in the Bose–Hubbard model require small lattice spacings to ensure desirable values of tunnel coupling and interaction strengths, these sub-micron optical potential structures are typically created by directly projecting a lithographically generated spatial light pattern onto the atom plane. This is in contrast to optical lattice setups that rely on the interference patterns generated by multiple laser beams and whose period is limited by the diffraction limit.

At first sight, then, probing these systems at the single atom level might appear to be an impossible task, since the size of the lattice sites is far smaller than the diffraction limit of optical microscopes. However, a recently developed quantum gas microscope [41] provides this capability through a remarkable combination of resolution and sensitivity that does enable the imaging of single atoms with near unit fidelity on individual sites of short-period optical lattices. This is a significant advance from the previously available site-resolved optical imaging of single atoms, which was limited to lattices with periods large compared to an optical wavelength [42]. The quantum gas microscope developed by W. S. Bakr and colleagues, and shortly thereafter by several other groups, is based on a high aperture optical system that simultaneously serves to generate the lattice potential and to detect single atoms with site-resolved resolution. By placing a two-dimensional quantum gas only a few microns away from the front surface of this microscope, an optical resolution of approximately 600 nm could be achieved. It was then possible to read out of up to tens of thousands lattice sites by imaging the light scattered by the atoms while enabling the detection of single atoms on each individual lattice site with near unity fidelity.

Figure 10.9 shows an example of the type of information that can be obtained with such quantum gas microscopes. As a result of the imaging process the many-body wave function of the quantum gas is projected onto number states on each lattice site. This provides a remarkable experimental view of the Mott insulator to superfluid transition as the depth of the lattice is changed. In the superfluid regime (a) and (b), sites can be occupied with odd or even atom numbers, which appear as full or empty sites, respectively, in the images. This is due to the fact that light-assisted collisions immediately eject atoms in pairs from individual lattice sites, leaving behind an atom on a site only if its initial occupation was odd. Deep in the Mott insulator phase of column (d), in contrast, site occupancies other than 1 are highly suppressed.

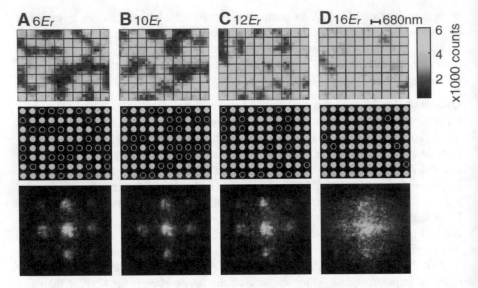

Fig. 10.9 Single-site imaging of atom number fluctuations across the superfluid–Mott insulator transition. (**a** to **d**) Images within each column are taken at the same final 2D lattice depth of (**a**) 6 E_{rec}; (**b**) $10E_{rec}$; (**c**), 12 E_{rec}; and (**d**) 16 E_{rec}. Top row: In situ fluorescence images from a region of 10×8 lattice sites within the $n = 1$ Mott shell that forms in a deep lattice. Middle row: Results of an atom detection algorithm for images in the top row. Solid and open circles indicate the presence and absence, respectively, of an atom on a site. The bottom row shows the corresponding time-of-flight fluorescence images (From Ref. [43])

Problems

Problem 10.1 A grand canonical ensemble is an ensemble of particles that can exchange both energy and particles with a reservoir, so that the expectation values of the energy and particle number are $E = \langle \hat{H} \rangle$ and $N = \langle \hat{N} \rangle$, with $\hat{N}$ the particle number operator. Much like the temperature is the Lagrange multiplier regulating the energy of the system as we saw in Sect. 2.3.1, the chemical potential μ (normalized to $-1/k_B T$) may be defined as the Lagrange multiplier regulating the particle number.

(a) Considering then a grand canonical ensemble of bosons with system Hamiltonian $\hat{H} = \sum_j E_j \hat{c}_j^\dagger \hat{c}_j$ show that the corresponding thermal equilibrium density operator is

$$\hat{\rho} = \frac{1}{Z} \prod_j e^{-\beta(E_j - \mu)\hat{c}_j^\dagger \hat{c}_j} ,$$

where Z is the partition function.

(b) Determine the corresponding mean particle number $\langle n_j \rangle$ in mode j.

Hint: Generalize the derivation of Eq. (2.85) to a multimode situation and to include the additional constraint on the mean number of particles.

Problem 10.2 Determine the density of states $\mathcal{D}(E)$ and the ground state population $\langle n_0 \rangle(T)$ of a condensate on non-interacting bosons in a trap with harmonic potential $V_{trap}(\mathbf{r}) = \frac{1}{2}m\left(\omega_1^2 x_1^2 + \omega_2^2 x_2^2 + \omega_3^2 x_3^2\right)$.

Problem 10.3 Show that independently of whether the particles are bosons or fermions the Schrödinger field operator $\hat{\Psi}(\mathbf{r}, t)$ satisfies the Heisenberg equation of motion

$$i\hbar \frac{d\hat{\Psi}(\mathbf{r}, t)}{dt} = \left(-\frac{\hbar^2}{2m}\nabla^2 + \hat{V}(\mathbf{r})\right)\hat{\Psi}(\mathbf{r}, t).$$

Problem 10.4 Prove the operator identity

$$[\hat{A}, \hat{B}_N \ldots \hat{B}_1] = \sum_{i=1}^{N} \hat{B}_N \ldots \hat{B}_{i+1}[\hat{A}, \hat{B}_i]\hat{B}_{i-1} \ldots \hat{B}_1 .$$

Problem 10.5

(a) Show that when applied on the N-particle wave function

$$|\phi_N\rangle = \frac{1}{\sqrt{N!}} \int d^3r_1 \ldots d^3r_N \phi_N(\mathbf{r}_1, \ldots, \mathbf{r}_N, t)\hat{\Psi}^\dagger(\mathbf{r}_N) \ldots \hat{\Psi}^\dagger(\mathbf{r}_1)|0\rangle$$

the number operator $\hat{N} \equiv \int d^3r\,\hat{\Psi}^\dagger(\mathbf{r})\hat{\Psi}(\mathbf{r})$ gives

$$\hat{N}|\phi_N\rangle = N|\phi_N\rangle .$$

(b) Show also that

$$\hat{\Psi}(\mathbf{r})|\phi_N\rangle = \frac{N}{\sqrt{N!}} \int d^3r_1 \ldots d^3r_{N-1}\phi_N(\mathbf{r}_1, \ldots \mathbf{r}_{N-1}, \mathbf{r})\hat{\Psi}^\dagger(\mathbf{r}_{N-1}) \ldots \hat{\Psi}^\dagger(\mathbf{r}_1)|0\rangle .$$

Problem 10.6 Derive the continuity equation (10.50)

$$\frac{d\hat{n}(\mathbf{r}, t)}{dt} = \nabla \cdot \mathbf{j},$$

where $\hat{n}(\mathbf{r}, t) = \hat{\Psi}^\dagger(\mathbf{r}, t)\hat{\Psi}(\mathbf{r}, t)$.

Problem 10.7 Show that for a Hamiltonian $\hat{\mathcal{H}}$ including in addition to the single-particle Hamiltonian $\hat{\mathcal{H}}_0$ a two-body interaction of the form (10.61) the time-dependent Hartree variational principle

$$\frac{\delta}{\delta \varphi_N^\star(\mathbf{r})} \left[\langle \phi_N | i\hbar \frac{\partial}{\partial t} - \hat{\mathcal{H}} | \phi_N \rangle \right] = 0$$

yields the equation of motion

$$i\hbar \frac{\partial \varphi_N(\mathbf{r})}{\partial t} = \hat{H}_0 \varphi_N(\mathbf{r}) + (N-1) \int d^3 r' \hat{V}(\mathbf{r}, \mathbf{r}') \varphi_N^\star(\mathbf{r}) \varphi_N^2(\mathbf{r}')$$

for the Hartree wave functions $\varphi_N(\mathbf{r}, t)$.

Problem 10.8 Carry out the explicit steps that lead from the BEC Hamiltonian

$$\hat{\mathcal{H}} = \sum_{\mathbf{k}} T_{\mathbf{k}} \hat{c}_{\mathbf{k}}^\dagger \hat{c}_{\mathbf{k}} + \frac{U_0}{2V} \sum_{\mathbf{k}, \mathbf{k}', \mathbf{q}} \hat{c}_{\mathbf{k}+\mathbf{q}}^\dagger \hat{c}_{\mathbf{k}'-\mathbf{q}}^\dagger \hat{c}_{\mathbf{k}'} \hat{c}_{\mathbf{k}},$$

to the linearized Hamiltonian

$$\hat{\mathcal{H}} = \sum_{\mathbf{k}} E_{\mathbf{k}} \hat{\alpha}_{\mathbf{k}}^\dagger \hat{\alpha}_{\mathbf{k}}$$

that characterizes its elementary excitations, and plot the resulting elementary excitation spectrum E_k as a function of the momentum k.

Problem 10.9 Evaluate the second-order correction to the Mott insulator energy for a Bose–Hubbard system with n atoms per site due to the mean-field correction resulting from intersite tunneling.

References

1. A. Einstein, Quantentheorie des einatomigen idealen Gases. Sitzungsber. Preuss. Akad. Wiss. **1924**, 261 (1924)
2. A. Einstein, Quantentheorie des einatomigen idealen Gases II. Sitzungsber. Preuss. Akad. Wiss. **1925**, 3 (1925)
3. S.N. Bose, Plancks Gesetz und Lichtquantenhypothese. Zeit. f. Phys. **26**, 178 (1924)
4. M.H. Anderson, J.R. Ensher, M.R. Matthews, C.E. Wieman, E.A. Cornell, Observation of Bose-Einstein condensation in a dilute atomic vapor. Science **269**, 198 (1995)
5. K.B. Davis, M.O. Mewes, M.R. Andrews, N.J. van Druten, D.S. Durfee, D.M. Kurn, W. Ketterle, Bose-Einstein condensation on a gas of Sodium atoms. Phys. Rev. Lett **75**, 3969 (1995)
6. L.D. Landau, E.M. Lifshitz, *Statistical Mechanics* (Pergamon, London, 1959)
7. K. Huang, *Statistical Mechanics* (Wiley, New York, 1987)
8. M. Toda, R. Kubo, N. Saito, *Statistical Physics I—Equilibrium Statistical Mechanics*, 2nd edn. (Springer, Berlin, 1992)
9. H. Politzer, Condensate fluctuations of a trapped ideal Bose gas. Phys. Rev. A **54**, 5048 (1996)

10. S. Grossman, M. Holthaus, Fluctuations of the particle number in a trapped Bose-Einstein condensate. Phys. Rev. Lett. **79**, 3557 (1997)
11. M. Gajda, K. Rzazewski, Fluctuations of Bose-Einstein condensate. Phys. Rev. Lett. **78**, 2686 (1997)
12. M. Wilkens, C. Weiss, Particle number fluctuations in an ideal Bose gas. J. Mod. Optics **44**, 1801 (1997)
13. C. Pethik, H. Smith, *Bose-Einstein Condensation in Dilute Gases*, 2nd edn. (Cambridge University, Cambridge, 2008)
14. N.N. Bogoliubov, On the theory of superfluidity. J. Phys. (USSR) **11**, 23 (1947)
15. B. Robertson, Introduction to field operators in quantum mechanics. Am. J. Phys. **41**, 678 (1973)
16. L.E. Ballentine, *Quantum Mechanics* (Prentice Hall, Englewood Cliffs, 1990)
17. A.K. Kerman, S.E. Koonin, Hamiltonian formulation of time-dependent variational principles for manybody systems. Ann. Physics (New York) **100**, 332 (1976)
18. J.W. Negele, The mean-field theory of nuclear structure and dynamics. Rev. Mod. Phys. **54**, 913 (1982)
19. L. Pitaevskii, S. Stringari, *Bose-Einstein Condensation* (Clarendon Press, Oxford, 2003)
20. Y.R. Shen, *Principles of Nonlinear Optics* (Wiley, New York, 1984)
21. R.W. Boyd, *Nonlinear Optics*, th edn. (Academic Press, San Diego, 2020)
22. A.C. Newell, J.V. Moloney, *Nonlinear Optics* (Addison-Wesley, Redwood City, 1992)
23. G. Lenz, P. Meystre, E.M. Wright, Nonlinear atom optics. Phys. Rev. Lett. **71**, 3271 (1993)
24. G. Lenz, P. Meystre, E.M. Wright, Nonlinear atom optics: General formalism and atomic solitons. Phys. Rev. A **50**, 1681 (1994)
25. W. Zhang, D.F. Walls, Quantum field theory of interaction of ultracold atoms with a light wave: Bragg scattering in nonlinear atom optics. Phys. Rev. A **49**, 3799 (1994)
26. Y. Castin, K. Mølmer, Maxwell-Bloch equations: A unified view of nonlinear optics and nonlinear atom optics. Phys. Rev. A **51**, R3426 (1995)
27. P. Meystre, *Atom Optics* (Springer, Berlin, 2001)
28. N.N. Bogoliubov, *Lectures in Quantum Statistics* (Gordon and Breach, New York, 1967)
29. A.L. Fetter, Nonuniform states of an imperfect Bose gas. Ann. Phys. (New York) **70**, 67 (1972)
30. E.M. Lifshitz, L.P. Pitaevskii, *Statistical Physics Part 2* (Pergamon Press, New York, 1980)
31. I. Bloch, Ultracold quantum gases in optical lattices. Nature Phys. **1**, 23 (2005)
32. G.-B. Jo, J. Guzman, C.K. Thomas, P. Hosur, A. Vishwanath, D.M. Stamper-Kurn, Untcold atoms in a tunable optical Kagome lattice. Phys. Rev. Lett **108**, 045305 (2005)
33. A. Celi, P. Massignan, J. Ruseckas, N. Goldman, I.B. Spielman, Synthetic gauge field in synthetic dimensions. Phys. Rev. Lett. **112**, 043001 (2014)
34. H.M. Price, O. Zilberberg, T. Ozawa, I. Carusotto, N. Goldman, Four-dimensional quantum Hall effect with ultracold atoms. Phys. Rev. Lett. **115**(19), 195303 (2015)
35. T.L. Nicholson, S.L. Campbell, R.B. Hutson, G.E. Marti, B.J. Bloom, R.L. McNally, W. Zhang, M.D. Barrett, M.S. Safronava, G.F. Strouse, W.L. Tew, J. Ye, Systematic evaluation of an atomic clock at 2×10^{-18} total uncertainty. Nature Commun. **6**, 6896 (2015)
36. S.L. Campbell, R.B. Hutson, G.E. Marti, A. Goban, N. Darkwah Oppong, R.N. McNally, L. Sonderhouse, J.M. Robinson, W. Zhang, B.J. Bloom, J. Ye, A Fermi-degenerate three-dimensional optical lattice clock. Science **358**, 90 (2017)
37. G.E. Marti, R.B. Hutson, A. Goban, S.L. Campbell, N. Poli, J. Ye, Imaging optical frequencies with 100 μHz precision and 1.1 μm resolution. Phys. Rev. Lett. **120**, 103201 (2018)
38. M. Greiner, O. Mandel, T. Essliner, T.W. Hänsch, I. Bloch, Quantum phase transition from a superfluid to a Mott insulator in a gas of ultracold atoms. Nature **415**, 39 (2002)
39. T. Esslinger, Fermi-Hubbard model with atoms in an optical lattice. Annual Rev. of Cond. Mat. Phys. **1**, 129 (2010)
40. C. Gross, I. Bloch, Quantum simulations with ultracold atoms in optical lattices. Science **357**, 995 (2017)
41. W.S. Bakr, J.I. Gillen, A. Peng, S. Fölling, M. Greiner, A quantum gas microscope for detecting single atoms in a Hubbard-regime optical lattice. Nature **462**, 74 (2009)

42. K. Nelson, X. Li, D. Weiss, Imaging single atoms in a three-dimensional array. Nature Phys. **3**, 556 (2007)
43. W.S. Bakr, A. Peng, M.E. Tai, R. Ma, J. Simon, J.I. Gillen, S. Fölling, L. Pollet, M. Greiner, Probing the superfluid to Mott insulator transition at the single-atom level. Science **329**, 548 (2010)

Chapter 11
Quantum Optomechanics

Quantum optomechanics extends the idea that light forces can achieve the quantum control of mechanical motion to mesoscopic and macroscopic systems. Following a semiclassical introduction to cold damping and the optical spring effect, we show how sideband cooling can bring those systems to their quantum mechanical ground state. We then discuss ways to prepare, manipulate, and characterize other quantum states of mechanical oscillators. This is followed by an analysis of the standard quantum limit of optomechanical interferometers that clarifies the roles of shot noise and radiation pressure noise. We finally return to ultracold atoms to show how their collective density excitations can likewise behave as optomechanical oscillators.

The previous three chapters discussed how to exploit the mechanical effects of light to trap and cool atoms or ions to extraordinarily low temperatures, even to their quantum ground state of motion, thereby providing a remarkable platform to address a wealth of questions at the boundary between AMO physics, condensed matter physics, and field theory. One may ask, then, whether similar advances can also be achieved in mesoscopic or macroscopic systems. Quantum optomechanics, the topic of this chapter, shows that this is indeed the case and that light forces provide a universal tool to achieve the quantum control of mechanical motion in devices spanning a vast range of parameters, with mechanical frequencies from a few Hertz to GHz and with masses from 10^{-20} g to several kilos. As such it offers a route to control the quantum state of truly macroscopic objects and opens the way to experimental and theoretical advances that may lead to a more profound understanding of the quantum world. And from the point of view of applications, quantum optomechanical techniques in both the optical and microwave regimes hold the promise of major advances in quantum metrology, in particular in motion and force detection near the fundamental limit imposed by quantum mechanics.

© The Author(s), under exclusive license to Springer Nature Switzerland AG 2021   325
P. Meystre, *Quantum Optics*, Graduate Texts in Physics,
https://doi.org/10.1007/978-3-030-76183-7_11

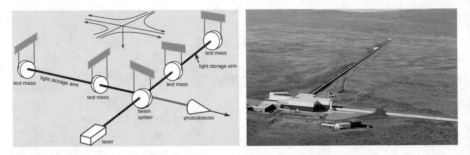

Fig. 11.1 Left: schematic of the LIGO laser interferometer gravitational wave antenna. Right: aerial view of the LIGO antenna in Hanford, in the state of Washington (From https://www.ligo.org/science/GW-Overview)

The underlying ideas of quantum optomechanics were largely driven by the developments in optical gravitational wave antennas spearheaded by V. Braginsky, K. Thorne, C. Caves, and others in the 1970s and 1980s [1–3]. These antennas operate by optically measuring changes in the positions of suspended kilogram-size test masses that serve as the end mirrors of large path length (kilometers long) Michelson interferometers as a result of the passage of gravitational waves, see Fig. 11.1. These waves produce time-dependent variations in the curvature of space-time, resulting in differential changes in the optical path length of the interferometer arms and a modulation of the optical transmission through it. It is in this context that researchers first understood a number of fundamental quantum optical effects on mechanics and mechanical detection, including the standard quantum limit of optical interferometers and the importance of back action evading measurements. They also recognized the importance of nonclassical states of light such as squeezed states in reducing the quantum noise in these interferometers.

Further progress in quantum optomechanics relied heavily on combining the detailed understanding of the mechanical effects of light that we introduced in Chaps. 8 and 9 with the availability of advanced micro- and nanomechanical devices. This opened a path to the realization of macroscopic mechanical systems that operate deep in the quantum regime, with no significant thermal noise remaining. This allows for the determination and control of their quantum state, resulting in the development of detectors of feeble forces and fields of increased sensitivity, precision, and accuracy.

Although this chapter is primarily about quantum effects, it will be useful to begin with a classical description of optomechanics, to introduce in simple terms the underlying phenomenology, most importantly the ideas of cold damping and optical spring effect. We will then turn to a full quantum description and show how a resolved sideband cooling method directly adapted from the technique discussed in Sect. 9.5 for trapped ions permits to cool these systems to their quantum mechanical ground state. This is an essential first step in eliminating the thermal fluctuations that normally mask quantum features.

However, the ground state is not particularly interesting by itself, so the next challenge is to prepare, manipulate, and characterize quantum states of mechanical oscillators directed at specific science or engineering goals. We then revisit the idea of standard quantum limit in the context of optomechanical detection and show that when light is used as the probe of mechanical motion, that limit arises from the balance between the uncertainty in photon number, or shot noise, and radiation pressure noise. The chapter finally returns to ultracold atoms and closes a loop by showing how collective density excitations of Bose condensates can behave precisely as optomechanical oscillators.

11.1 Classical Analysis

We have seen that resonant light-matter interactions can result in a very large enhancement of their coupling, but at the cost of being limited to narrow ranges of wavelengths. This is in contrast to non-resonant interactions, which are typically much weaker but largely wavelength independent. Cavity optomechanics exploits the best of both worlds by achieving resonant enhancement through engineered resonant structures rather than via the internal structure of materials. This permits to achieve optomechanical effects for a broad range of wavelengths, from the microwave to the optical regime, and in a vast range of platforms, from nanometer-sized devices with as little as 10^7 atoms to micromechanical structures of 10^{14} atoms and to the centimeter-sized mirrors used in gravitational wave detectors.

Generic Model A simple model system that displays the main features of cavity optomechanics consists of an optically driven Fabry–Pérot resonator with one fixed end mirror, effectively assumed to have infinite mass, and the other mirror, of mass m, harmonically bound and allowed to oscillate under the action of radiation pressure from the intracavity light field, as sketched in Fig. 11.2. V. B. Braginsky and A. B. Manukin recognized as early as 1967 that as the radiation pressure drives the mirror, it changes the cavity length and hence the intracavity field intensity and phase [4]. This results in two main effects: the *optical spring effect*, an optically induced change in the oscillation frequency of the mirror that can produce a significant stiffening of its effective frequency and even result in a form of radiation pressure driven optical bistability, and *optical damping*, or *cold damping*, whereby the optical field acts effectively as a viscous fluid that can damp the mirror oscillations and cool its center-of-mass motion.

Fig. 11.2 Generic cavity optomechanical system. The cavity consists of a highly reflective fixed input mirror and a small movable end mirror harmonically coupled to a support that acts as a thermal reservoir

One can readily understand how the optical spring effect can result in a more quantum behavior of the oscillator of frequency Ω_m by recalling that its mean number of thermal phonons $\langle n_m \rangle$ at temperature T is

$$\langle n_m \rangle = k_B T / \hbar \Omega_m . \tag{11.1}$$

For a given temperature, increasing Ω_m therefore automatically reduces $\langle n_m \rangle$ and the quantum regime can be approached without having to reduce T.

Cold damping, on the other hand, reduces the temperature of the oscillating mirror by opening up a dissipation channel to a reservoir that is effectively at zero temperature. To see how this works, we first recall that in the absence of an optical field, the average center- of-mass energy $\langle E \rangle$ of an oscillating mirror dissipatively coupled to a thermal bath at temperature T results from the balance between dissipation and heating,

$$\frac{d\langle E \rangle}{dt} = -\Gamma_m \langle E \rangle + \Gamma_m k_B T , \tag{11.2}$$

where Γ_m is the intrinsic mechanical damping rate and $k_B T$ the mean thermal energy at temperature T. When the oscillator is coupled in addition to an optical field, it is however possible to arrange their interaction in such a way that an additional dissipation channel with damping rate Γ_{opt} comes into play, so that

$$\begin{aligned}
\frac{d\langle E \rangle}{dt} &= -\Gamma_m \langle E \rangle + \Gamma_m k_B T - \Gamma_{opt} \langle E \rangle + \Gamma_{opt} k_B T_{opt} \\
&\approx -\Gamma_m \langle E \rangle + \Gamma_m k_B T - \Gamma_{opt} \langle E \rangle ,
\end{aligned} \tag{11.3}$$

where we have used the fact that the frequencies of optical fields are orders of magnitude higher than those of mechanical oscillators, and for a system at room temperature, the blackbody reservoir to which they are coupled can effectively be considered to be at zero temperature, $T_{opt} \approx 0$, as we have seen in Sect. 5.1.[1] In steady state, Eq. (11.3) gives $\langle E \rangle = \Gamma_m k_B T / (\Gamma_m + \Gamma_{opt})$, or

$$T_{eff} = \frac{\Gamma_m T}{\Gamma_m + \Gamma_{opt}} . \tag{11.4}$$

---

[1] Remember however that this is not the case for microwave fields.

This phenomenological picture predicts that the limit of cooling can approach $T \approx 0$ for $\Gamma_{\text{opt}} \gg \Gamma_m$, although a more detailed quantum analysis presented in the next section will yield a fundamental limit given by quantum noise, as expected. However, this is usually not a major limitation to cooling a mechanical mode arbitrarily close to the quantum ground state $\langle n_m \rangle \approx 0$.

More quantitatively, consider a single mode of the optical resonator of nominal frequency $\omega_c = \ell \pi c / L$, ℓ integer, and assume that radiation pressure causes a displacement $x(t)$ of the harmonically bound end mirror. This results in a change in the frequency of the optical mode to

$$\omega_c' = \omega_c - Gx(t),\tag{11.5}$$

where

$$G = -\partial \omega_c' / \partial x.\tag{11.6}$$

For a single-mode Fabry–Pérot resonator of length L and $x(t) \ll L$, this last expression becomes simply $G \approx \omega_c / L$.

Typical mechanical oscillator frequencies Ω_m are in the range of $2\pi \cdot 10\,\text{Hz}$ to $2\pi \cdot 10^9\,\text{Hz}$, and the mechanical quality factors of the mirrors are in the range of perhaps $Q_m \approx 10^3 - 10^7$, so that the mechanical damping rate $\Gamma_m = \Omega_m / Q_m$ of the oscillating mirror is typically much slower than the damping rate $\kappa/2$ of the intracavity field. One can therefore gain considerable intuition by first neglecting Γ_m altogether and assuming that the mirror motion is approximately harmonic,

$$x(t) \approx x_0 \sin(\Omega_m t).\tag{11.7}$$

For a classical monochromatic pump of frequency ω_L and amplitude α_{in}, the intracavity field obeys the equation of motion

$$\frac{d\alpha(t)}{dt} = [i\,(\Delta_c + Gx(t)) - \kappa/2]\,\alpha(t) + \sqrt{\kappa}\alpha_{\text{in}},\tag{11.8}$$

which is the classical limit of Eq. (5.159), and we introduced as in Sect. 9.4 the detuning

$$\Delta_c = \omega_L - \omega_c\tag{11.9}$$

between the frequency ω_L of the driving laser field and the cavity mode frequency ω_c.[2] The steady-state solution is

---

[2] Due to the number of relevant frequencies in optomechanics, we use ω_L instead of the more compact notation ω in this chapter to avoid possible confusion.

$$\alpha = \frac{\sqrt{\kappa}\alpha_{\rm in}}{-i(\Delta_c + Gx) + \kappa/2}, \tag{11.10}$$

where the intracavity field amplitude α is normalized in such a way that

$$|\alpha|^2 = \frac{\kappa}{(\Delta_c + Gx)^2 + (\kappa/2)^2} \left(\frac{P}{\hbar\omega_L}\right)$$

$$= \frac{\kappa}{(\Delta_c + \omega_c x/L)^2 + (\kappa/2)^2} \left(\frac{P}{\hbar\omega_L}\right), \tag{11.11}$$

so that

$$P \equiv \hbar\omega_L |\alpha_{\rm in}|^2 \tag{11.12}$$

is the input laser power driving the cavity mode. Remember that this normalization, which we already introduced in the discussion of cavity cooling of Sect. 9.4, is somewhat misleading since the analysis is completely classical at this point. Planck's constant $\hbar$ has been introduced "by hand" with the sole purpose of allowing for an easy generalization to the quantum description of Sect. 11.2. The amplitude α will then be interpreted as the square root of the mean number of intracavity photons $\alpha = \sqrt{\langle \hat{a}^\dagger \hat{a} \rangle}$, with $\hat{a}$ and $\hat{a}^\dagger$ the annihilation and creation operators of the intracavity field, and $|\alpha_{\rm in}|^2$ as an input flux having units of "photons per second."

For periodic mirror oscillations of the form

$$x(t) = x_0 \sin(\Omega_m t) \tag{11.13}$$

with x_0 small enough that $Gx_0 \ll \Omega_m$, the moving mirror boundary acts as a modulator, resulting in the generation of two sidebands at frequencies $\omega_L \pm \Omega_m$ [5]. Specifically, solving Eq. (11.8), for instance, in Fourier space, shows that the time-dependent complex field amplitude $\alpha(t)$ takes then the approximate form $\alpha(t) \simeq \alpha_0(t) + \alpha_1(t)$ with

$$\alpha_0(t) \simeq \frac{\sqrt{\kappa}\alpha_{\rm in}}{-i\Delta_c + \kappa/2},$$

$$\alpha_1(t) \simeq \left(\frac{Gx_0}{2}\right)\frac{\sqrt{\kappa}\alpha_{\rm in}}{-i\Delta_c + \kappa/2}$$

$$\times \left(\frac{e^{-i\Omega_m t}}{-i(\Delta_c + \Omega_m) + \kappa/2} - \frac{e^{+i\Omega_m t}}{-i(\Delta_c - \Omega_m) + \kappa/2}\right), \tag{11.14}$$

and

$$|\alpha(t)|^2 \approx |\alpha_0(t)|^2 + \alpha_0(t)\alpha_1^*(t) + \alpha_0^*(t)\alpha_1(t)$$

$$= \frac{\kappa|\alpha_{\text{in}}|^2}{\Delta_c^2 + \kappa^2/4}$$

$$\times \left[1 + Gx_0 \left(\frac{\Delta_c + \Omega_m}{(\Delta_c + \Omega_m)^2 + \kappa^2/4} + \frac{\Delta_c - \Omega_m}{(\Delta_c - \Omega_m)^2 + \kappa^2/4} \right) \sin(\Omega_m t) \right.$$

$$\left. + Gx_0 \left(\frac{\kappa/2}{(\Delta_c + \Omega_m)^2 + \kappa^2/4} + \frac{\kappa/2}{(\Delta_c - \Omega_m)^2 + \kappa^2/4} \right) \cos(\Omega_m t) \right].$$

$$(11.15)$$

The first sideband in Eq. (11.14) can be interpreted as an anti-Stokes line, with a resonance at $\omega_L = \omega_c - \Omega_m$, and the second one as a Stokes line at $\omega_L = \omega_c + \Omega_m$.[3] An important feature of these sidebands is that their amplitudes can be vastly different since they are determined by the cavity Lorentzian response function evaluated at $\omega_L - \Omega_m$ and $\omega_L + \Omega_m$, respectively. This asymmetry parallels a situation previously encountered in the sideband cooling of trapped ions of Sect. 9.5. It is therefore not surprising that this analogy can be exploited in the optomechanical cooling of oscillating membranes, as we will see in Sect. 11.1.2. First, however, we briefly return to the optical spring effect and show how in addition to increasing the oscillation frequency of the mechanical system, it can also result in radiation pressure induced optical bistability.

11.1.1 Static Phenomena: Optical Spring Effect

To properly describe the optomechanical system, it is of course not sufficient in general to consider only the intracavity field dynamics under the influence of periodic mirror oscillations, as we have done so far, but we must also include the back action of the field on the membrane motion. In a first step, we consider the limit where the damping rate $\kappa/2$ of the field is much faster than all other characteristic times of the system, in which case $\alpha(t)$ follows the membrane motion adiabatically.

We have seen that the oscillations of the mirror, which evolve under the combined effects of its harmonic restoring force and the radiation pressure force F_{rp} of the intracavity field $|\alpha|^2$, result in the generation of sidebands in that field, and these exert in turn a back action force on the mirror motion. Ignoring for a moment the restoring force, adiabatically eliminating $|\alpha|^2$ for large κ and with Eq. (11.5), we have that

---

[3] We recall that the Stokes and anti-Stokes nomenclature finds its origin in Raman scattering: Stokes scattering refers to the situation where the emitted radiation is of lower frequency than the incident radiation and anti-Stokes scattering to the case where it is of higher frequency.

$$F_{\text{rp}} = -\hbar \frac{\mathrm{d}}{\mathrm{d}x} \left(-G|\alpha|^2 x \right) = \hbar \frac{\omega_c}{L} |\alpha|^2, \tag{11.16}$$

where the first equality defines the optomechanical coupling "per photon"—again with the understanding that using the word photon is a stretch in the context of this classical description—and the second equality holds for a simple Fabry–Pérot, and $|\alpha|^2$ is then given by Eq. (11.11). One can easily show that the force F_{rp} can be derived from the potential

$$V_{\text{rp}} = -\frac{\hbar\kappa|\alpha_{\text{in}}|^2}{2} \arctan\left[2(\Delta_c + Gx)/\kappa\right], \tag{11.17}$$

and the mirror of mass m is therefore subject to the total potential

$$V(x) = \frac{1}{2} m\Omega_m^2 x^2 - \frac{\hbar\kappa|\alpha_{\text{in}}|^2}{2} \arctan\left[2(\Delta_c + Gx)/\kappa\right]. \tag{11.18}$$

The effect of V_{rp} is both to slightly shift the equilibrium position of the mirror to a position $x_0 \neq 0$, as would be intuitively expected, and to change its spring constant from its intrinsic value $k = m\Omega_m^2$ to

$$k_{\text{rp}} = m\Omega_m^2 + \left.\frac{\mathrm{d}^2 V_{\text{rp}}(x)}{\mathrm{d}x^2}\right|_{x=x_0}. \tag{11.19}$$

The second term in this expression is the static *optical spring effect*. For realistic parameters, it can increase the stiffness of the mechanical system by orders of magnitude.

An additional static effect of radiation pressure is that in general, there is a range of parameters for which the potential $V(x)$ can exhibit three extrema, see Fig. 11.3 and Problem 11.2. Two of them correspond to stable local minima of $V(x)$ and the third one to an unstable maximum. This results in the onset of radiation pressure induced optical bistability [6], the coexistence of two possible stable lengths x of the resonator for a given incident intensity $|\alpha_{\text{in}}|^2$. This effect is closely related to the more familiar form of bistability that can occur in Kerr nonlinear media. The difference is that in the latter case, it is the optical length of the resonator—the product of its physical length and the intensity dependent refractive index of the medium—that is changed, its physical length remaining unchanged. In contrast, in radiation pressure induced bistability, there is no nonlinear medium; it is x itself that is bistable.

11.1.2   Effects of Retardation: Cold Damping

Beside the adiabatic effects characteristic of the regime where the field damping rate κ dominates the system dynamics, cooling (or heating) of the mechanical motion

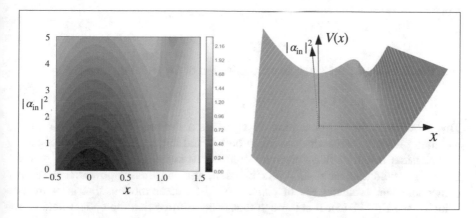

Fig. 11.3 Left: contour plot of the radiation pressure induced optical potential $V(x)$ of Eq. (11.18) resulting in radiation pressure induced optical bistability, in units of $\frac{1}{2}m\Omega_m^2$. Right: 3D rendition of the same potential as a function of the dimensionless position x of the moving mirror and the dimensional driving field intensity $|\alpha_{in}|^2$, scaled to $2/\hbar\kappa$. In this example, $|\alpha_{in}|^2$ is scaled to $2/\hbar\kappa$, $\Delta_c = -3\kappa$, and $G = 3\kappa$

becomes possible when this condition is no longer fulfilled. This is a consequence of the dispersive nature of the optomechanical interaction and of the resulting delayed response of the intracavity field to mechanical motion, which can produce under appropriate conditions an additional field-induced oscillator damping and cooling. We already encountered a similar mechanism in the description of cavity cooling of Sect. 9.4, although as we shall see a much closer analogy can be drawn with the sideband ion cooling of Sect. 9.5.

To analyze this regime, we proceed by assuming that the system is initially in equilibrium at some mirror position $\bar{x}$ with intracavity field $\bar{\alpha}$, taken to be real without loss of generality, and consider the linearized dynamics of small displacements $\delta x(t)$ and $\delta\alpha(t)$ from that state resulting from a small external perturbation $\delta F(t)$. To lowest order, Eq. (11.8) gives then

$$\frac{d}{dt}\delta\alpha = (i\Delta_c - \kappa/2)\delta\alpha + iG\bar{\alpha}\,\delta x\,, \tag{11.20}$$

with the resulting perturbation in the harmonically bound mirror position given by

$$\left[\frac{d^2}{dt^2} + \Gamma_m\frac{d}{dt} + \Omega_m^2\right]\delta x = \hbar G\bar{\alpha}\left(\delta\alpha + \delta\alpha^*\right)\,. \tag{11.21}$$

The first of these equations can be solved in Fourier space to give

$$\delta\alpha(\omega) = \left(\frac{iG\bar{\alpha}}{-i(\bar{\Delta}_c + \omega) + \kappa/2}\right)\delta x(\omega) \tag{11.22}$$

with

$$\bar{\Delta}_c = \Delta_c + G\bar{x}, \tag{11.23}$$

and from Eq. (11.21), the associated modification of the radiation pressure force is

$$\delta F_{\mathrm{rp}}(\omega) = -\hbar G\bar{\alpha}\left[\delta\alpha(\omega) + \delta\alpha^*(\omega)\right]. \tag{11.24}$$

This shows that the mirror motion $\delta x(t)$ exerts a *dynamical back action* on the radiation pressure force, which acquires both a real and an imaginary component. The physical origin of that imaginary component is the delayed response of the intracavity field to that motion. As a result, the intracavity power acquires a component that oscillates out of phase with the mirror motion, that is, with its velocity, see Eq. (11.15). It is through that friction force that the optical field acts as a viscous field for the mirror.

The net effect of the real and imaginary components of δF_{rp} can be conveniently cast in terms of a back action frequency shift $\delta\Omega_{\mathrm{opt}}$ and corresponding damping rate Γ_{opt}, with

$$\delta\Omega_{\mathrm{opt}} = \frac{\hbar G^2\bar{\alpha}^2}{2m\Omega_m}\left[\frac{\bar{\Delta}_c + \Omega_m}{(\bar{\Delta}_c + \Omega_m)^2 + \kappa^2/4} + \frac{\bar{\Delta}_c + \Omega_m}{(\bar{\Delta}_c - \Omega_m)^2 + \kappa^2/4}\right], \tag{11.25}$$

$$\Gamma_{\mathrm{opt}} = \frac{\hbar G^2\bar{\alpha}^2}{2m\Omega_m}\left[\frac{\kappa}{(\bar{\Delta}_c + \Omega_m)^2 + \kappa^2/4} - \frac{\kappa}{(\bar{\Delta}_c - \Omega_m)^2 + \kappa^2/4}\right], \tag{11.26}$$

as illustrated in Fig. 11.4.

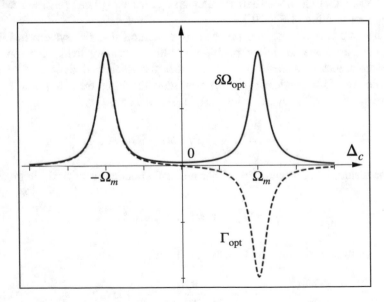

Fig. 11.4 Back action frequency shift $\delta\Omega_{\mathrm{opt}}(\Delta_c)$ (solid line) and damping rate $\Gamma_{\mathrm{opt}}(\Delta_c)$ (dashed line) as a function of Δ_c (arbitrary units)

For detunings $\bar{\Delta}_c \approx -\Omega_m$, the first term in Eq. (11.26) dominates over the second term, and the dynamical back action results in an increase in the damping of the mechanical oscillator and cooling, as already indicated in Eq. (11.4). It is therefore the asymmetry between the response function of the Fabry–Pérot at the frequencies of the two side modes that is responsible for cooling—or heating if one changes the sign of Δ_c and uses a blue-detuned instead of a red-detuned driving field. The existence of two sidebands, one associated with heating and the other with cooling, parallels closely the situation already encountered in Sect. 9.5. It can likewise be exploited to achieve sideband cooling, but now of a vibrating membrane rather than a trapped ion. Physically, this is a consequence of the fact that the optomechanical coupling between the intracavity field and the mirror results in the scattering of the driving field into an anti-Stokes line that is strongly damped due to the high density of states at the cavity resonance. Conversely, for the opposite detuning $\bar{\Delta}_c \approx -\Omega_m$, it is the Stokes line that is strongly damped, resulting in an anti-damping of the mirror motion that can lead to parametric oscillations and dynamical instabilities.

Together with Eq. (11.4), this analysis predicts that the cooling of the center-of-mass motion of the mirror can be arbitrarily close to $T_{\text{eff}} = 0$. More specifically, in the resolved sideband limit $\Omega_m \gg \kappa$, we find from Eq. (11.26)

$$\Gamma_{\text{opt}} \approx \left(\frac{2}{\kappa}\right)\frac{\hbar G^2 \bar{\alpha}^2}{m\Omega_m}, \tag{11.27}$$

which can become arbitrarily large for small optical damping rates.

The quantum description of the next section will show that cold damping and mirror cooling can also be interpreted in terms of the annihilation of phonons from the center-of-mass mode of oscillation when scattering the driving laser field into the anti-Stokes sideband, much like we interpreted the sideband cooling of trapped ions, and heating can be understood as resulting from the creation of phonons associated with the scattering of the driving field into the lower frequency Stokes side mode.

Optomechanical cooling was first observed in the microwave domain by D. G. Blair et al. [7] in a Niobium high-Q resonant mass gravitational radiation antenna and 10 years later in the optical domain in several laboratories around the world: first, via feedback cooling of a mechanical mirror by P. F. Cohadon et al. [8] and shortly thereafter in a broad range of mesoscopic and macroscopic systems ranging in weight from micrograms to kg scale, see e.g. Ref. [9–13], and most remarkably perhaps Ref. [14], which reported the cooling of a 10 kg mirror deep into the quantum regime, with about 10 thermal phonons of excitation left.

11.2 Quantum Theory

The classical prediction that one can in principle reach an arbitrarily large degree of cooling needs to be qualified to account for the effects of quantum noise. In particular, the open port of the resonator used to supply the optical drive of the oscillating mirror allows for the coupling of vacuum fluctuations into the resonator [15], as will be discussed in detail in Sect. 11.4. This leads to a fundamental limit to the degree of cooling that can be achieved.

Ignoring in a first step this coupling to the environment, the optomechanical Hamiltonian for a single optical mode of the Fabry–Pérot resonator and a single mode of oscillation of the suspended mirror is simply

$$\hat{H} = \hbar\omega(\hat{q})\hat{a}^\dagger\hat{a} + \frac{\hat{p}^2}{2m} + \frac{1}{2}m\Omega_m^2\hat{q}^2, \qquad (11.28)$$

where $\hat{a}$ and $\hat{a}^\dagger$ are bosonic annihilation and creation operators for the cavity mode of frequency $\omega(\hat{q})$, and $\hat{p}$ and $\hat{q}$ are the momentum and position of the oscillating mirror of mass m and frequency Ω_m. In reality, though, this Hamiltonian is more subtle than that may appear at first. This is because the mode frequency $\omega(\hat{q})$ depends on the length of the resonator, which in turn depends on the intracavity intensity. Stated differently, the boundary conditions for the quantization of the light field are changing in time and do so in a fashion that depends on the state of that field and its history. The rigorous quantization of this system is a far-from-trivial problem, but for most cases of interest in quantum optomechanics the situation is significantly simplified since the damping rate $\kappa/2$ of the optical field is much larger than the mechanical frequency Ω_m and the displacements considered are a small fraction of an optical wavelength. The intracavity field "learns" therefore about changes in its environment in times short compared to $1/\Omega_m$. Under these conditions, one can assume that the cavity frequency follows adiabatically any change in resonator length,

$$\omega(\hat{q}) = \frac{\ell\pi c}{L + \hat{q}} = \omega_c\left(\frac{1}{1 + \hat{q}/L}\right) \approx \omega_c(1 - \hat{q}/L), \qquad (11.29)$$

where ℓ is an integer that labels the mode of nominal frequency ω_c and L is the nominal resonator length, that is, its length in the absence of light. In the classical limit, we recover the result $G = \omega_c/L$ of Eq. (11.6) valid for a simple Fabry–Pérot. The Hamiltonian (11.28) reduces then to [17]

$$\hat{H} = \hbar\omega_c\hat{a}^\dagger\hat{a} + \frac{\hat{p}^2}{2m} + \frac{1}{2}m\Omega_m^2\hat{q}^2 - \hbar G\hat{a}^\dagger\hat{a}\hat{q}$$

$$= \hbar\omega_c\hat{a}^\dagger\hat{a} + \hbar\Omega_m\hat{b}^\dagger\hat{b} - \hbar g_0\hat{a}^\dagger\hat{a}(\hat{b} + \hat{b}^\dagger), \qquad (11.30)$$

where $\hat{b}$ and $\hat{b}^\dagger$ are the annihilation and creation operators for the mechanical oscillator,

$$\hat{q} = x_{zpf}(\hat{b} + \hat{b}^\dagger) \qquad (11.31)$$

with $x_{\text{zpf}} = \sqrt{\hbar/2m\Omega_m}$ the zero-point motion. We have also introduced the optomechanical coupling frequency

$$g_0 = x_{\text{zpf}}G = -x_{\text{zpf}}\partial\omega_c'/\partial x\,, \tag{11.32}$$

which scales the optomechanical displacement of the oscillator to x_{zpf}. The Hamiltonian (11.30) is the starting point for most quantum mechanical discussions of cavity optomechanics, with the system coupled in addition to two reservoirs, one for the intracavity field, with field decay rate $\kappa/2$, and the other for the mechanical oscillator, with decay rate Γ_m.

Cooling Limit We mentioned that the optical port through which the classical field driving the resonator is injected also admits vacuum and thermal noise. As we have seen in Sect. 5.3, this noise induces in the system transitions that are responsible for establishing its final equilibrium state and hence its final temperature. Splitting the field as $\hat{a} = \alpha + \hat{c}$, where α is its classical component and $\hat{c}$ the displaced annihilation operator that accounts for its fluctuations, their dominant contribution to the optomechanical interaction is

$$\hbar g_0(\alpha\hat{c}^\dagger + \alpha^*\hat{c})(\hat{b} + \hat{b}^\dagger) \equiv \hbar g_0\hat{F}(\hat{b} + \hat{b}^\dagger)\,. \tag{11.33}$$

For a field reservoir at effective temperature $T = 0$, the noise spectral density $S_{FF}(\omega)$ is, see Eq. (5.103) and Problem 11.3,

$$S_{FF}(\omega) = \int_{-\infty}^{\infty} d\tau\, e^{i\omega\tau}\langle\hat{F}(\tau)\hat{F}(0)\rangle = n\int_{-\infty}^{\infty} d\tau\, e^{i\omega\tau} e^{(i\Delta_c - \kappa/2)\tau}$$

$$= |\alpha|^2\frac{\kappa}{(\Delta_c + \omega)^2 + \kappa^2/4}\,. \tag{11.34}$$

Since the phononic levels of the mechanical oscillator are equidistant, the "cooling" and "heating" rates of noise-induced transitions between the states $|n_m - 1\rangle$ and $|n_m\rangle$ are simply $n_m A_{1\to0}$ and $n_m A_{0\to1}$, respectively, with

$$A_{0\to1} = g_0^2 S_{FF}(\omega = -\Omega_m)\quad;\quad A_{1\to0} = g_0^2 S_{FF}(\omega = \Omega_m)\,, \tag{11.35}$$

where we have used the result of Eqs. (5.106).

The asymmetry between these processes is analogous to the asymmetry between the damping and anti-damping contributions responsible for the resolved sideband cooling of trapped ions, see Eqs. (9.73), with the difference between cooling and heating rates maximized for $\Delta_c = -\Omega_m$, as illustrated in Fig. 11.5. In the absence of mechanical damping, we then have from detailed balance that the steady-state phonon number distribution is given by

$$\frac{\bar{n}_m^0 + 1}{\bar{n}_m^0} = \frac{S_{FF}(\Omega_m)}{S_{FF}(-\Omega_m)} = \exp[\hbar\Omega_m/k_B T_{\text{eff}}]\,, \tag{11.36}$$

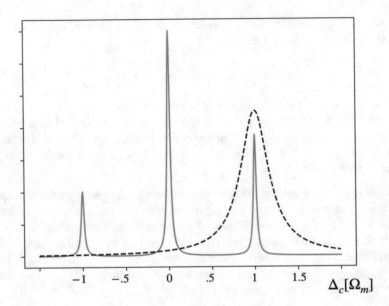

Fig. 11.5 Schematic of sideband cooling. A coherent electromagnetic field driving the optomechanical resonator acquires frequency sidebands due to the membrane oscillations, as discussed in Sect. 11.1. The solid line illustrates the corresponding spectrum of the intracavity field, with a peak at ω_L and two sidebands at $\omega_L \pm \Omega_m$. The dotted line illustrates the bandwidth κ of the cavity mode of frequency ω_c. Quantum mechanically, the origin of the high frequency sideband is the parametric transfer of phonons from the membrane to the microwave field and the lower sideband is due to the reverse process, see Sect. 11.2. Sideband cooling results when the upper sideband frequency $\omega_l + \Omega_m$ is resonant with the cavity frequency ω_c. The detuning Δ_c is in units of Ω_m, and the vertical axis is in arbitrary units

which gives in the resolved sideband limit $\kappa \ll \Omega_m$ [18, 19]

$$\bar{n}_m^0 = \left(\frac{\kappa}{4\Omega_m} \right)^2 . \tag{11.37}$$

The inclusion of mechanical damping changes this result to a cooling limit with minimum mean phonon number

$$\langle n_m \rangle = \frac{\Gamma_{\text{opt}} \bar{n}_m^0 + \Gamma_m \bar{n}_m^T}{\Gamma_m + \Gamma_{\text{opt}}}, \tag{11.38}$$

where $\bar{n}_m^T$ is the equilibrium phonon occupation determined by the mechanical bath temperature. This shows that in the presence of quantum fluctuations, the ground state can be approached, but not quite reached, as expected. For $\bar{n}_m^T \gg 0$, one recovers the classical result of Eq. (11.4), with $\langle n_m \rangle \to \bar{n}_m^0$ if the optical damping Γ_{opt} dominates over Γ_m.

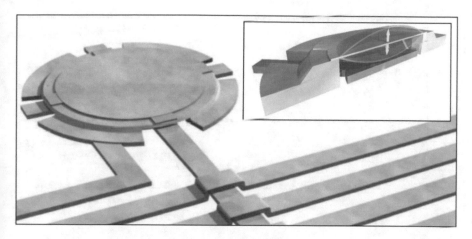

Fig. 11.6 Artist conception of the microwave optomechanical circuit of Ref. [21]. Capacitor element of the LC circuit is formed by a 15 μm diameter membrane lithographically suspended 50 nm above a lower electrode. Insert: cut through the capacitor showing the membrane oscillations. (Adapted from Refs. [21] and [24])

Experiments The cooling of the oscillator center-of-mass motion deep into the quantum regime has been demonstrated in a number of micromechanical systems, with a mean phonon number within a fraction of a phonon of their ground state of vibrational motion, $\langle n_m \rangle < 1$, see Refs. [20–22] for early experiments. In the first case, the frequency of the mechanical oscillator was high enough that cooling to the ground state could be achieved by conventional cryogenic refrigeration, while the other two experiments exploited resolved sideband cooling to approach the mechanical ground state of center-of-mass motion. In one case [21], the mechanical resonator was a suspended circular aluminum membrane tightly coupled to a superconducting lithographic microwave cavity, Fig. 11.6. That cavity was precooled to 20 mK, corresponding to an initial occupation of 40 phonons, and then further cooled by radiation pressure forces to an average phonon occupation of $\langle n_m \rangle \approx 0.3$, as shown in Fig. 11.7. In contrast, Ref. [22] utilized an optomechanical structure with co-located photonic and phononic band gaps in a suspended on-chip waveguide. The structure was precooled to 20 K, corresponding to about 100 thermal quanta, and then cooled via radiation pressure to $\langle n_m \rangle \approx 0.85$. Shortly thereafter, that same group also observed the motional sidebands generated on a second probe laser by a mechanical resonator optically cooled optically to near its vibrational ground state. They were able to detect the asymmetry in the sideband amplitudes between up-converted and down-converted photons, a smoking gun signature of the asymmetry between the quantum processes of emission and absorption of phonons, as we have seen [23].

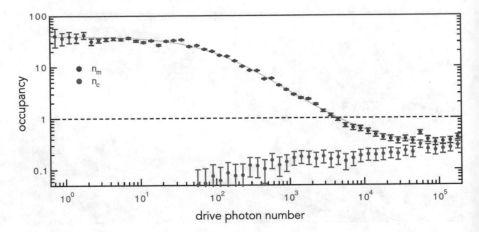

Fig. 11.7 Phonon occupancy (blue) and intracavity photon occupancy (red) as a function of the drive photon number. In this example, sideband cooling reduces the thermal occupancy of the mechanical mode from $n_m = 40$ into the quantum regime, reaching a minimum of $n_m = 0.34 \pm 0.05$. (From Ref. [21])

11.3 Beyond the Ground State

Cooling mechanical resonators to their ground state of motion is essential in eliminating the thermal fluctuations that normally mask quantum features. However, by itself that state is not particularly interesting, so the next challenge is to prepare, manipulate, and characterize quantum states of the mechanical resonator required for some specific science or engineering goal.

Like in cavity QED, the control of the quantum state of a mechanical oscillator requires typically that one operates in the strong coupling regime, where the energy exchange between the mechanical object and the system to which it is coupled is not adversely affected by dissipation and decoherence. This section reviews briefly selected examples of developments that have taken place along these lines. Substantially more details can be found in references [25–27].

Section (11.1) showed that the key frequencies that come into play in describing the interaction between the intracavity light field and the mechanical element are the detuning $\Delta_c = \omega_L - \omega_c$ between the driving field and cavity mode frequencies and the frequency Ω_m of the mechanical oscillator. To bring Δ_c more explicitly to the fore, it is useful to transform the optomechanical interaction (11.30) to a frame rotating at the frequency ω_L by applying the unitary transformation

$$\hat{U}(t) = e^{i\omega_L \hat{a}^\dagger \hat{a} t} , \tag{11.39}$$

resulting in the transformed Hamiltonian

$$\hat{H}' = \hat{U}(t)\hat{H}\hat{U}^\dagger(t) + i\hbar \left(\frac{\partial \hat{U}(t)}{\partial t} \right) \hat{U}^\dagger(t)$$

$$= -\hbar\Delta_c \hat{a}^\dagger \hat{a} + \hbar\Omega_m \hat{b}^\dagger \hat{b} - \hbar g_0 \hat{a}^\dagger \hat{a}(\hat{b} + \hat{b}^\dagger)$$

$$= \hat{H}_0 + \hat{H}_{\text{int}} . \tag{11.40}$$

11.3.1 Linearized Coupling

A number of quantum effects have been predicted in the regime when the radiation pressure of a single or very few photons displaces the mechanical oscillator by more than x_{zpf}. They include two-photon blockade and quantitative changes in the output spectrum and cavity response of the optomechanical system that can lead to the generation of non-Gaussian steady states of the oscillator [28–30]. Unfortunately, for realizable levels of the optomechanical coupling frequency, g_0 is typically much too small for these effects to dominate over the incoherent dynamics.

There is however a way around this difficulty, the trade-off being that the intrinsic nonlinear nature of the optomechanical interaction (11.40) is then replaced by an effective linear interaction. To achieve this goal, we proceed again by decomposing the electromagnetic field as the sum of a classical part and a small quantum mechanical component, $\hat{a} \rightarrow \alpha + \hat{c}$, with $\hat{c}$ interpreted as before as a displaced photon annihilation operator that accounts for quantum fluctuations. With $n = |\alpha|^2$ and introducing the effective optomechanical coupling frequency

$$g = g_0\sqrt{n} , \tag{11.41}$$

the optomechanical interaction in Eq. (11.40) becomes then to lowest order in $\hat{c}$ and $\hat{c}^\dagger$

$$\hat{H}_{\text{int}} \simeq -\hbar g_0 n(\hat{b} + \hat{b}^\dagger) - \hbar g \left(\hat{c} + \hat{c}^\dagger \right) (\hat{b} + \hat{b}^\dagger) . \tag{11.42}$$

The first term in this Hamiltonian describes a simple Kerr effect, with a change in resonator length proportional to the classically intracavity intensity. This is the term that leads to radiation pressure induced optical bistability. The second term can be reexpressed in an interaction picture with respect to $\hat{H}_0 = -\hbar\Delta_c \hat{a}^\dagger \hat{a} + \hbar\Omega_m \hat{b}^\dagger \hat{b}$ as

$$\hat{V} = -\hbar g \left[\hat{b}\hat{c}^\dagger e^{-i(\Delta_c + \Omega_m)t} + \text{h.c.} \right] - \hbar g \left[\hat{b}^\dagger \hat{c}^\dagger e^{-i(\Delta_c - \Omega_m)t} + \text{h.c.} \right] . \tag{11.43}$$

This interaction describes a linear coupling between the quantized component of the optical field and the mechanical oscillator, enhanced from the single-photon coupling frequency g_0 by a factor $\sqrt{n}$, which can be very substantial.

Beam Splitter Hamiltonian and State Transfer   On the red-detuned side of the Fabry–Pérot resonance, $\Delta_c = -\Omega_m$, and after invoking the rotating wave approximation, the interaction Hamiltonian (11.43) reduces to

$$\hat{V} \simeq -\hbar g \left(\hat{b}\hat{c}^{\dagger} + \text{h.c.} \right), \tag{11.44}$$

which is the beam splitter Hamiltonian (2.175) first encountered in the discussion of balanced homodyne detection of Sect. 2.4.2. In contrast, on the blue-detuned side of the resonance, $\Delta_c = +\Omega_m$, we have

$$\hat{V} \simeq -\hbar g \left(\hat{b}^{\dagger}\hat{c}^{\dagger} + \text{h.c.} \right). \tag{11.45}$$

This interaction describes the parametric amplification of the phonon mode and the optical field.

One of the remarkable properties of the beam splitter Hamiltonian (11.44) is that it provides a mechanism to precisely transfer the quantum state of the mechanical oscillator to the electromagnetic field, and conversely. This is easily seen by considering the Heisenberg equations of motion for the annihilation operators $\hat{b}$ and $\hat{c}$,

$$\hat{b}(t) = \hat{b}(0) \cos(\hbar g t) + i\hat{c}(0) \sin(\hbar g t)$$
$$\hat{c}(t) = \hat{c}(0) \cos(\hbar g t) + i\hat{b}(0) \sin(\hbar g t), \tag{11.46}$$

which we already encountered in Eqs. (2.180) in the context of balanced homodyne detection. Since the optomechanical interaction g can readily be made time dependent by pulsing the classical driving laser field intensity so that $n \rightarrow n(t)$, it follows that for an interaction time t_{int} and a driving laser pulse intensity such that

$$\hbar g_0 \int_0^{t_{\text{int}}} dt \sqrt{n(t)} = \pi/2,$$

we then have

$$\hat{b}(t_{\text{int}}) = \hat{c}(0) \quad ; \quad \hat{c}(t_{\text{int}}) = i\hat{b}(0), \tag{11.47}$$

indicative of a perfect state transfer between the optical and phonon modes—assuming of course that dissipation and decoherence can be ignored during that time interval.

The interest in high-fidelity state transfer between optical and acoustical fields is largely motivated by applications that would benefit from combining the best of their complementary features: the potentially slow decoherence rate of motional states in mechanical systems makes them well suited for information storage, with phonon lifetimes of 1.5 s or more and nano-acoustic resonators with quality factors Q as high as 5×10^{10} having been experimentally demonstrated [31]. However, these systems do not permit fast information transfer rates, in contrast to optical fields, which are ideal as information carriers but are typically subject to a fast decoherence that limits their interest for storage. *Hybrid systems* capable of combining the benefits of both subsystems are therefore of obvious benefit in quantum information applications. The coherent quantum mapping of phonon fields to optical modes also promises to be useful in quantum sensing applications, by combining the remarkable sensitivity of nanoscale cantilevers to feeble forces and fields with reliable and high-efficiency optical detection schemes. We will return to this point at some length in the next chapter.

In addition to standard state transfer between motional and optical states, phonon fields can also operate as high-fidelity transducers between optical fields of different wavelengths or even between optical and microwave fields. The first theoretical proposal that analyzed a scheme to transfer quantum states from a propagating light field to the vibrational state of a movable mirror by exploiting radiation pressure effects is due to J. Zhang and coworkers [32], and the first experimental demonstration of state transfer between a microwave field and a mechanical oscillator with amplitude at the single quantum level is due to T. A. Palomaki et al. [33].

Two-Mode Squeezing By explicitly introducing the (controllable) phase ϕ of the classical driving field so that

$$g = g_0 \sqrt{n} \rightarrow i g_0 \sqrt{n} \exp(i\phi),$$

the Hamiltonian (11.45) takes the form

$$\hat{V} = -i\hbar \left[g \hat{b}^\dagger \hat{c}^\dagger - g^* \hat{b} \hat{c} \right], \tag{11.48}$$

with associated evolution operator

$$\hat{S}_{ab}(t) = e^{(g^* \hat{b} \hat{c} - g \hat{b}^\dagger \hat{c}^\dagger)t}, \tag{11.49}$$

which we recognize as the two-mode squeezing operator (2.160).

We can demonstrate that $\hat{S}_{ab}(t)$ can indeed squeeze the state of a two-mode field by introducing the generalized two-mode quadrature operator

$$\hat{X} = \frac{1}{2^{3/2}} (\hat{a} + \hat{a}^\dagger + \hat{b} + \hat{b}^\dagger). \tag{11.50}$$

Problem 11.5 shows that the variance of a system initially in a two-mode vacuum state is then [34]

$$\sigma_X^2 = \frac{1}{4}\left[e^{-2|g|t}\cos^2(\phi/2) + e^{2|g|t}\sin^2(\phi/2)\right],\tag{11.51}$$

and taking for example $\phi = \pi/2$, one finds that σ_X^2 can be well below the standard quantum limit of $1/4$, a signature of two-mode squeezing. That same result also holds if the two modes are initially in coherent states, a direct consequence of the fact that the coherent state is a displaced vacuum.

Two-mode squeezed states are easily shown to be entangled, indicating that this form of interaction can result in quantum entanglement between the photon and phonon modes. As such this configuration represents a useful resource for the demonstration of fundamental quantum mechanical effects as well as to exploit cavity optomechanical devices in quantum information applications.

Back Action Evading Measurements   V. Braginsky and coworkers [1] showed in a work that was later on expanded upon by A. Clerk et al. [35] that it is possible to implement back action evading measurements of the membrane position. The way to do it is to drive it with an input field resonant with the cavity frequency ω_c, but modulated at the mirror frequency Ω_m. The mean-field amplitude of the intracavity field is then

$$\alpha(t) = \sqrt{n}\cos(\Omega_m t),\tag{11.52}$$

where, following an argument similar to the input–output analysis leading to Eq. (5.155),

$$\alpha = \frac{\sqrt{\kappa}\alpha_{in}}{i\Omega_m + \kappa/2}.\tag{11.53}$$

We proceed by introducing then the quadratures of the motional mode

$$\hat{X}_1(t) = \frac{1}{\sqrt{2}}\left(\hat{c}e^{i\Omega_m t} + \hat{c}^\dagger e^{-i\Omega_m t}\right),$$

$$\hat{X}_2(t) = -\frac{i}{\sqrt{2}}\left(\hat{c}e^{i\Omega_m t} - \hat{c}^\dagger e^{-i\Omega_m t}\right),\tag{11.54}$$

with $[\hat{X}_1(t), \hat{X}_2(t)] = i$ in terms of which the position operator is

$$\hat{x}(t) = \sqrt{2}x_{zpf}\left(\hat{X}_1(t)\cos\Omega_m t + \hat{X}_2(t)\sin\Omega_m t\right).\tag{11.55}$$

The term $-\hbar g \left(\hat{c} + \hat{c}^\dagger\right)(\hat{b} + \hat{b}^\dagger)$ of the interaction Hamiltonian (11.42) becomes then

$$\hat{V} = -\sqrt{2}\hbar g \left[\hat{X}_1(1 + \cos(2\Omega_m t)) + \hat{X}_2 \sin(2\Omega_m t)\right](\hat{b} + \hat{b}^\dagger), \qquad (11.56)$$

where $g = g_0\alpha = g_0\sqrt{n}$ as before, and its time-averaged form reduces to

$$\hat{V} \rightarrow -\sqrt{2}\hbar g \hat{X}_1(\hat{b} + \hat{b}^\dagger), \qquad (11.57)$$

an expression that commutes with $\hat{X}_1$. In that time-averaged sense, it is therefore possible to perform a back action evading measurement of the $\hat{d}_1$ quadrature of mirror motion.

11.3.2   Quadratic Coupling

So far we have considered geometries where the strong classical component of the optical field results in a quantum optomechanical coupling that is linear in the oscillator displacement. Other forms of coupling can however be considered, most interestingly perhaps a coupling quadratic in the displacement. This can be realized in a so-called membrane-in-the-middle geometry [36–39]. As implied by its name, this geometry involves an oscillating mechanical membrane placed inside a Fabry–Pérot with fixed end mirrors as sketched in Fig. 11.8.

An attractive feature of membrane-in-the-middle configurations is the ability to realize relatively easily either linear or quadratic optomechanical couplings, depending on the precise equilibrium position of the membrane. In case it is located

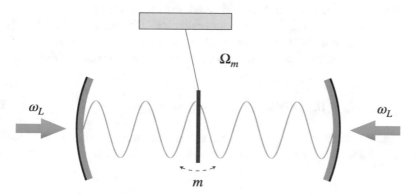

Fig. 11.8 Mirror-in-the-middle configuration

at an extremum of $\omega_c'(x)$, so that $G = -\partial \omega_c'/\partial x = 0$, see Eq. (11.6), we have, to lowest order,

$$\omega_c'(x) \approx \omega_c + \frac{1}{2} \frac{\partial^2 \omega_c}{\partial x^2} \tag{11.58}$$

so that the optomechanical Hamiltonian becomes

$$\hat{H} = \hbar \omega_c \hat{a}^\dagger \hat{a} + \hbar \Omega_M \hat{b}^\dagger \hat{b} + \frac{1}{2} \frac{\partial^2 \omega_c}{\partial x^2} x_{\text{zpt}}^2 (\hat{b} + \hat{b}^\dagger)^2 \hat{a}^\dagger \hat{a}. \tag{11.59}$$

In the rotating wave approximation, this reduces to

$$\hat{H} = \hbar \omega_c \hat{a}^\dagger \hat{a} + \hbar \Omega_M \hat{b}^\dagger \hat{b} + \hbar g_0^{(2)} \left(\hat{b}^\dagger \hat{b} + 1/2 \right) \hat{a}^\dagger \hat{a}, \tag{11.60}$$

where

$$g_0^{(2)} \equiv x_{\text{zpf}}^2 \frac{\partial^2 \omega_c}{\partial x^2}. \tag{11.61}$$

Since the quadratic coupling interaction commutes with $\hat{b}^\dagger \hat{b}$, it opens a number of interesting possibilities, including the direct quantum non-demolition measurement of energy eigenstates of the mechanical element.

11.3.3   Polariton Spectrum

A familiar characteristic of strongly coupled systems is the occurrence of normal mode splitting, which we first encountered in the dressed states discussion of Sects. 1.3 and 3.1. Similarly, the optomechanical Hamiltonian (11.40) can be diagonalized in terms of two bosonic normal modes with annihilation operators $\hat{A}$ and $\hat{B}$. These normal modes are often referred to as *optomechanical polariton modes*.

$$\hat{H}_0 \equiv \hat{H}_{(AB)} = \hbar \omega_A \hat{A}^\dagger \hat{A} + \hbar \omega_B \hat{B}^\dagger \hat{B} + \text{const.} \tag{11.62}$$

The diagonalization proceeds by expressing the polariton annihilation and creation operators $\hat{A}$ and $\hat{B}$ in terms of the bare modes via the Bogoliubov transformation[4]

---

[4]This is the two-mode extension

$$\hat{A}_i = \sum_{j=1,2} \left(u_{ij} \hat{a}_j + v_{ij} \hat{a}_j^\dagger \right)$$

$$\begin{pmatrix} \hat{A} \\ \hat{B} \\ \hat{A}^\dagger \\ \hat{B}^\dagger \end{pmatrix} = \begin{pmatrix} U^\dagger & -V^\dagger \\ -V^T & U^T \end{pmatrix} \begin{pmatrix} \hat{a} \\ \hat{b} \\ \hat{a}^\dagger \\ \hat{b}^\dagger \end{pmatrix}, \tag{11.63}$$

where U and V are 2×2 submatrices that satisfy the relationships

$$U^\dagger U - V^\dagger V = I, \tag{11.64}$$

$$U^T V - V^T U = 0, \tag{11.65}$$

with the inverse transformation

$$\begin{pmatrix} \hat{a} \\ \hat{b} \\ \hat{a}^\dagger \\ \hat{b}^\dagger \end{pmatrix} = \begin{pmatrix} U & V^* \\ V & U^* \end{pmatrix} \begin{pmatrix} \hat{A} \\ \hat{B} \\ \hat{A}^\dagger \\ \hat{B}^\dagger \end{pmatrix}. \tag{11.66}$$

Problem 11.5 carries out this diagonalization and shows that the resulting polariton mode frequencies are [40]

$$\omega_A(\Delta_c) = \frac{1}{\sqrt{2}} \left[\Delta_c^2 + \Omega_m^2 + \sqrt{(\Delta_c^2 - \Omega_m^2)^2 - 16g^2 \Delta_c \Omega_m} \right]^{1/2}, \tag{11.67}$$

$$\omega_B(\Delta_c) = \frac{1}{\sqrt{2}} \left[\Delta_c^2 + \Omega_m^2 - \sqrt{(\Delta_c^2 - \Omega_m^2)^2 - 16g^2 \Delta_c \Omega_m} \right]^{1/2}, \tag{11.68}$$

see Fig. 11.9. At the avoided crossing $\Delta_c = -\Omega_m$, they simplify to

$$\omega_{A,B}(-\Omega_m) = \Omega_m \sqrt{1 \pm \frac{2g}{\Omega_m}}, \tag{11.69}$$

indicating that the minimum frequency difference between the polariton branches "A" and "B" is proportional to g/Ω_m. Importantly, for $\Delta_c/\Omega_m \to -\infty$, we have $\omega_A(-\infty) \to -\Delta_c$ and $\omega_B(-\infty) \to \Omega_m$. In that limit, the branch "A" describes a

---

of the single-mode canonical Bogoliubov transformation $\hat{A} = u\hat{a} + v\hat{a}^\dagger$ that we already encountered in the discussions of the Rabi model in Sect. 3.6 and of quasiparticles in Sect. 10.3.2. We recall that the requirement that the transformation be canonical, $[\hat{A}, \hat{A}^\dagger] = 1$, resulted in the condition $|u|^2 - |v|^2 = 1$ or $u = \exp(i\theta_1)\cosh r$, $v = \exp(i\theta_2)\sinh r$.

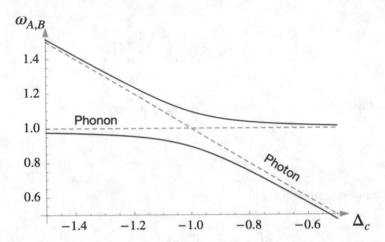

Fig. 11.9 Frequencies ω_A (red) and ω_B (blue) of the two polariton branches (normal modes) of the optomechanical system, in units of Ω_m, for $G/\Omega_m = 0.05$ in the red-detuned case $\Delta_c < 0$. The dashed curves correspond to the frequencies of the bare photon and phonon modes

photon-like excitation and the branch "B" a phonon-like excitation. In contrast, for $-\Omega_m < \Delta_c < 0$, we have

$$\omega_A(\Delta_c) \approx \Omega_m \left(1 + \frac{2g^2 \Delta_c}{(\Delta_c^2 - \Omega_m^2)\Omega_m} \right) ,$$

$$\omega_B(\Delta_c) \approx -\Delta_c \left(1 - \frac{2g^2 \Omega_m}{(\Delta_c^2 - \Omega_m^2)\Delta_c} \right) , \tag{11.70}$$

so that, or $\Delta_c \to 0^{(-)}$, the polariton "A" is phonon-like and "B" is photon-like. This is reminiscent of the situation for atom field coupling, where as we have seen, one of the dressed states becomes atom-like of the other one photon-like for large detunings.

For small dimensionless optomechanical couplings and detunings

$$g \equiv g/\Omega_m \ll 1 \quad ; \quad \delta \equiv \Delta_c/\Omega_m \ll -1, \tag{11.71}$$

the normal mode operators are given in terms of the bare mode operators by

$$\hat{A} = \left[1 + \frac{2\delta g^2}{(\delta - 1)^2} \right] \hat{a} - \frac{g}{1 + \delta} \hat{b} + \frac{g^2}{\delta(1 - \delta^2)} \hat{a}^\dagger + \frac{g}{1 - \delta} \hat{b}^\dagger, \tag{11.72}$$

$$\hat{B} = \frac{g}{1 + \delta} \hat{a} + \left[1 + \frac{2\delta g^2}{(\delta - 1)^2} \right] \hat{b} + \frac{g}{1 - \delta} \hat{a}^\dagger + \frac{g^2 \delta}{\delta^2 - 1} \hat{b}^\dagger, \tag{11.73}$$

and the polariton number operators $\hat{n}_A \equiv \hat{A}^\dagger \hat{A}$ and $\hat{n}_B \equiv \hat{B}^\dagger \hat{B}$ become

$$\hat{n}_A = \left[1 + \frac{4\delta g^2}{(\delta - 1)^2}\right]\hat{a}^\dagger \hat{a} + \frac{2(1 + \delta^2)g^2}{(\delta - 1)^2}\hat{b}^\dagger \hat{b} + \left(\frac{g}{1 - \delta}\right)^2$$

$$- \frac{g}{1 + \delta}(\hat{a}^\dagger \hat{b} + \hat{b}^\dagger \hat{a}) + \frac{g^2}{\delta(1 - \delta^2)}(\hat{a}^2 + \hat{a}^{\dagger 2})$$

$$+ \frac{g}{1 - \delta}(\hat{a}\hat{b} + \hat{b}^\dagger \hat{a}^\dagger) - \frac{g^2}{1 - \delta^2}(\hat{b}^2 + \hat{b}^{\dagger 2}), \tag{11.74a}$$

$$\hat{n}_B = \left[1 + \frac{4\delta g^2}{(\delta - 1)^2}\right]\hat{b}^\dagger \hat{b} + \frac{2(1 + \delta^2)g^2}{(\delta - 1)^2}\hat{a}^\dagger \hat{a} + \left(\frac{g}{1 - \delta}\right)^2$$

$$+ \frac{g}{1 + \delta}(\hat{a}^\dagger \hat{b} + \hat{b}^\dagger \hat{a}) + \frac{g^2}{1 - \delta^2}(\hat{a}^2 + \hat{a}^{\dagger 2})$$

$$+ \frac{g}{1 - \delta}(\hat{a}\hat{b} + \hat{b}^\dagger \hat{a}^\dagger) + \frac{g^2\delta}{\delta^2 - 1}(\hat{b}^2 + \hat{b}^{\dagger 2}). \tag{11.74b}$$

To the lowest order, one can neglect the intermode correlations and squeezing terms appearing in these expressions. This approximates the mean polariton numbers by the mean thermal occupation $\bar{n}_a$ of the optical reservoir and $\bar{n}_b$ of the mechanical reservoir, respectively. When the optomechanical coupling is small but finite, though, all terms in Eqs. (11.74a) and (11.74b) contribute, and the steady-state polariton populations deviate from thermal equilibrium. For $-1 < \delta < 0$, the expressions for $\hat{A}$ and $\hat{B}$ are simply interchanged.

11.4 Standard Quantum Limit of Optomechanical Detection

The next chapter will show examples of optomechanical measurements of extraordinarily feeble displacements and forces and discuss situations where detectors of even greater sensitivity would be highly desirable. But efforts to increase the precision of measurements are constrained by the Heisenberg uncertainty principle, as we have seen in Chap. 6. This section revisits this issue in the context of optomechanical detection and shows that when light is used as the probe of mechanical motion, the standard quantum limit arises from the balance between the uncertainties in photon number, the shot noise, and the back action of the radiation pressure applied to the object.

The most remarkable optomechanical detectors to date are without a doubt the LIGO and VIRGO gravitational wave antennas. They work by optically measuring changes in the positions of two suspended masses that serve as the end mirrors of kilometers-long Michelson interferometers, see Fig. 11.1. Gravitational waves, resulting, for example, from the merger of two black holes, cause minute changes in the curvature of space-time and hence in the lengths of the interferometer arms. Because the changes caused by such collisions are so small, the sensitivity of these antennas must be truly extraordinary. At its most sensitive state, the Advanced LIGO system will be able to detect a change in distance between its mirrors 1/10,000th the width of a proton, equivalent to measuring the distance to the nearest star, some 4.2 light-years away, to an accuracy smaller than the width of a human hair [41]. It should therefore come as no great surprise that the development of these systems has played, and continues to play, a central role in developing a more profound understanding of the role of quantum noise and quantum back action in optomechanical detectors. For this reason, we focus in the following on the specific case of a LIGO-type Michelson interferometer antenna.

Intuitively, one might argue that since the photon statistics of single mode lasers are Poissonian, their intensity fluctuations, or *shot noise*, scale as the square root of the intensity. One might therefore contend that interferometer noise can be reduced simply by increasing that intensity. This, however, is not the case, as discussed by C. Caves and colleagues in a series of classic papers [15, 16]. Optomechanical interferometers are subject to two fundamental sources of quantum noise: the shot noise that we just mentioned, which is a sensing noise, and *radiation pressure noise*, which is a back action noise. It is their combined effect that results in the standard quantum limit of optomechanical interferometers.[5]

Michelson interferometers are characterized by two input modes, labeled a_1 and a_2 in Fig. 11.10. A laser field a_1 of wavelength λ is injected through the first port and distributed to the two interferometer arms by a beam splitter. The mode a_2 is the field entering the interferometer through the detection port. Typically, it is in a vacuum state. The output signal is obtained by a photodetector measuring the power in the exit mode c_2.

We saw in the discussion of homodyne detection that the fields b_1 and b_2 that enter the interferometer arms after the beam splitter are related to the input modes a_1 and a_2 by the input–output matrix relation (2.174)

$$U(\theta, \phi) = \begin{pmatrix} \cos(\theta/2) & i\sin(\theta/2) \\ i\sin(\theta/2) & \cos(\theta/2) \end{pmatrix}, \tag{11.75}$$

where we have set $\phi = 0$. To eliminate unimportant phase factors, we assume that the beam splitter is complemented by two phase shifters acting on the mode a_1, thereby imposing on the two fields exiting the beam splitter opposite phase shifts

---

[5]We ignore here all technical noise, which can at least in principle be reduced to an arbitrarily low level.

Fig. 11.10 Schematic of a
Michelson interferometer,
with the various fields
considered in the text labeled.
Typically (but not always),
the input field $\hat{a}_1$ is a laser
field, and $\hat{a}_2$ is a vacuum field

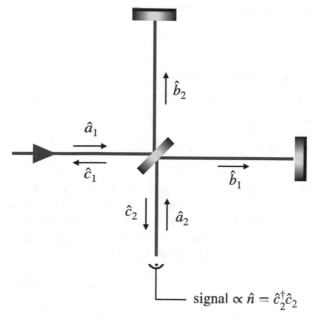

signal $\propto \hat{n} = \hat{c}_2^\dagger \hat{c}_2$

$\pm\varphi$, that is,

$$U(\theta, \phi) \rightarrow \begin{pmatrix} e^{i\varphi} & 0 \\ 0 & 1 \end{pmatrix} \begin{pmatrix} \cos(\theta/2) & i\sin(\theta/2) \\ i\sin(\theta/2) & \cos(\theta/2) \end{pmatrix} \begin{pmatrix} e^{-i\varphi} & 0 \\ 0 & 1 \end{pmatrix}$$

$$= \begin{pmatrix} \cos(\theta/2) & ie^{i\varphi}\sin(\theta/2) \\ ie^{-i\varphi}\sin(\theta/2) & \cos(\theta/2) \end{pmatrix}. \tag{11.76}$$

For a 50/50 beam splitter, we have $\theta = \pi/2$, and with $\phi = -\pi/2$, this gives finally

$$U = \frac{1}{\sqrt{2}} \begin{pmatrix} 1 & -1 \\ 1 & 1 \end{pmatrix}. \tag{11.77}$$

The annihilation operators of the interferometer modes are then related to those of
the input modes by the unitary transformation

$$\begin{pmatrix} \hat{b}_1 \\ \hat{b}_2 \end{pmatrix} = \frac{1}{\sqrt{2}} \begin{pmatrix} 1 & -1 \\ 1 & 1 \end{pmatrix} \begin{pmatrix} \hat{a}_1 \\ \hat{a}_2 \end{pmatrix} = \frac{1}{\sqrt{2}} \begin{pmatrix} \hat{a}_1 - \hat{a}_2 \\ \hat{a}_1 + \hat{a}_2 \end{pmatrix}. \tag{11.78}$$

Similarly, those of the output modes are given by

$$\begin{pmatrix} \hat{c}_1 \\ \hat{c}_2 \end{pmatrix} = \frac{1}{\sqrt{2}} \begin{pmatrix} e^{i\zeta/2}\hat{b}_1 - e^{-i\zeta/2}\hat{b}_2 \\ e^{i\zeta/2}\hat{b}_1 + e^{-i\zeta/2}\hat{b}_2 \end{pmatrix} = \begin{pmatrix} i\sin(\zeta/2)\hat{a}_1 - \cos(\zeta/2)\hat{a}_2 \\ \cos(\zeta/2)\hat{a}_1 - i\sin(\zeta/2)\hat{a}_2 \end{pmatrix}, \tag{11.79}$$

where the phase

$$\zeta = 4\pi \Delta z/\lambda \tag{11.80}$$

accounts for the difference in length Δz between the two arms of the interferometer.

Shot Noise The power measured at the output of the interferometer is proportional to

$$\hat{n} = \hat{c}_2^\dagger \hat{c}_2 = \cos^2(\zeta/2)\hat{a}_1^\dagger \hat{a}_1 + \sin^2(\zeta/2)\hat{a}_2^\dagger \hat{a}_2 + \frac{i}{2}\sin(\zeta)\left[a_2^\dagger \hat{a}_1 - \hat{a}_1^\dagger \hat{a}_2\right]. \tag{11.81}$$

We assume that the field injected in mode a_1 is a strong coherent field $|\alpha\rangle$, with $\hat{a}_1|\alpha\rangle = \alpha|\alpha\rangle$ and mean photon number α^2, with α taken real for convenience and without loss of generality. In contrast, the field injected in mode a_2 is very weak, a reasonable assumption since it is in most (but not all) cases the vacuum field. Equation (11.81) can then be approximated by

$$\hat{n} \approx \cos^2(\zeta/2)\hat{a}_1^\dagger \hat{a}_1 + \alpha \sin(\zeta)\hat{d}_2(a_2), \tag{11.82}$$

where α is the amplitude of the coherent state and

$$\hat{d}_2(a_2) = \frac{1}{2i}\left(\hat{a}_2 - \hat{a}_2^\dagger\right) \tag{11.83}$$

is the quadrature of the input mode a_2, see Eq. (2.122). The average photon number at the detector is therefore

$$\langle \hat{n} \rangle = \alpha^2 \cos^2(\zeta/2), \tag{11.84}$$

and its variance is

$$\sigma_n^2 = \langle \hat{n}^2 \rangle - \langle \hat{n} \rangle^2 = \alpha^2 \left[\cos^4(\zeta/2) + \sin^2(\zeta/2)\sigma_{d_2}^2(a_2)\right], \tag{11.85}$$

where $\sigma_{d_2}^2(a_2)$ is the variance of the quadrature $\hat{d}_2(a_2)$, and we have accounted for the fact that the photon number variance of a coherent state $|\alpha\rangle$ is $|\alpha|^2$. This shows that the shot noise of the interferometer is not due only to the shot noise of the input laser field. Rather, it also comprises a component proportional to the variance of the quadrature d_2 of the weak-field injected in its second port, typically the vacuum field.

The measurement of small relative phase changes $\delta\zeta$, and hence of small differences Δz in the lengths of the interferometer arms, can be inferred from the detected mean photon number $\langle \hat{n} \rangle$ since from Eq. (11.84), we have

$$\frac{d\langle \hat{n} \rangle}{d\zeta} = -\tfrac{1}{2}\alpha^2 \sin(\zeta/2)\cos(\zeta/2) = -\tfrac{1}{2}\alpha^2 \sin(\zeta), \tag{11.86}$$

or, with $\zeta = (4\pi/\lambda)\Delta z$,

$$\frac{d\langle \hat{n} \rangle}{d(\Delta z)} = -\frac{2\pi}{\lambda} \alpha^2 \sin(\zeta). \tag{11.87}$$

With Eq. (11.85), this gives for the variance $\sigma_{sn}^2(\Delta z)$ of the random fluctuations of Δz resulting from the shot noise in the input fields

$$\sigma_{sn}^2(\Delta z) = \frac{\lambda^2}{4\pi^2 \alpha^2 \sin^2(\zeta)} \left[\cos^4(\zeta/2) + \sin^2(\zeta/2)\sigma_{d_2}^2(a_2) \right]$$

$$= \frac{\lambda^2}{16\pi^2 \alpha^2} \left[\cot^2(\zeta/2) + 4\sigma_{d_2}^2(a_2) \right]. \tag{11.88}$$

The first term in this expression can be eliminated by operating the interferometer near a dark fringe, in which case $\cot^2(\zeta/2) = 0$. The variance in signal resulting from the shot noise reduces then to

$$\sigma_{sn}^2(\Delta z) = \frac{\lambda^2}{4\pi^2 \alpha^2} \sigma_{d_2}^2(a_2). \tag{11.89}$$

Remarkably, it is entirely due to the fluctuations in the quadrature $\hat{d}_2$ of the (typically vacuum) input mode a_2. Because it scales as $1/\alpha^2$, the shot noise decreases with increased laser intensity as expected. However, we still have to account for the second unavoidable source of quantum noise, radiation pressure noise, and as we now show, it becomes increasingly dominant as α^2 increases.

Radiation Pressure Noise In optomechanical interferometers such as the LIGO antennas, the end mirrors of the two interferometer arms are suspended and can therefore move under the influence of radiation pressure from the intracavity field. Differential changes in the lengths of the two arms result in changes in the corresponding intracavity field intensities and hence in a differential radiation pressure force between the two mirrors proportional to the difference in photon numbers of these two fields, $\hat{b}_2^\dagger \hat{b}_2 - \hat{b}_1^\dagger \hat{b}_1$. This back action force imparts a change in the relative momenta of the mirrors

$$\Delta p = 2\hbar k (\hat{b}_2^\dagger \hat{b}_2 - \hat{b}_1^\dagger \hat{b}_1), \tag{11.90}$$

and an associated additional noise source. A derivation that parallels the analysis of balanced homodyne detection of Sect. 2.4.2 and is the topic of Problem 11.8 shows that the variance in momentum resulting from radiation pressure fluctuations is

$$\sigma_{rp}^2(\Delta p) = (4\hbar k)^2 \alpha^2 \sigma_{d_1}^2(a_2), \tag{11.91}$$

where $\sigma^2_{d_1}(a_2)$ is the variance of the quadrature

$$\hat{d}_1 = \frac{1}{2}(\hat{a}_2 + \hat{a}_2^\dagger) \tag{11.92}$$

of the input mode a_2. For a measurement time τ, the corresponding variance in Δz will be

$$\sigma^2_{rp}(\Delta z) = \left(\frac{\tau}{2m}\right)^2 \sigma^2_{rp}(\Delta p) = \left(\frac{4\pi\hbar\tau}{m\lambda}\right)^2 \alpha^2 \sigma^2_{d_1}(a_2). \tag{11.93}$$

Standard Quantum Limit The total variance of z is the sum of its shot noise and radiation pressure noise contributions,

$$\sigma^2(\Delta z) = \sigma^2_{sn}(\Delta z) + \sigma^2_{rp}(\Delta z)$$

$$= \frac{1}{\alpha^2}\left(\frac{\lambda}{2\pi}\right)^2 \sigma^2_{d_2}(a_2) + \alpha^2\left(\frac{4\pi\hbar\tau}{m\lambda}\right)^2 \sigma^2_{d_1}(a_2). \tag{11.94}$$

Importantly, in contrast to the shot noise contribution, which scales as $1/\alpha^2$, the radiation pressure contribution scales linearly with α^2, as illustrated in Fig. 11.11. As already indicated, then, the intuitively appealing idea that increasing the laser

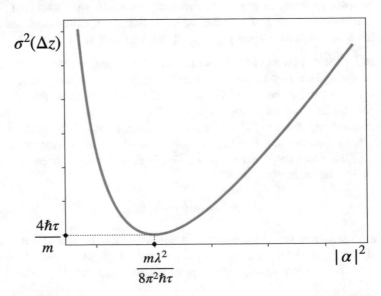

Fig. 11.11 Schematic of the dependence of the variance in the differential lengths of the optomechanical interferometer arms $\sigma^2(\Delta z)$ on the dimensionless input laser power α^2, illustrating the transition from shot noise dominated to radiation pressure dominated fluctuations as the intensity is increased. Arbitrary units

power will reduce the noise of the interferometer is therefore not correct, as it fails to account for the effects of radiation pressure.

From the Heisenberg uncertainty relation (2.126) for the variances of the single-mode field quadratures, we have

$$\sigma_{d_1}(a_2)\,\sigma_{d_2}(a_2) \geq \frac{1}{4}, \tag{11.95}$$

with $\sigma_{d_1}(a_2) = \sigma_{d_2}(a_2) = \frac{1}{2}$ for coherent states, and in particular for the input field mode a_2 is in a vacuum. In that case, Eq. (11.94) reduces to

$$\sigma^2(\Delta z) = \frac{1}{\alpha^2}\left(\frac{\lambda}{2\pi}\right)^2 + \alpha^2\left(\frac{4\pi\hbar\tau}{m\lambda}\right)^2. \tag{11.96}$$

Because the shot noise and radiation pressure noise contributions scale as $1/\alpha^2$ and α^2, respectively, the variance $\sigma^2(\Delta z)$ is minimized for the optimum mean photon number of the input laser

$$\alpha^2 = \frac{m\lambda^2}{8\pi^2\hbar\tau}, \tag{11.97}$$

in which case shot noise and radiation pressure noise contributions are equal and

$$\sigma^2(\Delta z) = 4\hbar\tau/m. \tag{11.98}$$

This is the standard quantum limit of optomechanical interferometers. But as already pointed out in Sect. 6.2, it is important to remember that the standard quantum limit is not a fundamental measurement precision limit and that it is sometimes possible to circumvent it. In the particular example considered here, this can be achieved by injecting a squeezed state with an appropriately chosen squeezed variance in the a_2 input mode of the interferometer, so as to reduce either the shot noise or the radiation pressure noise. The impact of the remaining variance can then be further reduced with an appropriate choice of laser power α^2. This technique has now been implemented in the advanced LIGO gravitational wave antenna.

11.5 Ultracold Atoms

We conclude this chapter by drawing a few additional connections and parallels between mesoscopic and macroscopic systems and atomic ensembles made to behave much like "moving atomic mirrors." For this brief overview, we limit ourselves to neutral atomic samples cooled well below their recoil temperature and trapped inside a single-mode Fabry–Pérot resonator. These could be, for example, a nearly homogeneous and collisionless Bose–Einstein condensate at $T \approx 0$ or an atomic sample cooled near the vibrational ground state of an optical trap. Side-mode excitations of the condensate in the first case, and the vibrational motion of thermal atoms in the second case, provide formal analogs of one or several moving mirrors.

Bose–Einstein Condensate   To see how this works, we consider a generic model consisting of a collisionless Bose condensate at $T = 0$ trapped inside a Fabry–Pérot cavity of length L and mode frequency ω_c. The atoms of mass m are driven by a pump laser of frequency ω_L and wave number k. When ω_L is far detuned from the atomic transition frequency ω_0, the excited electronic state of the atoms can be adiabatically eliminated, as we have seen, and the atoms interact dispersively with the cavity field, see e.g. Eq. (3.13). In a frame rotating at the laser frequency ω, the system is therefore described by the Hamiltonian

$$\hat{H} = \hat{H}_{\text{bec}} + \hat{H}_{\text{field}} + \hat{H}_\kappa , \tag{11.99}$$

where

$$\hat{H}_{\text{field}} = i\hbar\eta(\hat{a}^\dagger - \hat{a}) - \hbar\Delta_c\hat{a}^\dagger\hat{a} \tag{11.100}$$

describes the driven cavity field mode,

$$\hat{H}_{\text{bec}} = \int dx\,\hat{\Psi}^\dagger(x)\left[\frac{\hat{p}_x^2}{2m} + \frac{g^2}{\Delta}\cos^2(k\hat{x})\hat{a}^\dagger\hat{a}\right]\hat{\Psi}(x), \tag{11.101}$$

which accounts for the (one-dimensional) condensate and its off-resonant interaction with the intracavity field, and $\hat{H}_\kappa$ depicts the coupling of the resonator field to the outside world, leading to its dissipation at rate $\kappa/2$. Here, $\hat{\Psi}(x)$ is the bosonic Schrödinger field operator of the atoms, g is the vacuum Rabi frequency, and $\Delta = \omega_0 - \omega$ is the atom–field detuning.

When the light field can be approximated by a plane wave, the atomic field operator can likewise be expanded in terms of plane waves as

$$\hat{\Psi}(x) = \frac{1}{\sqrt{L}}\sum_q \hat{b}_k e^{iqx} , \tag{11.102}$$

where $\hat{b}_q$ and $\hat{b}_q^\dagger$ are annihilation and creation operators for atomic bosons with momentum k, satisfying the bosonic commutation relations $[\hat{b}_q, \hat{b}_{q'}^\dagger] = \delta_{q,q'}$ and $[\hat{b}_q, \hat{b}_{q'}] = 0$.

In the absence of light and at $T = 0$, the ground state of the atomic sample is a condensate of N atoms with zero momentum in the state

$$|\Psi_0\rangle = (\hat{b}_0^\dagger)^N|0\rangle . \tag{11.103}$$

However, as a result of virtual transitions induced by the intracavity field, the atoms can acquire a recoil momentum $\pm 2\ell\hbar k$, where $\ell = 0, 1, 2, \ldots$, as we have seen in Chap. 8 in the context of atomic diffraction. In the limit of low photon numbers, it is sufficient to consider the lowest diffraction order, $\ell = 1$. The atomic field operator can therefore be expressed in terms of a zero-momentum and a "sine mode" component as

$$\hat{\Psi} \sim \hat{b}_0 \phi_0(x) + \hat{b}_2 \phi_2(x), \tag{11.104}$$

where $\phi_0(x) = 1$ is the condensate wave function and $\phi_2(x) = \sqrt{2}\cos(2kx)$, with the $\sqrt{2}$ factor required for normalization of the mode function.

For very weak optical fields, the occupation of the sine mode remains much smaller than the zero-momentum mode, so that $\hat{b}_0 \simeq \sqrt{N}$ and $\langle \hat{b}_2^\dagger \hat{b}_2 \rangle \ll N$. Substituting then Eq. (11.104) into the Hamiltonian (11.101) and in a frame rotating at the pump laser frequency, the Hamiltonian $\hat{H}_{\text{om, bec}}$ becomes [42]

$$\hat{H}_{\text{om,bec}} = 4\hbar\omega_{\text{rec}}\hat{b}_2^\dagger \hat{b}_2 + \hbar\hat{a}^\dagger\hat{a}\left[\Delta' + g_2(\hat{b}_2 + \hat{b}_2^\dagger)\right], \tag{11.105}$$

where

$$g_2 = \frac{g^2\sqrt{N/2}}{\Delta_s}, \tag{11.106}$$

$\omega_{\text{rec}} = \hbar k^2/2m$ is the recoil frequency, and $\Delta' = \Delta + g^2 N/2\Delta$ is an effective Stark-shifted detuning.

The reduced Hamiltonian (11.105) describes the coupling of two oscillators, the cavity mode $\hat{a}$ and the momentum side mode $\hat{b}_2$, via the optomechanical coupling $g_2\hat{a}^\dagger\hat{a}(\hat{b}_2 + \hat{b}_2^\dagger)$. In this sense, the momentum side mode of the condensate behaves formally just like a moving mirror driven by the radiation pressure of the intracavity field, in complete analogy with the situation described by Eq. (11.30).

F. Brenneke et al. [42] studied the dynamics of a Bose condensate of ^{87}Rb atoms trapped inside a high-finesse Fabry–Pérot and driven by a feeble optical field. This experiment demonstrated the optomechanical coupling of a collective density excitation of the condensate, showing that it behaves precisely as a mechanical oscillator coupled to the cavity field, in quantitative agreement with a cavity optomechanical model of Eq. (11.105).

Ultracold Atoms A similar analogy can be established when considering a sample of ultracold two-level atoms with transition frequency ω_0 tightly confined in a harmonic trap of frequency ω_z centered at the location z_0 along the axis of an optical resonator and driven far off resonance by field at frequency ω. The vacuum Rabi frequency of the intracavity field at the location $z_i = z_0 + \delta z_i$ of atom i is

$$g(z_i) = g_0 \sin(kz_0 + 2k\delta z_i), \tag{11.107}$$

so that far off resonance, the atom–field Hamiltonian is

$$\hat{H} = \hbar\omega\hat{a}^\dagger\hat{a} + \hbar\omega_z\hat{b}_i^\dagger\hat{b}_i - \hbar\sum_i \frac{|g(z_i)|^2}{\Delta}\hat{a}^\dagger\hat{a}, \tag{11.108}$$

where $\hbar\omega_z\hat{b}_i^\dagger\hat{b}_i$ accounts for the harmonic center-of-mass motion of that atom and $\Delta = \omega_0 - \omega$.

Summing over all atoms in the sample and expanding the far off-resonant atom–field interaction to the lowest order in $k\delta z_i$, one finds, see Problem 11.10,

$$\hat{H} \approx \hbar\left(\omega - \frac{Ng_0^2}{\Delta}\sin^2 kz_0\right)\hat{a}^\dagger\hat{a} + \hbar\omega_z\sum_i\hat{b}_i^\dagger\hat{b}_i - \frac{\hbar g_0^2}{\Delta}\sin(2kz_0)\hat{a}^\dagger\hat{a}\left[\sum_i k\,\delta z_i\right], \tag{11.109}$$

where N is the number of atoms and the operator $\hat{b}_i$ describes the annihilation of a phonon from the center-of-mass motion of atom i.

The last term of this Hamiltonian describes the optomechanical coupling of the intracavity field to the collective atomic variable

$$\hat{Z}_{cm} = \frac{1}{N}\sum_i\hat{\delta z}_i, \tag{11.110}$$

which is nothing but the center-of-mass position of the sample, that is, its normal mode. For small displacements, that mode can be described as a harmonic oscillator of frequency ω_z and mass Nm. In this picture, the atom–field system is therefore modeled by the optomechanical Hamiltonian

$$\hat{H}_{om, at} = \hbar\omega_c'\hat{a}^\dagger\hat{a} + \hbar\omega_z\hat{b}_{cm}^\dagger\hat{b}_{cm} - \hbar g_N(\hat{b}_{cm} + \hat{b}_{cm}^\dagger)\hat{a}^\dagger\hat{a}, \tag{11.111}$$

where $\omega' = \omega - (Ng_0^2/\Delta)\sin^2 kz_0$, $\hat{b}_{cm}$ and $\hat{b}_{cm}^\dagger$ are bosonic annihilation and creation operators for the center-of-mass mode of motion of the atomic ensemble, $z_{zpf} = \sqrt{\hbar/2Nm\omega_z}$, and

$$g_N = Nz_{zpf}\frac{\hbar kg_0^2}{\Delta}\sin(2kz_0). \tag{11.112}$$

In a trailblazing experiment [43], T. Purdy and colleagues positioned a sample of cold atoms with sub-wavelength accuracy at various positions in a Fabry–Pérot cavity and succeeded in particular in demonstrating the tuning from linear to quadratic optomechanical coupling regime. A comprehensive review of cavity optomechanics with cold atoms can be found in Ref. [44].

11.6 Functionalization and Hybrid Systems

The rapid progress witnessed by quantum optomechanics makes it increasingly realistic to consider the use of mechanical systems operating in the quantum regime to make precise and accurate measurements of feeble forces and fields. In many cases, these measurements amount to the detection of exceedingly small displacements, and in that context, the remarkable potential for functionalization of optomechanical devices is particularly attractive. Their motional degree(s) of freedom can be coupled to a broad range of other physical systems, including photons via radiation pressure from a reflecting surface, spin(s) via coupling to a magnetic material, electric charges via the interaction with a conducting surface, etc. In that way, the mechanical element can serve as a universal transducer or intermediary that enables the coupling between otherwise incompatible systems. The potential of these hybrid systems for functionalization also suggests that quantum optomechanical systems have the potential to play an important role in classical and quantum information processing, where transduction between different information-carrying physical systems is crucial.

Much potential for the functionalization of optomechanical devices is offered by interfacing them with a single quantum object. This could be, for example, an atom, a molecule, or a Bose–Einstein condensate [45]. For example, S. Camerer and coworkers [46] realized a hybrid optomechanical system by coupling ultracold atoms trapped in an optical lattice to a micromechanical membrane, the coupling being mediated by the light field, and observed both the effect of the membrane motion on the atoms and the back action of the atomic motion on the membrane.

Alternatively, one can also couple optomechanical systems to artificial atoms, for example, to superconducting qubits [20, 47, 48] or to another type of artificial atoms called nitrogen vacancy centers or NV centers. NV centers consist of a nearest neighbor pair of a nitrogen atom, which substitutes for a carbon atom and a lattice vacancy in a diamond crystal. Their ground state is a spin triplet that can be optically initialized, manipulated, and read out by a combination of optical and microwave fields, and they are characterized by remarkably long room temperature coherence times for solid-state systems. Due to their attractive combination of optical and electronic spin properties, they are of particular interest for hybrid optomechanical systems [49] and offer much promise for applications in quantum information processing and in ultrasensitive magnetometry, where the spin is used as an atomic-sized magnetic sensor, see e.g. Refs. [50, 51].

Even more remarkably perhaps, micromechanical oscillators in the quantum regime also offer a possible route toward new tests of the quantum mechanics and gravitation at unprecedented size and mass scales, as well as to the exploration of the "dark sector" and its yet unidentified constituents of Dark Matter and Dark Energy, topics to which we turn in the final chapter.

Problems

Problem 11.1 Show that the field equation of motion

$$\frac{d\alpha(t)}{dt} = [i\left(\Delta_c + Gx(t)\right) - \kappa/2]\,\alpha(t) + \sqrt{\kappa}\alpha_{in}\,,$$

with $x(t) = x_0 \sin(\Omega_m t)$, has the approximate solution $\alpha(t) \simeq \alpha_0(t) + \alpha_1(t)$, with $\alpha_0(t)$ and $\alpha_1(t)$ given by Eq. (11.14).

Problem 11.2 Determine the conditions on $|\alpha_{in}|^2$, Δ_c, and G under which the potential

$$V(x) = \frac{1}{2}m\Omega_m^2 x^2 - \frac{\hbar\kappa|\alpha_{in}|^2}{2}\arctan\left[2(\Delta_c + Gx)/\kappa\right]$$

results in optical bistability in the steady-state mirror displacement x. Plot the potential, as well as the steady-state resonator length x as a function of $|\alpha_{in}|^2$ for parameters that satisfy this condition.

Problem 11.3 Consider an optomechanical resonator where the cavity field is coupled with damping rate $\kappa/2$ to a reservoir at temperature $T = 0$. Decomposing the intracavity field operator as $\hat{a} = \alpha + \hat{c}$, show that to the lowest order in the creation and annihilation operators $\hat{c}^\dagger$ and $\hat{c}$ the associated noise power spectrum is

$$S_{nn}(\omega) = n\frac{\kappa}{(\omega + \Delta_c)^2 + (\kappa/2)^2}\,,$$

where $n = |\alpha|^2$.

Problem 11.4 Carry out the steps that lead to the polariton mode frequencies $\omega_A(\Delta_c)$ and $\omega_B(\Delta_c)$ of Eq. (11.68).

Problem 11.5 Determine the polariton modes and polariton spectrum for the simple optomechanical Hamiltonian (11.40), $\hbar\Delta_c\hat{a}^\dagger\hat{a} + \hbar\Omega_m\hat{b}^\dagger\hat{b} - \hbar g_0\hat{a}^\dagger\hat{a}(\hat{b}+\hat{b}^\dagger)$.

Problem 11.6 Find the evolution of the operators $\hat{a}(t)$ and $\hat{b}(t)$ for the beam splitter Hamiltonian $\hat{V} = -\hbar g(\hat{a}\hat{b}^\dagger + \text{h.c.})$ *Hint: The Baker–Hausdorff relation may prove useful.*

Problem 11.7 Show that under the action of the two-mode squeezing operator

$$\hat{S}_{ab}(t) = e^{\left[g^*\hat{a}\hat{b} - g\hat{a}^\dagger\hat{b}^\dagger\right]t}$$

the variance of the generalized two-mode quadrature operator $\hat{X} = \frac{1}{2^{3/2}}(\hat{a} + \hat{a}^\dagger + \hat{b} + \hat{b}^\dagger)$ of a system initially in a two-mode vacuum state becomes

$$\sigma_X^2 = \frac{1}{4}\left[\exp(-2|g|t)\cos^2(\phi/2) + \exp(2|g|t)\sin^2(\phi/2)\right].$$

Determine also the variance of a system initially when the two modes are initially in coherent states. Hint: Use the fact that the coherent state is a displaced vacuum.

Problem 11.8 Show that in a LIGO-type two-arm interferometer, the variance in momentum resulting from radiation pressure fluctuations is given by Eq. (11.91),

$$\sigma_{rp}^2(\Delta p) = (4\hbar k)^2 \alpha^2 \sigma_{d_1}^2(a_2),$$

where $\sigma_{d_1}^2(a_2)$ is the variance of the quadrature $\hat{d}_1 = (1/2)(\hat{a}_2 + \hat{a}_2^\dagger)$ of the input mode a_2. Show also that for a measurement time τ, the corresponding variance in Δz will be

$$\sigma_{rp}^2(\Delta z) = \left(\frac{4\pi\hbar\tau}{m\lambda}\right)^2 \alpha^2 \sigma_{d_1}^2(a_2).$$

Problem 11.9 Show that in the limit where the Schrödinger field $\hat{\Psi}(x) \approx \phi_0(x)\hat{b}_0 + \sqrt{2}\cos(2kx)\hat{b}_2(x)$, the atom–field Hamiltonian

$$\hat{H} = \hbar\omega\hat{a}^\dagger\hat{a} + \int dx\, \hat{\Psi}^\dagger(x)\left[\frac{\hat{p}_x^2}{2m} - \frac{g^2}{\Delta}\cos^2(k\hat{x})\hat{a}^\dagger\hat{a}\right]\hat{\Psi}(x)$$

reduces to the form

$$\hat{H}_{om,bec} = 4\hbar\omega_{rec}\hat{b}_2^\dagger\hat{b}_2 + \hbar\hat{a}^\dagger\hat{a}\left[\Delta' + g_2(\hat{b}_2 + \hat{b}_2^\dagger)\right]$$

of Eq. (11.105), with $g_2 = \frac{g^2\sqrt{N/2}}{\Delta_s}$, $\omega_{rec} = \hbar k^2/2m$, and $\Delta' = \Delta + g^2 N/2\Delta$.

Problem 11.10 Carry out the steps that lead to the Hamiltonian (11.111)

$$\hat{H}_{om,\,at} = \hbar\omega_c'\hat{a}^\dagger\hat{a} + \hbar\omega_z\hat{b}_{cm}^\dagger\hat{b}_{cm} - \hbar g_N(\hat{b}_{cm} + \hat{b}_{cm}^\dagger)\hat{a}^\dagger\hat{a},$$

which describes a sample of ultracold atoms tightly confined in a harmonic trap of frequency ω_z centered at some location z_0 along the axis of an optical resonator and driven far off resonance by a field at frequency ω, in the limit where the positions of the individual atoms are $z_i = z_0 + \delta z_i$, with δz_i much smaller than an optical wavelength.

References

1. V. Braginsky, Y.I. Voronstsov, K.S. Thorne, Quantum nondemolition measurements. Science **209**, 547 (1980)
2. V. Braginski, F.Ya. Khalili, *Quantum Measurement* (Cambridge University, Cambridge, 1992), ed. by K. Thorne
3. P. Meystre, M.O. Scully, *Quantum Optics, Experimental Gravitation, and Measurement Theory* (Plenum Press, New York, 1983)
4. V.B. Braginski, A.B. Manukin, Ponderomotive effects of electromagnetic radiation. JETP **25**, 653 (1967)
5. A. Schliesser, *Cavity Optomechanics and Optical Frequency Comb Generation with Silica Whispering Gallery Mode Generators*, PhD thesis (Ludwig Maximillians University Munich, Munich, 2005)
6. A. Dorsel, J.D. McCullen, P. Meystre, E. Vignes, H. Walther, Optical bistability and mirror confinement induced by radiation pressure. Phys. Rev. Lett. **51**, 1550 (1983)
7. D.G. Blair, E.N. Ivanov, M.E. Tobar, P.J. Turner, F. van Kann, I.S. Heng, High sensitivity gravitational wave antenna with parametric transducer readout. Phys. Rev. Lett. **74**, 1908 (1995)
8. P.F. Cohadon, A. Heidmann, M. Pinard, Cooling of a mirror by radiation pressure. Phys. Rev. Lett. **83**, 3174 (1999)
9. S. Gigan, H.R. Böhm, M. Paternostro, F. Blaser, G. Langer, J.B. Hertzberg, K.C. Schwab, D. Bäuerle, M. Aspelmeyer, A. Zeilinger, Self-cooling of a micromirror by radiation pressure. Nature **444**, 67 (2006)
10. O. Arcizet, P.F. Cohadon, T. Briant, M. Pinard, A. Heidmann, Radiation-pressure cooling and optomechanical instability of a micromirror. Nature **444**, 71 (2006)
11. D. Kleckner, D. Bouwmeester, Sub-kelvin optical cooling of a micromechanical resonator. Nature **444**, 75 (2006)
12. A. Schliesser, P. Del'Haye, N. Nooshi, K.J. Valhala, T.J. Kippenberg, Radiation pressure cooling of a micromechanical oscillator using dynamical backaction. Phys. Rev. Lett. **97**, 243905 (2006)
13. T. Corbitt, C. Wipf, T. Bodiya, D. Ottaway, D. Sigg, N. Smith, S. Whitcomb, N. Mavalvala, Optical dilution and feedback cooling of a gram-scale mirror to 6.9 mK. Phys. Rev. Lett. **99**, 160801 (2007)
14. C. Whittle et al., Approaching the motional ground state of a 10 kg object. Science **372**, 1333 (2021)
15. C.M. Caves, Quantum-mechanical radiation-pressure fluctuations in an interferometer. Phys. Rev. Lett **45**, 75 (1980)
16. C. M. Caves, K. S. Thorne, R. W. P. Drever, V. D. Sandberg, M. Zimmerman, On the measurement of a weak classical force coupled to a quantum mechanical oscillator, Rev. Mod. Phys. **52**, 341 (1980).
17. C.K. Law, Interaction between a moving mirror and radiation pressure: A Hamiltonian formulation. Phys. Rev. A **51**, 2537 (1995)
18. F. Marquardt, J.P. Chen, A.A. Clerk, S.M. Girvin, Quantum theory of cavity-assisted sideband cooling of mechanical motion. Phys. Rev. Lett **99**, 093902 (2007)
19. I. Wilson-Rae, N. Nooshi, W. Zwerger, T.J. Kippenberg, Theory of ground state cooling of a mechanical oscillator using dynamical backaction. Phys. Rev. Lett. **99**, 093901 (2007)
20. A.D. O'Connell, M. Hofheinz, M. Ansmann, R.C. Bialczak, M. Lenander, E. Lucero, M. Neeley, D. Sank, H. Wang, M. Weides, J. Wenner, J.M. Martinis, A.N. Cleland, Quantum ground state and single-phonon control of a mechanical resonator. Nature **464**, 697 (2010)
21. J.D. Teufel, T. Donner, D. Li, J. W. Harlow, M.S. Allman, K. Cicak, J.D. Whittaker, K.W. Lehnert, R.W. Simmonds, Sideband cooling of micromechanical motion to the quantum ground state. Nature **475**, 359 (2011)

22. J. Chan, T.P. Mayer Alegre, A.H. Safavi-Naeini, J.T. Hill, A. Krause, S. Gröblacher, M. Aspelmeyer, O. Painter, Laser cooling of a nanomechanical oscillator into its quantum ground state. Nature **478**, 89 (2011)

23. A.H. Safavi-Naeini, J. Chan, J.T. Hill, T.P. Mayer Alegre, A. Krause, O. Painter, Observation of quantum motion of a nanomechanical resonator. Phys. Rev. Lett. **108**, 033602 (2012)

24. P. Meystre, Cool vibrations. Science **333**, 832 (2011)

25. M. Aspelmeyer, T.J. Kippenberg, F. Marquardt, Cavity optomechanics. Rev. Mod. Phys. **86**, 1391 (2014)

26. P. Meystre, A short walk through quantum optomechanics. Ann. Phys. (Berlin) **525**, 215 (2013)

27. P.-F. Cohadon, J. Harris, F. Marquardt, L. Cugliandolo, *Quantum Optomechanics and Nanomechanics: Lecture Notes of the Les Houches Summer School, August 2015*, vol. 105 (Oxford University Press, Oxford, 2020)

28. U. Akram, N. Kiesel, M. Aspelmeyer, G.J. Milburn, Single-photon opto-mechanics in the strong coupling regime. New J. Phys. **12**, 083030 (2010)

29. P. Rabl, Photon blockade effect in optomechanical systems. Phys. Rev. Lett. **107**, 063601 (2011)

30. A. Nunnenkamp, K. Børkje, S.M. Girvin, Single-photon optomechanics. Phys. Rev. Lett. **107**, 063602 (2011)

31. G.S. MacCabe, H. Ren, J. Luo, J.D. Cohen, H. Zhou, A. Sipahigi, M. Mirhosseini, O. Painter, Nano-acoustic resonator with ultralong phonon lifetime. Science **370**, 840 (2020)

32. J. Zhang, K. Pen, S.L. Braunstein, Quantum-state transfer from light to macroscopic oscillators. Phys. Rev. A **68**, 013808 (2003)

33. T.A. Palomaki, J.W. Harlow, J.D. Teufel, R.W. Simmonds, K.W. Lehnert, Coherent state transfer between itinerant microwave fields and a mechanical oscillator. Nature **495**, 210 (2013)

34. R. Loudon, P.L. Knight, Squeezed light. J. Modern Optics **34**, 709 (1987)

35. A.A. Clerk, F. Marquardt, K. Jacobs, Back-action evasion and squeezing of a mechanical resonator using a cavity detector. New J. Phys. **10**, 1 (2008)

36. P. Meystre, E.M. Wright, J.D. McCullen, E. Vignes, Theory of radiation-pressure-driven interferometers. J. Opt. Soc. Am. **2**, 1830 (1985)

37. J.D. Thompson, B.M. Zwickl, A.M. Jayich, F. Marquardt, S.M. Girvin, J.G.E. Harris, Strong dispersive coupling of a high-finesse cavity to a micromechanical membrane. Nature **452**, 72 (2008)

38. A.M. Jayich, J.C. Sankey, B.M. Zwickl, C. Yang, J.D. Thompson, S.M. Girvin, A.A. Clerk, F. Marquardt, J.G.E. Harris, Dispersive optomechanics: a membrane inside a cavity. New J. Phys. **10**, 095008 (2008)

39. M. Bhattacharya, H. Uys, P. Meystre, Optomechanical trapping and cooling of partially reflecting mirrors. Phys. Rev. A **77**, 033819 (2008)

40. K. Zhang, F. Bariani, P. Meystre, Theory of an optomechanical quantum heat engine. Phys. Rev. A **90**, 023819 (2014)

41. LIGO Caltech website. https://www.ligo.caltech.edu (2020). Accessed November 24, 2020

42. F. Brennecke, S. Ritter, T. Donner, T. Esslinger, Cavity optomechanics with a Bose-Einstein condensate. Science **322**, 235 (2008)

43. T. Purdy, D.W.C. Brooks, N. Brahms, Z.-Y. Ma, D.M. Stamper-Kurn, Tunable cavity optomechanics with ultracold atoms. Phys. Rev. Lett. **105**, 133602 (2010)

44. D.M. Stamper-Kurn, Cavity Optomechanics with Cold Atoms, in *Cavity Optomechanics*, ed. by M. Aspelmeyer, T. Kippenberg, F. Marquardt (Springer, Berlin, 2015), p. 283

45. P. Treutlein, D. Hunger, S. Camerer, T.W. Hänsch, J. Reichel, Bose-Einstein condensate coupled to a nanomechanical resonator on an atom chip. Phys. Rev. Lett. **99**, 140403 (2007)

46. S. Camerer, M. Korppi, A. Jöckel, D. Hunger, T.W. Hänsch, P. Treutlein, Realization of an optomechanical interface between ultracold atoms and a membrane. Phys. Rev. Lett. **107**, 223001 (2011)

47. A.D. Armour, M.P. Blencowe, K.C. Schwab, Entanglement and decoherence of a micromechanical resonator via coupling to a Cooper box. Phys. Rev. Lett. **88**, 148301 (2002)

48. M.D. LaHaye, J. Suh, P.M. Echternach, K.C. Schwab, M.L. Roukes, Nanomechanical measurements of a superconducting qubit. Nature **459**, 960 (2009)
49. P. Rabl, P. Cappellaro, M.V. Gurudev Dutt, L. Jiiang, J.R. Maze, M.D. Lukin, Strong magnetic coupling between an electronic spin qubit and a mechanical resonator. Phys. Rev. B **79**, 041302(R) (2009)
50. J.M. Taylor, P. Cappellaro, L. Childress, L. Jiang, D. Budker, P.R. Hemmer, A. Yacoby, R. Walsworth, M.D. Lukin, High-sensitivity diamond magnetometer with nanoscale resolution. Nature Phys. **4**, 810 (2011)
51. J.R. Maze, P.L. Stanwix, J.S. Hodges, S. Hong, J.M. Taylor, P. Cappellaro, L. Jiang, M.V. Gurudev Dutt, E. Togan, A.S. Zibrov, A. Yacoby, R.L. Walsworth, M.D. Lukin, Nanoscale magnetic sensing with an individual electronic spin in diamond. Nature **455**, 644 (2008)

Chapter 12
Outlook

In addition to being a field of research in its own right, quantum optics is of considerable value for basic and applied science, as well as for engineering and technology. But even more remarkably perhaps, it is also exceptionally positioned to help shed light on aspects of the physical world that is still a profound mystery to us. This final chapter elaborates on that point with a brief overview of some contributions of quantum optics to tests of the fundamental laws of nature and to searches for the particles and fields populating the Dark Sector, these 95% of the physical world that we still do not understand.

If there is one main lesson to be learned from the previous chapters, it is that over the years quantum optics has perfected a remarkable set of tools to tame the quantum. They range from manipulating the precise quantum state of optical or microwave fields to isolating and trapping atoms, molecules, and ions in controlled environments; from being able to address these particles individually to tailoring their interactions in many-body systems; and from bringing objects of increasing mass deep into the quantum regime to integrating them in powerful hybrid systems.

Combined with our increased understanding, at least at an operational level, of quantum measurements and of the importance and implications of quantum entanglement, these developments have uniquely positioned quantum optics to tackle a remarkable spectrum of problems in fundamental and applied science and engineering—most notably perhaps in quantum metrology and quantum information science. They also create important synergies at the boundary between quantum optics and a broad spectrum of physics subfields from AMO and condensed matter physics to astrophysics and high-energy physics and provide new opportunities in areas ranging from chemistry to biology and from engineering to health science as well. These scientific and technical advances, sometimes referred to as the second quantum revolution, bring to the fore the extraordinary potential of quantum mechanics, a potential that was initially largely unforeseen. Interested readers are encouraged to consult the timely and eye-opening report of the National Academy of Sciences, Engineering and Medicine *Manipulating Quantum Systems:*

© The Author(s), under exclusive license to Springer Nature Switzerland AG 2021
P. Meystre, *Quantum Optics*, Graduate Texts in Physics,
https://doi.org/10.1007/978-3-030-76183-7_12

An Assessment of Atomic, Molecular and Optical Physics in the United States, which covers many of these points in much detail [1].

Despite all these successes, however, we should try not to be too complacent: in many ways, the current state of physics is not completely unlike the situation faced by classical physics at the turn of the twentieth century, when it had reached the superb degree of maturity that was central to the industrial revolution. Now as then, things are not quite as fully understood as one would wish. In 1900, Lord Kelvin gave a celebrated talk entitled "Nineteenth Century Clouds over the Dynamical Theory of Heat and Light" [2] in which he stated with remarkable insight that "The beauty and clearness of the dynamical theory, which asserts heat and light to be modes of motion, is at present obscured by two clouds." He went on to explain that the first of these two clouds was the inability to experimentally detect the luminous ether—the medium that was thought to be vibrating to create light waves; and the second was the so-called ultraviolet catastrophe of blackbody radiation—the fact that Maxwell's theory utterly failed to predict the amount of ultraviolet radiation emitted by objects as a function of their temperature. As we know, these two clouds led to two earthshaking revolutions in physics: relativity and quantum mechanics.

The parallel with the current state of physics is actually quite striking. Despite the extraordinary successes of the Standard Model, one might argue that it is not just two clouds that obscure our full understanding of the physical world, but something more like a thick fog. Indeed, it is now well established that we only understand only about 5% of the Universe. The particles included in the Standard Model, quarks, leptons and the associated force carriers of the weak, strong, and electromagnetic interaction, plus the Higgs boson, properly describe ordinary matter. But they account neither for dark matter, which accounts for 26% of the Universe composition, nor for dark energy, which accounts for another 69%. Dark matter and dark energy are jointly referred to as the *dark sector*. And in addition, despite continuing attempts the Standard Model still fails to unify gravity with the other fundamental interactions.

Understanding the dark sector and its hypothetical collection of yet unobserved quantum fields and corresponding particles is arguably the greatest current challenge in fundamental physics. Not surprisingly, its experimental component relies heavily on the use of large particle colliders and/or astronomical observations—after all, the existence of dark matter has been inferred for decades from the observation of its gravitational effects such as the rotation of spiral galaxies [3, 4], and dark energy somewhat more recently from the accelerating expansion of the Universe [5, 6]. But as we shall see, quantum optics also contributes essential tools and expertise to the cross-disciplinary efforts aimed at solving this deepest puzzle of physics.[1]

The next two sections give a brief overview of some of the ways in which quantum optics contributes to this enterprise. It does not go into any detail and

---

[1]This is of course not a new state of affairs. Optics has a long and distinguished history of doing so, having for example been at the heart of the development of fields from astronomy to quantum mechanics and to relativity, to mention just three examples.

is by no means meant to be comprehensive. Rather, it is limited to a mention of a few selected examples that rely on concepts and approaches that we have become familiar with, most importantly interferometry, atom interferometry, and optomechanics. The goal here is simply to give a taste of the kind of questions and challenges that can be addressed with these tools. Because so little is known about the dark sector, it is hard to go beyond this elementary level without diving into some of the theories and scenarios that have been proposed to describe it. Unfortunately the current lack of experimental guidance—none of the proposed dark particles and fields have been observed yet—makes it challenging to decide which of these theories to focus on. A very thorough review of AMO and quantum optics contributions to recent developments in tests of fundamental physics, including parity violation, searches for permanent electric dipole moments, tests of general relativity and of the equivalence principle, and searches for dark matter and dark energy can be found in Ref. [7].

12.1 Gravitation

Guided by the fact that we are quite familiar with gravity, which we understand reasonably well—at least at the classical level since we are constantly confronted by its effects in everyday life, we focus on that topic first. We then turn to the much more elusive dark sector in the following section, with the caveat that this is a somewhat arbitrary and not very satisfactory division, since as we shall see there is a considerable overlap between these two topics, which really address the same fundamental questions.

It has been known since Newton's 1687 publication of the law of universal gravitation in Philosophiæ Naturalis Principia Mathematica that the gravitational attraction between two point bodies is proportional to the product of their masses M and m and inversely proportional to the square of their distance r, $F = GMm/r^2$, where $G = 6.674 \times 10^{-11}$ m^{3}kg^{-1}s^{-2} is Newton's gravitational constant. Einstein's 1915 general relativity theory improved significantly on that law by providing a description of gravity as a geometric property of space and time, or four-dimensional space-time, and predicted new effects such as gravitational waves, gravitational lensing, and gravitational time dilation. But despite all its successes, general relativity is largely expected not to be the final word, as it cannot so far be unified with the other known forces of nature, the electromagnetic, weak, and strong forces, which are themselves unified in the Standard Model. The quest to unify all known forces of nature remains a major challenge, and precise tests of the laws of gravity may provide important hints toward the solution of this problem.

Consider for example the equivalence principle, the fundamental principle that forms the basis of Einstein's theory of general relativity. It states that all objects fall with the same acceleration under the influence of gravity, that is, that all forms of matter or energy respond to gravity in the same way. So far the equivalence principle has been experimentally verified at a level of accuracy such that the accelerations of two falling objects have been shown to differ by no more than one part in 10^{13}. But it is widely believed that general relativity and the Standard Model are both low energy limits of a more complete theory that unifies all forces at high energies. We simply do not know if the equivalence principle will hold for that unified theory, or whether it will be violated at some level. Higher precision experiments are therefore needed to determine if, when, and how it ceases to hold. A similar motivation also lies behind improved tests of the $1/r^2$ dependence of the law of universal gravitation. As we shall see, such precision experiments also provide powerful tools in the search for new physics, such as candidates for dark matter and dark energy.

We will turn to these topics in the following subsections but first focus on a prediction of general relativity that has been spectacularly confirmed in the last few years, the existence of gravitational waves.

12.1.1 Gravitational Wave Detection

Four centuries after Galileo Galilei used telescopes to study and revolutionize our understanding of the Universe they remain our most powerful tool to learn about it, whether they detect radio waves, submillimeter waves, infrared radiation, visible light, ultraviolet radiation, or X-rays. However, extremely significant additional information is provided by the detection and characterization of other "messenger signals" from outer space such as neutrinos, cosmic rays, and, since 2015, gravitational waves. These disturbances of space-time are produced by the motion of massive objects, in particular by closely orbiting compact massive objects such as neutron stars or black holes binaries, merging supermassive black holes, collapsing supernovae, or pulsars, and they propagate outward from their source at the speed of light. Gravitational waves also hold the promise to provide information on the processes that took place in the early Universe, shortly after the Big Bang.

While gravitational waves were predicted by Einstein in 1916, it is not until 1974 that R. A. Hulse and J. H. Taylor obtained the first indirect evidence of their existence through the observation of the gradual decline of the orbit period of a binary pulsar that matched the loss of energy and angular momentum by gravitational radiation predicted by general relativity [8–10]. The first direct detection of gravitational waves finally occurred on September 14, 2015, when LIGO (which stands for Laser Interferometer Gravitational Wave Observatory) physically sensed the undulations in space-time caused by gravitational waves generated by two colliding black holes 1.3 billion light-years away [11], see Fig. 12.1. Without a doubt, LIGO's discovery will go down in history as one of the humanity's greatest scientific achievements.

Fig. 12.1 Signals of the first gravitational waves detected by the twin LIGO observatories in Livingston, Louisiana, and Hanford, Washington. The signals come from two merging black holes, each about 30 times the mass of the sun and 1.3 billion light-years away. The top two plots show data received at Livingston and Hanford, with the thin lines showing the predicted shapes for the waveform. Time is plotted on the horizontal axis and strain on the vertical axis. The strain represents the fractional amount by which distances are distorted. The bottom plot compares data from both detectors, where the Hanford data have been inverted for comparison, due to the differences in orientation of the detectors at the two sites. (Image Credit: Caltech-MIT-LIGO Lab)

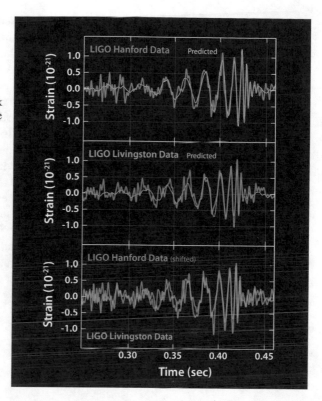

LIGO We mentioned in Chap. 6 and again in Sect. 11.4 that the LIGO antennas use large-scale Michelson interferometers[2] with free standing end mirrors to detect the changes in space-time curvature resulting from the passage of gravitational waves, see Fig. 11.1. Gravitational waves passing through the interferometer produce a metric disturbance that results in an effective differential change in the arm lengths, imparting in turn a phase shift to the optical fields circulating in the arms.

This seems straightforward enough in principle, so one might ask why it took a century after Einstein's prediction to directly detect these waves? The simple reason is that while the processes that generate gravitational waves can be extremely violent and destructive, by the time the waves reach the Earth they are thousands of billions of times weaker. In fact, by the time gravitational waves from LIGO's first detection reached us, the amount of space-time wobbling they generated was a 1000 times smaller than the nucleus of an atom! These inconceivably small perturbations in space-time are what LIGO is designed to measure.

---

[2]The arms of the LIGO interferometer are 4 km long, but each arm consists of a Fabry-Pérot interferometer such that the light beams effectively traverse the cavity about 450 times, resulting in an effective arm length of 1800 km, see Fig. 12.2.

Achieving the seemingly impossible task of detecting such minute length changes required to gain a deeper understanding of quantum measurements and of the standard quantum limit. This led as we have seen to the development of quantum non-demolition measurements and to the use of squeezed states to reduce shot noise in interferometric gravitational wave detectors [12–15]. As pointed out in the third of these references, during the first LIGO observing run, which took place from September 2015 to January 2016, three binary black hole detections were made, and the second observing run, which ran from November 2016 to August 2017, detected seven binary black hole mergers, and one binary neutron star merger. In contrast, the improved performance of the detectors, combined with the permanent addition of the Italy-based Advanced Virgo antenna in the third observing run, which took place from April 1 to September 30, 2019 and from November 1, 2019 until March 27, 2020, resulted in the observation of 56 candidate gravitational wave signals, including at least one new compact binary coalescence in the binary neutron star mass range. In particular, as discussed in Sect. 11.4 and shown in Fig. 12.2, shot noise was reduced by the use of squeezed vacuum injected in the second input of the interferometer.

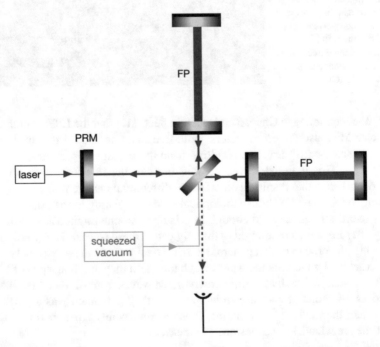

Fig. 12.2 Schematic of the LIGO antenna, illustrating the presence of Fabry–Pérot interferometers (FP) in the two arms of the Michelson interferometer, thereby increasing their effective length by a factor of 450, as well as the power recycling mirror (PRM) whose goal is to effectively increase the laser power inside the antenna, and the squeezed vacuum injection in the otherwise "empty" interferometer input

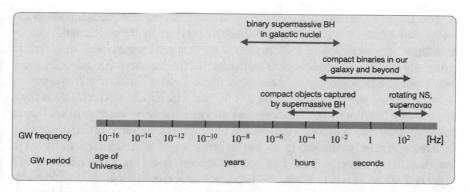

Fig. 12.3 Spectrum of the expected frequency ranges of the gravitational waves emitted in selected cosmic events. BH stands for black hole and NS for neutron star. (Adapted from https://LISA.NASA.gov)

LISA The LIGO gravitational wave detectors and their cousin, the Virgo interferometer, detect gravitational waves in the hertz to kilohertz range, a consequence of the effective length of the interferometer arms. There is also considerable interest in detecting gravitational waves of much lower frequencies, in the millihertz to hertz range, corresponding to binaries that orbit each other with periods of seconds to hours [16]. Examples of systems emitting such waves include white dwarf binaries, as well as stellar-size black holes with large separations. They initially lose energy in the form of low frequency gravitational waves, leading to a slow decrease of their orbital separation before their orbital frequency increases in the last instants preceding their merger. The detection of gravitational waves in the millihertz frequency range would therefore permit to observe the evolution of these systems long before their final merger.

A third class of objects that could be observed in a low frequency observatory is the merger of massive black holes of tens of thousands of solar masses. Because these systems have a larger effective radius than solar-mass black holes, they cannot approach one another sufficiently closely to reach the high orbital frequencies and merge instead at the low frequencies, see Fig. 12.3. It is expected that the merger of binary supermassive black holes located at arbitrary distances will be observable with such an antenna, thus making it a deep Universe observatory where sources detected with good sky localization would help chart the cosmic expansion history up to high cosmological redshifts.[3] Finally, it may be that at much lower frequencies the primordial gravitational wave background analog of the cosmic microwave background might also be detectable [17].

---

[3]The cosmological redshift refers to the lengthening of a radiation wavelength as it travels through expanding space. It should not be confused with the Doppler shift, which depends on the motion of the radiating object relative to an observer.

The length of the interferometer arms needs to be at least comparable to the wavelength of the signal, as its sensitivity decreases rapidly for wavelengths shorter than them. The detection of low frequency gravitational waves leads therefore to extremely challenging requirements on the size of the antennas. For example, since gravitational waves propagate at the velocity of light a signal at a frequency of $\omega \approx 6s^{-1}$ has a wavelength of about 3×10^8 m, which quite obviously calls for a space-based interferometer. The proposed gravitational wave antenna LISA (Laser Interferometer Space Antenna), expected to be placed on orbit in 2030, consists of three spacecrafts arranged in an equilateral triangle with sides 2.5 million km long and flying along an Earth-like heliocentric orbit, see Fig. 12.4. All three will carry two telescopes, two lasers, and two test masses, each a 46 mm, roughly 2 kg gold-coated cube of gold/platinum, pointed at the other two spacecrafts and forming Michelson-like interferometers, each centered on one of the spacecrafts [18]. The entire arrangement will be ten times larger than the orbit of the Moon, and the mean linear distance between the formation and the Earth will be 50 million kilometers. The arms of the interferometer will be 2.5×10^6 km long, corresponding to a frequency of $0.75 \, s^{-1}$.

An Earth orbiting system, the Gravity Recovery and Climate Experiment Follow-on mission (GRACE-FO), was developed in part to test some of the elements of the future LISA space antenna. It consists of two satellites separated by about 220 km, a distance constantly measured by an extremely precise microwave ranging

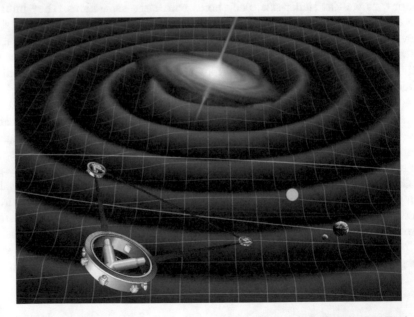

Fig. 12.4 Artist illustration of LISA. The three LISA spacecrafts will be in individual Earth-like orbits around the sun to create the triangle. The test masses are free-falling and shielded by the enclosing spacecraft from disturbances of the solar wind and radiation pressure. (Image credit: http://lisa.jpl.nasa.gov/gallery/lisa-waves.html)

system (Fig. 12.5). Since areas of slightly stronger gravity affect the lead satellite first and pull it away from the trailing satellite, distance variations amount to a monitoring of small changes in Earth gravity. This information provides in particular global measurements on seasonal river basin water storage variations, human influences on water storage changes, continental aquifer changes, and more. This is a spectacular and powerful example of the beneficial cross-fertilization that can arise between basic sciences and applications, in that case gravitational waves detection and the study of climate changes and human impact on the environment. More details on this remarkable project can be found on the GRACE-FO website https://gracefo.jpl.nasa.gov.

Atom Interferometer Antennas Atom interferometers continue to witness remarkable experimental developments and can now achieve coherence times of seconds and wave packet separations of tens of centimeters. Of particular relevance in the context of gravitational wave detection are "large momentum transfer clock atom interferometers" such as the system recently reported in Ref. [19]. In contrast with most atom interferometers, which make use of two-photon interactions to minimize the negative effects of spontaneous emission as we mentioned in Sect. 8.6, they rely on single-photon transitions on the $^1S_0$-$^3P_1$ intercombination line of ^{88}Sr, a transition characterized by a very narrow natural linewidth of just 7.4 KHz. Using that transition J. Rudolph and coworkers have demonstrated an atomic momentum

Fig. 12.5 Illustration of GRACE-FO in orbit. (Image credit: NASA/JPL-Caltech)

separation of 141 $\hbar k$ between the two arms of the interferometer. At the same time, as briefly mentioned in Sect. 10.4, progress in atomic clocks has been just as spectacular, with the best atomic clocks having now a stability better than 1 s over the age of the Universe [20, 21].[4]

These developments have lead M. Kasevich and his coworkers to propose an alternative gravitational wave detector that combines optical methods and atom interferometry [24]. Their proposal draws on the use of atomic clocks to measure the differential acceleration of two spatially separated, free-falling atom interferometers, as discussed in Sect. 8.6, Problems 8.7–8.9. As such it expands on the same basic idea as LIGO or LISA, but with the important difference that the free standing mirrors are now replaced by atom interferometers.

We outlined in Problem 8.9 how atom interferometers can be used to measure the local acceleration due to gravity. This results from the fact that if the excited and ground state atoms have momenta $\mathbf{p}$ and $\mathbf{p} + \hbar \mathbf{k}_{\text{eff}}$, once they are recombined the phase difference between the two paths of the interferometer is given by

$$\Delta \Phi = \mathbf{k}_{\text{eff}} \cdot \mathbf{g} T^2 , \tag{12.1}$$

where T is the time interval between the $\pi/2$ and π optical pulses that act as atomic "beam splitter" and "mirror," respectively, $\mathbf{k}_{\text{eff}}$ is the effective momentum difference imparted on the excited and ground electronic states of the atom, and $\mathbf{g}$ the acceleration of gravity, see Refs. [25, 26] for more details. Furthermore, using two spatially separated atom interferometers allows to measure gravity gradients as well. If they are separated by a large distance, and with the help of high precision clocks, comparing the matter-wave interference fringes in the two interferometers provides a record of the effect of gravitational waves on the travel time of a laser pulse linking the two atom interferometers. It is argued in particular that using atoms instead of mirrors as test masses could reduce a number of systematic errors.

This technique is also central to the proposed Mid-band Atomic Gravitational wave Interferometric Sensor (MAGIS) proposal [27], which will be designed to detect gravitational waves in the frequency band of 0.03 Hz–10 Hz complementary to existing detectors. MAGIS-100, a 100-m tall atomic sensor now being constructed at Fermilab, will serve as a prototype of such a gravitational wave detector. Interestingly, it is also expected to be sensitive to proposed ultralight dark matter candidates,[5] see Sect. 12.2.2, as well as to provide several other opportunities in fundamental physics, such as testing the equivalence principle and the inverse square law of gravity, to which we now turn.

---

[4]The $^1S_0$-$^3P_0$ clock transition of ^{88}Sr has an extremely narrow linewidth of just 1 mHz, corresponding to an excited state lifetime in excess of 100 s. Currently, strontium clocks equipped with the best ultra-stable lasers show a fractional stability at the low 10^{-18} total uncertainty level after preparing ultracold strontium atoms in an optical lattice and probing the $^1S_0$-$^3P_0$ mHz clock transition with a 698-nm laser stabilized to a linewidth of 26 mHz [20, 22, 23].

[5]As we will see in Sect. 12.2 ultralight dark matter is a proposed class of dark matter consisting of bosons with masses ranging from 10^{-22}eV $< m < 1$eV [28].

12.1.2 Tests of the Equivalence Principle

The equivalence principle states the equivalence of the inertial and gravitational masses of an object. That is, when objects are in free fall, the trajectories that they follow are entirely independent of their masses. This principle can be tested by measuring the differential acceleration Δa of two test masses of different composition relative to each other while falling in the gravitational field of a source body of mass M, for example the Earth. Combining the law of gravitational attraction

$$F = Gm_g M/r^2 \tag{12.2}$$

for the force F on a gravitational mass m_g due to the gravitational field of an attractor of mass M with the definition of the inertial mass m_i

$$F = m_i a \tag{12.3}$$

gives for the acceleration a of that mass

$$a = \frac{GM}{r^2}\left(\frac{m_g}{m_i}\right), \tag{12.4}$$

which is independent of mass if the equivalence principle holds and $m_g = m_i$. Deviations from the universality of free fall can be quantified by the Eötvös parameter

$$\eta = \frac{a_1 - a_2}{(a_1 + a_2)/2} = 2\frac{(m_g/m_i)_1 - (m_g/m_i)_2}{(m_g/m_i)_1 + (m_g/m_i)_2}, \tag{12.5}$$

where the subscripts $\ell = 1, 2$ refer to the two test masses 1 and 2, and a_ℓ is their free-fall acceleration. The second equality results from the assumption that the ratio of gravitational to inertial masses $(m_g)_\ell$ and $(m_i)_\ell$ depends on their composition. The finding of a value $\eta \neq 0$ would indicate a violation of the equivalence principle.

We mentioned that general relativity and the Standard Model are believed to be low energy limits of a unified theory. It is hoped that testing fundamental laws such as the equivalence principle with ever increasing accuracy and under a broad range of conditions will provide valuable hints toward achieving this unification. The most accurate laboratory measurements of the Eötvös parameter to date were carried out by the Eöt-Wash group [29, 30] using torsion balances and found that $\eta < 10^{-13}$ for the case of Be–Ti test masses with Earth as the attractor.[6] Space experiments

---

[6]Note that strictly speaking, in order to truly test the equivalence principle and constrain the Eötvös parameter to the $\eta < 10^{-13}$ level would require to repeat the experiment for every possible combination of materials.

carried out by the MICROSCOPE satellite, which measured the force required to maintain two test masses of titanium and platinum alloys exactly in the same orbit [31], found an upper limit of $\eta < 1.3 \times 10^{-14}$.

Atom Interferometer Tests Because of their remarkable potential sensitivity as gravimeters and gravity gradiometers, atom interferometers are an alternative tool of choice to test the equivalence principle, an important distinction with the torsion pendulum or space-based experiments being that they deal with single atoms. Atom clouds are well-suited test masses because they spend 99.9% of the interrogation time in free fall and the remainder in precisely controlled interactions with the lasers serving as beam splitters and mirrors. In addition, atoms have uniform and well-characterized physical properties.

The group of M. Kasevich at Stanford performs tests of the equivalence principle by interferometrically measuring the relative acceleration of freely falling clouds of ^{85}Rb and ^{87}Rb atoms in a 8.9-meter-long dual species atom interferometer located in a 10-meter-high drop tower, see Fig. 12.6—an arrangement that is the inspiration for the proposed MAGIS detector mentioned earlier. The long drift time of that interferometer enables it to have an acceleration sensitivity of 7×10^{-12} g, which is roughly the same as the gravitational attraction that you would feel toward a person 10 m away from you. Their latest results [32] report an atom-interferometric Eötvös parameter measurement of $\eta = [1.6 \pm 1.8(\text{stat}) \pm 3.4(\text{sys})] \times 10^{-12}$, consistent with $\eta = 0$. Here "sys" stands for systematic errors and "stat" for statistical error.

Fig. 12.6 Stanford 10-m tower used for tests of the equivalence principle. (Courtesy Mark Kasevich, Stanford University)

12.1.3 Testing the Inverse Square Law

Testing gravity at the shortest possible distances is important as proposed theories to unify gravity and the Standard Model, as well as hypothesized dark matter particles and dark energy candidates, involve features that could imply short-range deviations from the gravitational inverse square law [33].

The results of experimental tests of that law are usually formulated in the form of constraints on the parameters α and λ that characterize a possible Yukawa interaction of the form

$$V(r) = \frac{GmM}{r^2}\left[1 + \alpha e^{-r/\lambda}\right].\tag{12.6}$$

Below $10\,\mu$m the tests of the inverse square law rely often on measurements of the Casimir force of Sect. 7.5. S. Lamoreaux [34] used a torsion balance to measure that force in the range 0.6 to $6\,\mu$m and was able to probe the Casimir force with about 5% accuracy. These measurements permitted to set a limit on the violation of the inverse square law, excluding Yukawa-type forces with $\alpha > 10^9$ in the range $100\,\text{nm} \le \lambda < 10\,\mu$m [35]. More recent studies using a variety of alternative experimental setups that largely avoid the need to rely on the Casimir effect have further enlarged the excluded region of the $\{\lambda, \alpha\}$ parameter space, see Fig. 12.7. These include the observation of the oscillations of a test mass attached to a micro-electromechanical oscillator and gravitationally coupled a rotating source mass [36], as well as the study of the harmonic torque exerted on a detector pendulum by a rotating attractor [37, 38].

12.1.4 Gravitationally Induced Decoherence

In Chap. 6 we adopted an operational approach to the problem of quantum measurement based on the von Neumann projection postulate. We motivated it phenomenologically in terms of the coupling of the measuring apparatus to a reservoir with a very large number of degrees of freedom where quantum information is irreversibly dispersed. We also adopted the view that the state vector describes our knowledge of the system and as such provides us with a means to calculate probabilities when a measurement is performed. Since each time we make a measurement our knowledge of the system is changed, it is not surprising that the state vector should make a "quantum jump" as a result. This is essentially the so-called Copenhagen interpretation of quantum mechanics. This, however, is not necessarily a satisfying state of affairs, and there has been a lively and ongoing discussion of this problem, including most famously perhaps the "many-worlds" interpretation of H. Everett [39], as well as a number of other interpretations, although these proposals do not normally result in falsifiable deviations from standard quantum mechanics.

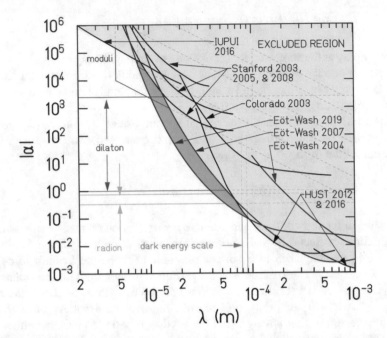

Fig. 12.7 The colored area shows the experimental constraints on a Yukawa violation of the Newtonian $1/r^2$ law, with that region excluded at the 90% level. IUPUI2016 refers to the results of Ref. [36]. The green area shows the improvement reported in Ref. [37]. The purple and green lines indicate the expected domains of impact on the $1/r^2$ law of two dark matter candidates, the dilaton and the radion. (From Ref. [37], which also gives the full journal references of the results summarized in the figure)

The belief that general relativity and the Standard Model are low energy limits of a more complete theory that unifies all forces at high energies raises the question of whether the measurement problem is related to that issue, and more specifically whether the collapse of the wave function might be a gravitational effect. In contrast to the view that it is a mechanism such as environment-induced decoherence that rapidly destroys quantum superpositions in massive objects and establishes the transition from the quantum to the classical world, several authors [40–43] have advanced collapse models associated with more fundamental mechanisms and the appearance of new physical principles. Specifically, they propose that gravitational effects play a role in the reduction of the state vector, based on the idea that macroscopic quantum superpositions of two differing mass distributions are unstable.

D. Bouwmeester et al. [44] pioneered the idea that quantum optomechanics experiments may shed light on this issue and on possible unconventional decoherence processes by attempting to generate spatial quantum superpositions of massive objects. O. Romero-Isart has analyzed the experimental requirements to test some of these models and discussed the feasibility of a quantum optomechanical implementation using levitating dielectric nanospheres [45]. He estimates that combined with cryogenic, extreme high vacuum, and low-vibration environments it should be experimentally feasible to prepare the center of mass of a micrometer-sized object in a spatial quantum superposition comparable to its size. In such a hitherto unexplored parameter regime gravitationally induced decoherence could be unambiguously falsified. An important step in this direction was recently achieved with the cooling of a levitated 150-nm silica microsphere to its quantum mechanical ground state [46] by M. Aspelmeyer's group in Vienna. Such levitated optomechanical systems are a promising playground to study the interface between quantum physics and gravity using quantum optical control over the motion of increasingly massive solids and provide a viable route toward experiments in which macroscopic systems deep in the quantum regime act as gravitational source masses.

12.2 The Dark Sector

A number of astronomical observations point to the existence of dark matter, which constitutes the majority of the mass of galaxies, and of an unknown force field, called dark energy, which is responsible for the accelerated expansion of the Universe. Together they constitute 95% of the Universe, and understanding them remains a major challenge.

The ΛCDM (Lambda Cold Dark Matter) model, see e.g. Ref. [47], is the model that provides our best current understanding of the origin of the Cosmos. In this cosmological model the Universe contains three major components: a cosmological constant Λ associated with dark energy, cold dark matter (abbreviated CDM), and ordinary matter. It is frequently referred to as the Standard Model of cosmology because it accounts in particular for the existence and structure of the cosmic microwave background, the large-scale structure in the distribution of galaxies, and the observed accelerating expansion of the Universe.

Although it has not been directly detected so far, cosmologists believe that cold dark matter is comprised of cold, slow moving particles that do not absorb or scatter electromagnetic radiation, so that it appears "dark," or perhaps more accurately "invisible." Astronomical observations indicate that the dark matter density in the solar neighborhood is of the order of[7] $\rho = 0.3\,\text{GeV}/\text{cm}^3 \approx 4.8 \cdot 10^{-25}\,\text{g/cm}^3$, so that assuming that it is comprised of a single component of particles of mass m_{DM}, the local number density of these particles would be

$$n_{\text{DM}} = 0.3 \left(\frac{1\text{GeV}}{m_{\text{DM}}} \right) \text{cm}^{-3} . \tag{12.7}$$

---

[7]To put the density ρ in perspective, if the Earth consisted of dark matter, with its volume of approximately $V = 10^{21}\,\text{m}^3$ its mass would be about 0.1 kg.

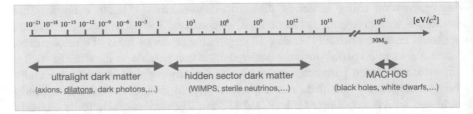

Fig. 12.8 Dark matter candidates as a function of their expected mass, in units of eV/c^2, which covers a range of over 80 orders of magnitude. For reference we note that $10^{62}eV/c^2$ corresponds approximately to 30 solar masses $M_\odot$, while $1eV/c^2$ is the mass of the neutrino. The axion is a proposed dark particle that would solve the so-called strong CP problem. It is characterized as something like a cousin of the photon, but with a small mass. Dilatons are scalar particles that appear in particular in the Brans–Dicke theory of gravity, where Newton's constant G is no longer presumed to be constant, but replaced by a scalar field ϕ and its associated particle, the dilaton. The dark photons and the axions are vector and pseudo-scalar candidate particles, respectively. The hypothetical WIMPs, or weakly interacting massive particles, are another proposed candidate for dark matter

This indicates that dark matter candidates can be separated into objects with mass greater than about $m_{DM} \approx 1\,eV/c^2$, which would appear as distinct particles, and ultralight dark matter particles with masses below $1\,eV$, which would be characterized by enormous occupation numbers, see Fig. 12.8. Dark matter particles of the first type and with a velocity v_{DM} would impart a momentum $p = m_{DM}v_{DM}$ when colliding with a detector. In the second case, on the other hand, they would have enormous occupation numbers, indicating that they are bosonic. They would behave as a background of oscillating waves at frequency $\omega_{DM} = m_{DM}c^2/\hbar$ and would produce extremely weak, coherent, and persistent signals. Searching for these two classes of signals calls therefore for different measurement techniques.

While much of the past focus has been on "weakly interacting massive particles," or WIMPs, which are usually theorized to be heavier than the proton, the absence of new particles beyond the Standard Model being discovered at the LHC, combined with the failure by large-scale detectors to find them after decades of effort, see e.g. Refs. [48, 49], has led to a plethora of new ideas. Many of them are ideally suited for AMO and quantum optics experiments and their broad range of tools including atomic clocks, interferometers, magnetometers, atom interferometers, and optomechanical detectors.

Cold dark matter is currently thought to be mostly non-baryonic, that is, to consist of matter other than protons and neutrons (and electrons). Since at this point we are aware of its existence only through its gravitational effects, dark matter particles are assumed to interact with each other and other particles primarily through gravity— and possibly the weak force. It is thought that they may also couple weakly to photons, exert minute torques on magnetic spins, be associated to new forces, or lead in small variations of the fundamental constants. However, there is no experimental evidence for any of these speculated manifestations.

Since gravitational effects are the only certain influence of dark matter, it is not surprising that experiments aimed at testing the laws of gravity of the previous section are also intimately related to investigations of the dark sector. It is however important not to forget that our level of ignorance of the dark sector is such that speculations about the nature of dark matter cover over 80 orders of magnitude. Needless to say, this makes their search extraordinarily challenging, and it is therefore essential that all possible tools that might provide any light on their nature should be considered.

What quantum optics brings to the table is the availability of exquisitely controlled normal matter such as ultracold atoms and mesoscopic mechanical oscillators, optical and atom interferometry measurement techniques of distances and accelerations at or below the standard quantum limit, and more. Combined with spectacular advances in precision time and frequency metrology, these techniques complement and enhance significantly the powerful tools offered by high energy physics and astronomy. They allow for tests of the fundamental laws of physics of unprecedented sensitivity and are likely to contribute in multiple and important ways to the search for an understanding of the dark sector.

12.2.1 Coupling to Photons

While dark matter is believed not to absorb or scatter light, P. Sikivie [50, 51] proposed that a modification of the Maxwell equations might arise from a light, stable axion, which could be one of the dark matter particles. This hypothetical elementary particle has been proposed to resolve the so-called strong charge-parity (CP) problem in particle physics, the puzzling fact that Quantum Chromodynamics seems to preserve CP symmetry, although that symmetry can in principle be violated in strong interactions [52]. Incorporating the axion into Maxwell's equations has the effect of rotating the electric and magnetic fields into each other,

$$\begin{pmatrix} \mathbf{E'} \\ c\mathbf{B'} \end{pmatrix} = \frac{1}{\cos \xi} \begin{pmatrix} \cos \xi & \sin \xi \\ -\sin \xi & \cos \xi \end{pmatrix} \begin{pmatrix} \mathbf{E} \\ c\mathbf{B} \end{pmatrix} , \tag{12.8}$$

where the mixing angle ξ depends on a coupling constant κ and the axion field strength θ, $\tan \xi = -\kappa\theta$. If this is correct, then these axions could be detected on Earth by converting them into photons with the help of strong magnetic fields. This proposal has led to several experiments, in particular the Axion Dark Matter eXperiment (ADMX) [53]. We saw in Chap. 7 in the context of cavity QED that resonant electromagnetic cavities provide a powerful platform for detecting particles that couple to the electromagnetic field. This is the general approach used by this experiment, which began in the 1990s. It combines a strong (8 T) magnetic field with a cryogenically cooled, tunable high-Q electromagnetic resonator and relies on the proposed electromagnetic coupling whereby ultralight axions in a magnetic field are converted into detectable microwave photons. Another experiment, set up at CERN and dubbed OSQAR for "Optical Search for QED Vacuum Birefringence, Axions and Photon Regenerations," looks for axions and axion-like particles by exposing a laser beam to a 9 T magnetic field, with the idea that that field will cause some of the photons in the laser to turn into axions. The experiment takes place in a vacuum chamber containing a barrier that stops the laser beam but lets axions pass through. If light is observed on the other side of the barrier, they would be evidence that axions have travelled through the barrier and turned back into detectable photons on the other side.

The signals that can be expected from these experiments are extraordinarily weak, and so far no evidence for the existence of the axions has been found [53, 54]. It is however possible that the advances in the preparation and measurement of quantum states of microwave frequency electrical circuits hinted at in Sect. 7.4 could prove valuable in improving these detectors, as would be the ability to search for axions by performing quantum non-demolition measurements of the microwave photon number.

12.2.2   Atom-Interferometric Searches

We mentioned briefly in Sect. 12.1 that precision tests of the law of universal gravitation also provide powerful tools in the search for dark matter candidates. This is not surprising since gravitational effects are the most important signature of dark matter, and the only one so far.

One such example is the 100-meter-long Matter-wave Atomic Gradiometer Interferometric Sensor MAGIS-100 already mentioned in the context of tests of the equivalence principle (Fig. 12.9). In this system, atomic clouds located at either end of the MAGIS-100 shaft are manipulated by counter-propagating sequences of optical pulses that are used to divide, redirect, and recombine the atomic de Broglie waves, resulting in the formation of two atom interferometers at elevations x_1 and x_2, as illustrated in Fig. 12.10. The phase difference between these interferometers produces the gradiometer signal that is sensitive to the relative acceleration between the atoms. MAGIS-100 will look for ultralight dark matter candidates that weigh much less than a WIMP, see Fig. 12.8. An example of such a particle is again the axion, which would weigh less than the mass of a neutrino. Axions are expected to cause quantities thought of as fundamental constants to oscillate with time, and these oscillations, for example in the gravitational constant G, could then be detected by the gradiometer.

12.2.3   Cavity Optomechanical Searches

Optomechanical systems that are read out interferometrically below the standard quantum limit have been demonstrated across a wide range of mass scales, with

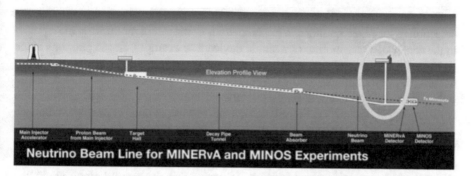

Fig. 12.9 Proposed site for MAGIS-100: Elevation profile view of the existing Neutrino Main Injector tunnel at Fermilab. MAGIS-100 will be located in the 100-m-deep access shaft (yellow circle). (Figure from the U.S. Dept. of Energy Proposal P-110: Matter-wave Atomic Gradiometer Interferometric Sensor (MAGIS-100) [55], courtesy of J. Hogan, Stanford University)

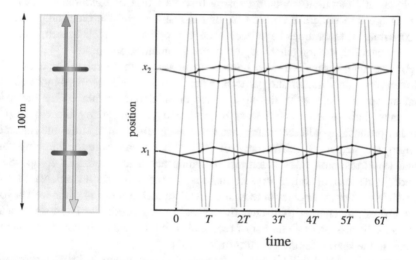

Fig. 12.10 Left: Atomic clouds (blue) are located at either end of the MAGIS-100 shaft, and counter-propagating sequences of optical pulses are used to divide, redirect, and recombine the atomic de Broglie waves, resulting in the formation of two atom interferometers at elevations x_1 and x_2. The phase difference between these interferometers produces the gradiometer signal. Right: Space-time diagram showing the resonant atom interferometry scheme in the MAGIS gradiometer. The thin black lines represent the counter-propagating pulses of light, the blue lines are the ground state, and the red lines the excited electronic state of the atoms. (Adapted from Figs. 6 and 7 of the U.S. Dept. of Energy Proposal P-110: Matter-wave Atomic Gradiometer Interferometric Sensor (MAGIS-100) [55], courtesy J. Hogan, Stanford University)

natural frequencies ranging from millihertz to terahertz [56]. As such they open promising avenues to search for dark matter over a large range of energy scales. In particular, monitoring solid objects allows for the coherent integration of long-wavelength interactions, as well as for the integration of small cross-sections over large numbers of target atoms.

To detect heavy (particle-like) dark matter candidates on can consider either the detection of localized phonons in bulk materials or the direct monitoring of impulses to the center-of-mass motion of a single device. For example, single phonons at the μeV level can be read out in micromechanical oscillators [57, 58] or superfluid helium [59]. Alternatively, one can monitor the center-of-mass motion of an entire object. This technique could be particularly advantageous in the setting where the collisions act coherently on the entire mechanical component, for example when the dark matter couples to the sensor through a long-range force. As a long-term goal, mechanical sensing might open the possibility of direct detection of particle dark matter purely through its gravitational interaction with visible matter, a coupling that is the only one guaranteed to exist. Achieving this goal would require reducing noise levels to well below the standard quantum limit of impulse sensing [60, 61].

Ultralight dark matter, in contrast, is characterized by extremely high population numbers, so that it behaves as classical fields rather than individual particles. Several ultralight field candidates would cause a time variation of fundamental constants such as the fine structure constant α or the mass m_e of the electron. A variety of experimental techniques have been used or proposed for their search, including resonant cavities, torsion balances, atom interferometers, and more.

For example, optical cavities have been proposed in searches for a coupling between axion dark matter and photons, whereby axions are absorbed by or emitted from an optical field, with the appearance of sidebands displaced by the axion frequency. This process could then be resonantly enhanced by the cavity [62]. Another proposal would search for these effects via differential strain measurement of rigid and suspended-mirror cavities [63]. Other ideas include searching for a signal using resonant mass antennas, including acoustic resonators composed of superfluid helium or single crystal materials, producing displacements that are accessible with opto- or electromechanical readout techniques [64]. It is well beyond the scope of this chapter to discuss these ideas in any depth, but interested readers will find a discussion of a number of recent ideas on the use of mechanical quantum sensing in the search for dark matter in Ref. [65].

For now, though, it will be sufficient to conclude by underscoring one more time the dual role of quantum optics as a research field in its own right and as an enabling science, with an impact on science and technology second to none. A partial list would include astronomy, astrophysics and cosmology, physics broadly understood, life and health sciences, earth sciences, quantum information science and quantum metrology, engineering, and industrial development. And remarkably, quantum optics also stands to help shed significant light on the 95% of the physical world that is still a deep mystery to us. All of this is what makes it a uniquely attractive and fulfilling research field. There is something in it for just about everybody!

"There is a crack in everything
that's how the light gets in"
Leonhard Cohen – Anthem (1992)

References

1. National Academy on Science, Engineering and Medicine. *Manipulating Quantum Systems: An Assessment of Atomic, Molecular, and Optical Physics in the United States (2020)* (The National Academies Press, Washington, 2020). https://doi.org/10.17226/25613
2. L. Kelvin, Royal Institution Lecture: Nineteenth-Century clouds over the dynamical theory of heat and light. Phil. Mag. S. 6 **2**, 1 (1901)
3. V. Rubin, W.K. Ford Jr, Rotation of the Andromeda nebula from a spectroscopic survey of emission regions. Astrophys. J. **159**, 379 (1970)
4. V. Rubin, W.K. Ford Jr., N. Thonnard, Rotational properties of 21 SC galaxies with a large range of luminosities and radii, from NGC 4605 (R=44 kpc) to UGC 2885 (R=122 kpc). Astrophys. J. **238**, 471 (1980)
5. A.G. Riess, A.V. Fillipenko, P. Challis, A.A. Clocchiatti, A. Diercks, P.M. Garnavich, R.L. Gilliand, C.J. Hogan, S. Jha, R.P. Kirshner, B. Leibundgut, M.M. Phillips, D. Reiss, B.P. Schmidt, R.A. Schommer, R.C. Smith, J. Spyromilio, C. Stubbs, N.B. Suntzeff, J. Tonry, Observational evidence from supernovae for an accelerating Universe and a cosmological constant. Astronomical J. **116**, 1009 (1998)
6. S. Perlmutter, G. Aldering, G. Goldhaber, R.A. Knop, P. Nugent, P.G. Castro, S. Deustua, S. Fabbro, A. Goobar, D.E. Groom, I.M. Hook, A.G. Kim, M.Y. Kim, J.C. Lee, N.J. Nunes, R. Pain, C.R. Pennypacker, R. Quimby, C. Lidman, R.S. Ellis, M. Irwin, R.G. McMahon, P. Ruiz-Lapuente, N. Walton, B. Schaefer, B.J. Boyle, A.V. Filippenko, T. Matheson, A.S. Fruchter, N. Panagia, H.J.M. Newberg, W.J. Couch, Measurements of Ω and Λ from 42 high-redshift supernovae. Astrophysical J. **517**, 565 (1999)
7. M.S. Safronava, D. Budker, D. DeMille, D.F. Jackson Kimball, A. Derevianko, C.W. Clark, Search for new physics with atoms and molecules. Rev. Mod. Phys. **90**, 025008 (2018)
8. R.A. Hulse, J.H. Taylor, Discovery of a pulsar in a binary system. Astrophys J. **195**, L51 (1975)
9. R.A. Hulse, The discovery of the binary pulsar. Rev. Mod. Phys. **66**, 699 (1994)
10. J.H. Taylor, Binary pulsars and relativistic gravity. Rev. Mod. Phys. **66**, 711 (1994)
11. B.P. Abbott et al., LIGO Scientific Collaboration, Virgo Collaboration, Observation of gravitational waves from a binary black hole merger. Phys. Rev. Lett. **116**, 061102 (2016)
12. LIGO Scientific Collaboration, A gravitational wave observatory operating beyond the quantum shot-noise limit. Nature Phys. **7**, 962 (2011)
13. M. Tse et al., Quantum-enhanced Advanced LIGO detectors in the era of gravitational-wave astronomy. Phys. Rev. Lett. **123**, 231107 (2019)
14. A. Buikema, et al. Sensitivity and performance of the Advanced LIGO detectors in the third observing run. Phys. Rev. D **102**, 062003 (2020)
15. F. Acernese et al. (The Virgo Collaboration), Quantum backaction on kg-scale mirrors: observation of radiation pressure noise in the Advanced Virgo detector. Phys. Rev. Lett. **125**, 131101 (2020)
16. P. Amaro-Seoane, S. Aoudia, S. Babak, P. Binétruy, E. Berti, A. Bohé, C. Caprini, M. Colpi, N.J. Cornish, K. Danzmann, J.-F. Dufaux, J. Gair, O. Jentrich, Ph. Jetzer, A. Klein, R.N. Lang, A. Lobo, T. Littenberg, S.T. McWilliams, G. Nenemans, A. Petiteau, E.K. Porter, B.F. Schutz, A. Sesana, R. Stebbins, T. Sumner, M. Vallisneri, M. Volonteri, H. Ward, eLISA: astrophysics and cosmology in the millihertz regime. arXiv:1201.3621 (2012)
17. C.J. Moore, R.H. Cole, C.P.L. Berry, Gravitational-wave sensitivity curves. Classical Quantum Gravity **32**, 015014 (2015)
18. K. Danzmann et al., *LISA Laser Interferometer Space Antenna, A proposal in Response to the ESA Call for L3 Mission Concepts* (2017). https://www.elisascience.org/files/publications/LISA-L3-20170120.pdf. Accessed December 21, 2020
19. J. Rudolph, T. Wilkason, M. Nantel, H. Swan, C.M. Holland, Y. Jiang, B.E. Garber, S.P. Carman, J.S. Hogan, Large momentum transfer clock atom interferometry on the 689 nm intercombination line of strontium. Phys. Rev. Lett. **124**, 083604 (2020)

20. G.E. Marti, R.B. Hutson, A. Goban, S.L. Campbell, N. Poli, J. Ye, Imaging optical frequencies with $100\,\mu$Hz precision and $1.1\,\mu$m resolution. Phys. Rev. Lett. **120**, 103201 (2018)

21. S.M. Brewer, J.-S. Chen, A.M. Hankin, E.R. Clemenrs, C.W. Chou, D.J. Wineland, D.B. Hume, D.R. Leidbrandt, ^{27}Al$^+$ quantum-logic clock with a systematic uncertainty below 10^{-18}. Phys. Rev. Lett. **123**, 033201 (2019)

22. T.L. Nicholson, S.L. Campbell, R.B. Hutson, G.E. Marti, B.J. Bloom, R.L. McNally, W. Zhang, M.D. Barrett, M.S. Safronava, G.F. Strouse, W.L. Tew, J. Ye, Systematic evaluation of an atomic clock at 2×10^{-18} total uncertainty. Nature Commun. **6**, 6896 (2015)

23. S.L. Campbell, R.B. Hutson, G.E. Marti, A. Goban, N. Darkwah Oppong, R.N. McNally, L. Sonderhouse, J.M. Robinson, W. Zhang, B.J. Bloom, J. Ye, A Fermi-degenerate three-dimensional optical lattice clock. Science **358**, 90 (2017)

24. P.W. Graham, J.M. Hogan, M.A. Kasevich, S. Rajendran, New method for gravitational wave detection with atomic sensors. Phys. Rev. Lett. **110**, 171102 (2013)

25. M. Kasevich, S. Chu, Measurement of the gravitational acceleration of an atom with a light-pulse atom interferometer. Appl. Phys. B **54**, 321 (1992)

26. B. Young, M. Kasevich, S. Chu, in *Atom Interferometry*, ed. by P. Berman (Academic Press, San Diego, 1997), pp. 363–406

27. P.W. Graham, J. Hogan, M. Kasevich, S. Rajendran, R. Romani, Mid-band gravitational wave detection with precision atomic sensors. arXiv:1711.02225v1 (2017)

28. E.G.M. Ferreira, Ultra-light dark matter. arXiv:2005.03254v1 (2020)

29. S. Schlamminger, K.Y. Choi, T.A. Wagner, J.H. Grundlach, E.G. Adelberger, Test of the equivalence principle using a rotation torsion balance. Phys. Rev. Lett. **100**, 041101 (2008)

30. T.A. Wagner, S. Schlamminger, J.H. Gundlach, E.G. Adelberger, Torsion-balance tests of the weak equivalence principle. Class. Quantum Gravity **29**, 184002 (2012)

31. P. Touboul, G. Métris, M. Rodrigues, Y. André, Q. Baghi, J. Bergé, D. Boulanger, S. Bremer, P. Carle, R. Chhun, B. Christophe, V. Cipolla, T. Damour, P. Danto, H. Dittus, P. Fayet, B. Foulon, G. Gageant, P.-Y. Guidotti, D. Hagedorn, E. Hardy, P.-A. Huynh, H. Inchauspe, P. Kayser, S. Lala, C. Lämmerzahl, V. Lebat, P. Leseur, F. Liorzou, M. List, F. Löffler, I. Panet, B. Pouilloux, P. Prieur, A. Rebray, S. Reynaud, B. Rievers, A. Robert, H. Selig, L. Serron, T. Sumner, N. Tanguy, P. Visser, MICROSCOPE Mission: First results of a space test of the equivalence principle. Phys. Rev. Lett. **119**, 231101 (2017)

32. P. Asenbaum, C. Overstreet, M. Kim, J. Curti, M.A. Kasevich, Atom-interferometric test of the Equivalence Principle at the 10^{-12} level. Phys.. Rev. Lett. **125**, 191101 (2020)

33. E.G. Adelberger, B.R. Heckel, A.E. Nelson, Tests of the gravitational inverse square law. Annu. Rev. Nucl. Part. Sci **53**(1), 77–121 (2003)

34. S.K. Lamoreaux, Demonstration of the Casimir force in the 0.6–6μm range. Phys. Rev. Lett. **78**, 5 (1997)

35. J.C. Long, H.W. Chan, J.C. Price, Experimental status of gravitational-strength forces in the sub-centimeter regime. Nucl. Phys. B **539**, 23 (1999)

36. Y.J. Chen, W.K. Tham, D.E. Krause, D. López, E. Fischbach, R.S. Decca, Stronger limits of hypothetical Yukawa interactions in the 30–8000 nm range. Phys. Rev. Lett **116**, 221102 (2016)

37. J.G. Lee, E.G. Adelberger, T.S. Cook, S.M. Fleischer, B.R. Eckel, New test of the gravitational $1/r^2$ law at separations down to 55 μm. Phys. Rev. Lett. **124**, 101101 (2020)

38. W.H. Tan, A.B. Du, W.C. Dong, S.W.Q. Yang, C.G. Shao, S.G. Guan, Q.L. Wang, B.F. Zhan, P.S. Luo, L.C. Tu, J. Luo, Improvement to testing the gravitational inverse-square law at the submillimeter range. Phys. Rev. Lett. **124**, 051301 (2020)

39. H. Everett, 'Relative state' formulation of quantum mechanics. Rev. Mod. Phys. **29**, 454 (1957)

40. R. Penrose, On gravity's role in quantum state reduction. Gen. Relativ. Gravitation **28**, 581 (1996)

41. G.C. Ghirardi, A. Rimini, T. Weber, Unified dynamics for microscopic and macroscopic systems. Phys. Rev. D **34**, 470 (1986)

42. L. Diósi, Models for universal reduction of macroscopic quantum fluctuations. Phys. Rev. A **40**, 1165 (1989)

43. G.C. Ghirardi, R. Grassi, A. Rimini, Continuous-spontaneous-reduction model involving gravity. Phys. Rev. A **42**, 1057 (1990)
44. W. Marshall, C. Simon, R. Penrose, D. Bouwmeester, Towards quantum superpositions of a mirror. Phys. Rev. Lett **91**, 159903 (2003)
45. O. Romero-Isart, Quantum superposition of massive objects and collapse models. Phys. Rev. A **84**, 052121 (2012)
46. U. Delić, M. Reisenbauer, K. Dare, D. Grass, V. Vuletić snf, M. Aspelmeyer, Cooling of a levitated nanoparticle to the motional quantum ground state. Science **367**, 892 (2020)
47. J. Frieman, M. Turner, D. Huterer, Dark energy and the accelerating universe. Ann. Rev. Astron. Astrophys. **65**, 385 (2008)
48. J.H. Davis, The past and future of light dark matter direct detection. Int. J. Mod. Phys. A **30**, 1530038 (2015)
49. E. Gibney, Last chance for WIMPs: physicists launch all-out hunt for dark-matter candidate. Nature **586**, 344 (2020)
50. P. Sikivie, Experimental tests of the "invisible" axion. Phys. Rev. Lett **51**, 1415 (1983)
51. P. Sikivie, Detection rates for "invisible" axion searches. Phys. Rev. D **32**, 2988 (1985)
52. J.E. Kim, G. Carosi, Axions and the strong CP problem. Phys. Rev. Lett **82**, 557 (2010)
53. T. Braine, R. Cervantes, N. Crisosto, N. Du, S. Kimes, L.J. Rosenberg, G. Rybka, J. Yang, D. Bowring, A.S. Chou, R. Khatiwada, A. sonnenschein, W. Wester, G. Carosi, N. Woollett, L.D. Duffy, R. Bradley, C. Boutan, M. Jones, B.H. LaRoque, N.S. Oblath, M.S. Taubman, J. Clarke, A. Dove, A. Eddins, S. R. O'Kelley, S. Nawaz, I. Siddiqi, N. Stevenson, A. Agrawal, A.V. Dixit, J.R. Gleason, S. Jois, P. Sikivie, J.A. Solomon, N.S. Sullivan, D.B. Tanner, E. Lentz, E.J. Daw, J.H. Buckley, P.M. Harrington, E.A. Henriksen, K.M. Murch, Extended search of the invisible axion in the Axion Dark Matter eXperiment. Phys. Rev. Lett **124**, 101303 (2020)
54. P. Pugnat, R. Ballou, M. Schott, T. Husek, M. Sulc, G. Deferne, L. Duvillaret, M. Finger Jr., M. Finger, L. Flekova nd J. Hosek, V. Jary, R. Jost, M. Kral, S. Kunc, K. Macuchova, K.A. Meissner, J. Morville, D. Romanini, A. Siemko, M. Slunewcka, G. Vitrant, J. Zicha, Search for weakly inteacting sub-eV particles with the OSQAR laser-based experiment: results and perspectives. European Phys. J. **74**, 3027 (2014)
55. P. Adamson, S. Chattopadhyay, J. Coleman, P. Graham, S. Geer, R. Harnik, S. Hahn, J. Hogan, M. Kasevich, T. Kovachy, J. Mitchell, R. Plunkett, S. Rajendran, L. Vaerio, A. Vaspmos, *Proposal P-1101 Matter-wave Atomic Gradiometer Interferometric Sensor (MAGIS-100)* (2018). https://www.osti.gov/biblio/1605586
56. M. Aspelmeyer, T.J. Kippenberg, F. Marquardt, Cavity optomechanics. Rev. Mod. Phys. **86**, 1391 (2014)
57. J.D. Cohen, S.M. Meenehan, G.S. McCabe, S. Gröblacher, A.H. Safavi-Naemi, F. Marsili, M.D. Shaw, O. Painter, Phonon counting and intensity interferometry of a nanomechanical resonator. Nature **520**, 522 (2015)
58. R. Riedinger, S. Hong, R.A. Norte, J.A. Slater, J. Shang, A.G. Krause, V. Anant, M. Aspelmeyer, S. Gröblacher, Non-classical correlations between single photons and phonons from a mechanical oscillator. Nature **530**, 313–316 (2015)
59. A. Shkarin, A. Kashkanova, C. Brown, S. Garcia, K. Ott, J. Reichel, J. Harris, Quantum optomechanics in a liquid. Phys. Rev. Lett. **122**, 153601 (2019)
60. E.D. Hall, R.X. Adhikari, V.V. Frolov, H. Müller, M. Pospelov, Laser interferometers as dark matter detectors. Phys. Rev. D **98**, 083019 (2018)
61. A. Kawasaki, Search for kilogram-scale dark matter with precision displacement sensors. Phys. Rev. D **99**, 023005 (2019)
62. A.C. Mellisinos, Proposal for a search for cosmic axions using an optical cavity. Phys. Rev. Lett. **102**, 202001 (2009)
63. A.A. Geraci, C. Bradley, D. Gao, J. Weinstein, A. Derevienko, Searching for ultralight dark matter with optical cavities. Phys. Rev. Lett. **123**, 031304 (2019)
64. J. Manley, D.J. Wilson, R. Stump, D. Grin, S. Singh, Searching for scalat dark matter with compact mechanical resonators. Phys. Rev. Lett. **124**, 151301 (2020)

65. D. Carney, G. Krnjaic, D.C. Moore, C.A. Regal, G. Afek, S. Bhave, B. Brubaker, T. CVorbitt, J. Cripe, N. Cristosto, A. Geraci, S. Ghosh, J.G.E. Harris, A. Hook, E.W. Kolb, J. Kunjummen, R.F. Lang, T. Li, Z. Liu, J. Luykken, L. Magrini, J. Manley, N. Matsumoto, A. Monte, F. Monteiro, T. Purdy, C.J. Riedel, R. Singh, S. Singh, K. Sinha, J.M. Taylor, J. Qin, D.J. Wilson, Y. Zhao, Mechanical quantum sensing in the search for dark matter. Quantum Sci. Technol. **6**, 024002 (2021)

Index

© The Author(s), under exclusive license to Springer Nature Switzerland AG 2021
P. Meystre, *Quantum Optics*, Graduate Texts in Physics,
https://doi.org/10.1007/978-3-030-76183-7

Printed in the United States
by Baker & Taylor Publisher Services